Springer Proceedings in Physics 41

Springer Proceedings in Physics

Managing Editor: H. K. V. Lotsch

Volume 30 *Short-Wavelength Lasers and Their Applications*
Editor: C. Yamanaka

Volume 31 *Quantum String Theory*
Editors: N. Kawamoto and T. Kugo

Volume 32 *Universalities in Condensed Matter*
Editors: R. Jullien, L. Peliti, R. Rammal, and N. Boccara

Volume 33 *Computer Simulation Studies in Condensed Matter Physics: Recent Developments*
Editors: D. P. Landau, K. K. Mon, and H.-B. Schüttler

Volume 34 *Amorphous and Crystalline Silicon Carbide*
Editors: G. L. Harris and C. Y.-W. Yang

Volume 35 *Polycrystalline Semiconductors*
Editors: H. J. Möller, H. P. Strunk, and J. H. Werner

Volume 36 *Nonlinear Optics of Organics and Semiconductors*
Editor: T. Kobayashi

Volume 37 *Dynamics of Disordered Materials*
Editors: D. Richter, A. J. Dianoux, W. Petry, and J. Teixeira

Volume 38 *Electroluminescence*
Editors: S. Shionoya and H. Kobayashi

Volume 39 *Disorder and Nonlinearity*
Editors: A. R. Bishop, D. Campbell, and S. Pnevmatikos

Volume 40 *Static and Dynamic Properties of Liquids*
Editors: M. Davidović and A. K. Soper

Volume 41 *Quantum Optics V*
Editors: J. D. Harvey and D. F. Walls

Volumes 1–29 are listed on the back inside cover

Quantum Optics V

Proceedings of the Fifth International Symposium
Rotorua, New Zealand, February 13–17, 1989

Editors: J. D. Harvey and D. F. Walls

With 103 Figures

Springer-Verlag Berlin Heidelberg New York
London Paris Tokyo Hong Kong

Professor John D. Harvey
Professor Daniel F. Walls
University of Auckland, Private Bag, Auckland, New Zealand

ISBN-13: 978-3-642-74953-7 e-ISBN-13: 978-3-642-74951-3
DOI: 10.1007/978-3-642-74951-3

Softcover reprint of the hardcover 1st edition 1989

2154/3150-543210 – Printed on acid-free paper

Preface

This volume contains contributions based on the lectures delivered at the Fifth International Symposium on Quantum Optics. This Conference, the fifth in a triennial series hosted in New Zealand, was held in Rotorua, 13–17 February 1989. The Conference was attended by 75 participants from New Zealand, Australia, Japan, USA, France, Italy and Germany. There was also a high level of participation from graduate students from New Zealand and Australia, who greatly benefitted from the opportunity to attend world-class conferences. The participants were housed in the Hyatt Hotel and surrounding motels and all enjoyed the relaxed atmosphere offered by Rotorua in the Southern Hemisphere summer.

There were 24 invited papers, given as oral presentations of 40 minutes, and 22 poster papers. The major topics covered at the Conference were new experimental and theoretical results in nonclassical light, including sub-shot-noise light sources. We were fortunate in that all major experimental groups in the world working in this area were represented. The latest experimental results from AT & T Laboratories, NT & T Laboratories, IBM Laboratories, Ecole Normale Supérieure and the Californian Institute of Technology were reported. New theoretical results from Southern Hemisphere participants included a true phase operator for quantum fields derived by Professor David Pegg of Griffiths University and a general treatment of lasers pumped without shot noise by Professor D.F. Walls of Auckland University. There was also the first extensive report of very short pulses produced in an argon ion laser by swept gain superfluorescence in a talk by Professor J.D. Harvey of Auckland.

The Conference was notable for the active audience participation evident in the time reserved for questions after each talk (and at other times). A number of graduate students were able to present their research in the poster sessions. Sessions were conducted in the morning and the evening and this left the afternoons free for informal discussions and exploring the mountains and geysers of the local environment. This format worked extremely well.

The editors would like to thank the participants for providing detailed notes for publication shortly after the meeting, and the various organisations that have provided financial and other support. These include:

University of Auckland
International Union of Pure and Applied Physics
Royal Society of New Zealand
Fletcher Challenge
Laserex

Spectra Physics
Institute of Physics (UK)
Springer-Verlag
New Zealand Institute of Physics

We look forward to another stimulating meeting in 1992.

Auckland, February 1989

John D. Harvey
Daniel F. Walls

Contents

Part I

Quantum Noise Reduction and Squeezing

Photon Noise Reduction in Lasers by Suppression of Pump Fluctuations

D.F. Walls[1], *F. Haake*[2], *and M.J. Collett*[1]

[1]Physics Department, University of Auckland, Auckland, New Zealand
[2] Fachbereich Physik, Universität-Gesamthochschule Essen, D-4300 Essen, Fed. Rep. of Germany

1. Introduction

We wish to address the problem of reducing the photon number or intensity fluctuations in the output of a laser. Usual incoherently pumped lasers exhibit photon statistics which tend to Poissonian well above threshold. Ways to reduce the intensity fluctuations in laser light have recently been discussed by a number of authors [1-4].

GOLUBEV AND SOKOLOV [1] have demonstrated theoretically that regular pump excitation will result in subpoissonian photon statistics of the output light. A theoretical analysis by YAMAMOTO *et al.* [2] showed that the photon number fluctuations of the laser output can be reduced compared to the usual incoherently pumped laser, if the amplitude fluctuations of the pump are reduced. MACHIDA and coworkers [1,3] have recently demonstrated the operation of a semiconductor laser pumped with an electron beam with reduced fluctuations and shown that the output light has a 7.3% reduction in number fluctuations below the vacuum level. MARTE *et al.* [4] have considered pumping lasers with squeezed light. If the phase of the squeezing is such that the amplitude fluctuations of the pump are reduced, then subpoissonian photon statistics will result. The role of pump fluctuations has also been considered in the micromaser [5] where they manifest themselves in the irregularity of the injected atoms. Regular atomic injection has been shown to yield subpoissonian photon statistics.

In this paper we will demonstrate using the method of GOLUBEV AND SOKOLOV [1] how the SCULLY LAMB [6] derivation of the laser master equation may be modified for regular pump excitation. We shall also describe how the model of MARTE *et al.* [4] for a squeezed pump laser may be realised in practice.

2. Regular Pump Excitation

The first model we shall consider is based on a modification of the Scully Lamb laser master equation and is due to GOLUBEV AND SOKOLOV [1]. We shall follow their method but work out results explicitly using the results of the Scully Lamb treatment. We shall consider the situation where the atoms are regularly excited. A possible mechanism for such a regular excitation is described by Golubev and Sokolov. In their scheme the atoms were excited by periodic pulses of duration T_0. It was possible to collect all the atoms in the upper working level during the pulse duration which was much less than the atomic lifetime ($T_0 \ll \tau_a$). The field energy was modulated at the pulse sequence period T_1. Under the condition $T_1 \ll \kappa^{-1}$ (where κ^{-1} is the lifetime of the cavity field), the modulation depth was small.

We consider the following model for a laser. The infinitesimal change in the field density operator after interaction with a single excited atom is described by the operator $u\rho$. The density operator at time $t + \Delta t$ is then

$$\rho(t + \Delta t) = (1 + u)\,\rho(t) \tag{2.1}$$

Springer Proceedings in Physics, Vol. 41 **Quantum Optics V**
Editors: J.D. Harvey and D.F. Walls

where $\Delta t \ll \kappa^{-1}$ the cavity decay time.

If n atoms are excited during the time interval Δt

$$\rho(t + \Delta t) = (1 + u)^n \rho(t) \tag{2.2}$$

assuming that each atom finds the field in a state prepared by the preceding atom.

If the excitation process obeys Poisson statistics, we must average (2.2) with the weight function $\exp(- r\Delta t) \frac{(r\Delta t)^n}{n!}$ where r is the pumping rate of the atoms. This yields

$$\rho(t + \Delta t) = \exp(r\Delta t u)\, \rho(t) \tag{2.3}$$

Upon differentiating this gives

$$\dot{\rho} = ru\rho \tag{2.4}$$

When we include the cavity damping of the field mode, this yields the usual Scully Lamb master equation for the laser

$$\dot{\rho} = ru\rho + \Lambda\rho \tag{2.5}$$

where in the number state representation [7]

$$(u\rho)_m = - a_{m+1}\, \rho_m + a_m\, \rho_{m-1} \tag{2.6}$$

where

$$a_m = \frac{2g^2\, \gamma_b\, \gamma_{ab} m}{\gamma_a\, \gamma_b\, \gamma_{ab}^2 + 4\gamma_{ab}^2\, g^2 m}$$

where g is the atom field dipole coupling constant, and γ_a, $\gamma_b \left(\gamma_{ab} = \frac{\gamma_a + \gamma_b}{2}\right)$ are the decay rates of the upper and lower atomic levels (see Fig.1) and

$$(\Lambda\rho)_m = \kappa\left[-m\rho_m + (m + 1)\, \rho_{m+1}\right] \tag{2.7}$$

where κ is the cavity decay rate.

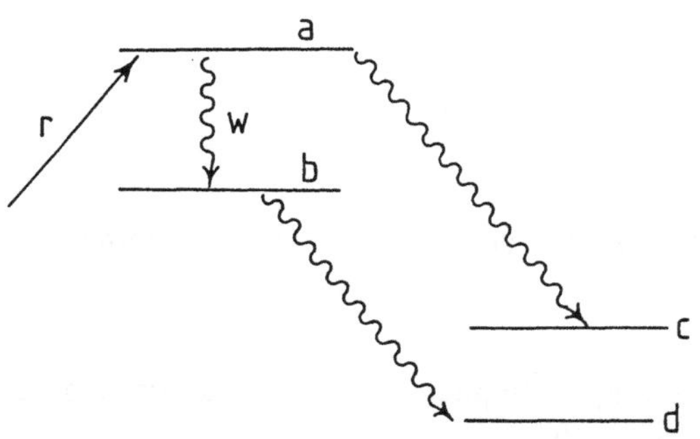

Figure 1 Atomic level scheme. Laser action takes place between levels a and b which decay to levels c and d with decay constants γ_a and γ_b respectively. The excitation rate to the upper level a is r.

If, instead of a Poissonian excitation, there is a regular excitation, we set $n = r\Delta t$ which yields

$$\dot{\rho} = r\ell n(1+u)\rho \tag{2.8}$$

Including the cavity damping the complete laser master equation is

$$\dot{\rho} = r\ell n(1+u)\rho + \Lambda\rho \tag{2.9}$$

Since u represents a very small change we shall expand the logarithm $\ell n(1 + u) \approx u - \frac{u^2}{2}$. This gives the following master equation for a regularly pumped laser in the number state representation.

$$\dot{\rho}_m = \kappa\left[- m\rho_m + (m + 1)\rho_{m+1}\right] + r\left[- a_{m+1}\, \rho_m + a_m\, \rho_{m-1}\right]$$

$$+ \frac{r}{2}\left[- a_{m+1}^2\, \rho_m + a_m(a_m + a_{m+1})\rho_{m-1} - a_m a_{m-1}\, \rho_{m-2}\right] \tag{2.10}$$

We shall now seek a solution for the steady state mean and variance in the photon number. We shall use a technique developed by GÖRTZ AND WALLS [8] and introduce the following slowly varying function of m

$$q_m = \frac{\rho_{m+1}}{\rho_m} \tag{2.11}$$

Since $q_m = q_{m\pm j}$ for $j = 1,2$. We also set $a_{m\pm1} = a_m$. Then dividing the steady state version of (10) by ρ_m we get an algebraic equation for q_m. Eliminating the unphysical root $q_m = 1$, we arrive at the quadratic equation

$$q_m^2 - \frac{ra_m}{m\kappa}\left(1 + \frac{1}{2}a_m\right)q_m + \frac{ra_m^2}{2m\kappa} = 0 \tag{2.12}$$

The mean number $\bar{n}$ is found by setting $q_m = 1$ (which gives the most likely photon number).

We find

$$\bar{n} = \frac{r}{2\kappa}\left(\frac{\gamma_b}{\gamma_{ab}}\right) - n_s \tag{2.13}$$

where

$$n_s = \frac{\gamma_a\gamma_b}{4g^2}$$

Since we are looking for a low noise regime, we shall be working well above threshold where $\frac{r}{2\kappa}\left(\frac{\gamma_b}{\gamma_{ab}}\right) \gg n_s$.

It is shown by Görtz and Walls that when the steady state photon number is approximated by a Gaussian, the photon number variance is given by

$$\sigma^2 = -\left(\frac{\partial q_m}{\partial m}\right)^{-1}_{q=1} \qquad (2.14)$$

We find

$$\sigma^2 = \bar{n}\left(1 - \frac{\gamma_b}{2(\gamma_a + \gamma_b)}\right) + n_s \qquad (2.15)$$

We may neglect the n_s term sufficiently far above threshold.

We consider two limits of (2.15)

(a) $\gamma_a \ll \gamma_b$ where there is no spontaneous emission is minimized in which case $\sigma^2 = \frac{\bar{n}}{2}$.

(b) $\gamma_a = \gamma_b$ where the spontaneous emission rate equals the loss rate from the lower level, in which case $\sigma^2 = \frac{3}{4}\bar{n}$.

These results agree with those found by GOLUBEV AND SOKOLOV [1] and recently by SCULLY and coworkers [9].

Spectrum of Intensity Fluctuations

The work in the previous section refers to the number fluctuations of the light inside the cavity. However what is experimentally observable is the intensity fluctuations of the light which emerges from the cavity.

The spectral variance of intensity correlations is given by

$$\langle \delta i^2 \rangle_\omega = \int_{-\infty}^{\infty} dt\, e^{i\omega t} \langle i(t), i(0) \rangle \qquad (2.16)$$

where $\langle i(t), i(0)\rangle = \langle i(t) \; i(0)\rangle - \langle i(t)\rangle\langle i(0)\rangle$

The intensity correlations of the light emerging from the cavity are related to the internal photon number correlations by

$$\langle i(t)\, i(0)\rangle = \kappa\bar{n}\delta(t) + \kappa^2 g(t) \qquad (2.17)$$

where κ is the cavity loss rate

and $g(t) = \langle a^+(0)a^+(t)\, a(t)\, a(0)\rangle$

$$= \mathrm{Tr}(a^+(t)\, a(t)\, a(0)\rho\, a^+(0)\rangle \qquad (2.18)$$

By writing the master equation for the density operator $\rho(t)$ of the field as $\dot{\rho} = L\rho$ we have

$$g(t) = \mathrm{Tr}(a^+ a e^{Lt} a\rho\, a^+) \qquad (2.19)$$

Evaluating the trace using the photon number representation

$$g(t) = \sum_m m \langle m|e^{Lt} a\rho\, a^+|m\rangle$$

$$= \sum_{m,n} m\, (e^{Lt})_{mn} \langle n|\, a\rho\, a^+|n\rangle$$

$$= \sum_{m,n} m\, (e^{Lt})_{mn}\, (n+1)\rho_{n+1}$$

$$= \sum_{m,n} m\, (e^{Lt})_{m,n-1}\, n\, \rho_n \tag{2.20}$$

where (e^{Lt}) is the solution of the master equation (2.10) reducing to the sharp distribution δ_{mn} initially. Since the stationary distribution ρ_n is smooth, we may approximate

$$\left(e^{Lt}\right)_{m,n-1} = (e^{Lt})_{mn} - \frac{\partial}{\partial n}\, (e^{Lt})_{mn} \tag{2.21}$$

Finally reverting to the continuum description appropriate to the high excitation regime, $\left(e^{Lt}\right)mn$ goes to the conditional probability density $P(mt|n)$ and

$$g(t) = \langle m(t)\, m(0)\rangle - \int dmdn\ m \frac{\partial}{\partial n}\left[P(mt|n)\right] n\, \rho_n \tag{2.22}$$

which on integrating by parts becomes

$$g(t) = \langle m(t)\, m(0)\rangle + \int dmdn\ m\, P(mt|n)\, \left(\rho(n) + n\frac{\partial}{\partial n}\rho(n)\right) \tag{2.23}$$

Approximating the stationary distribution as a gaussian of the form

$$\rho_{\bar{n}+\delta m} = \rho_{\bar{n}}\, e^{-(\delta m)^2/2\sigma^2} \tag{2.24}$$

we find

$$g(t) = \langle m(t)m(0)\rangle \left(1 + \frac{\bar{n}}{\sigma^2}\right) + \bar{n} - \langle m(t)\, m^2(0)\rangle/\sigma^2 \tag{2.25}$$

Finally by using the Gaussian and Markovian properties of the random process we find

$$\langle m(t)\, m(0)\rangle = \sigma^2 e^{-\delta t} + \bar{n}^2 \tag{2.26}$$

where $\delta = \dfrac{\kappa\frac{\bar{n}}{n_s}}{1 + \frac{\bar{n}}{n_s}}$

and

$$g(t) = \bar{n}^2 + (\sigma^2 - \bar{n})\, e^{-\delta t} \tag{2.27}$$

Thus the spectral variance of the intensity correlations is given by

$$\langle \delta i^2\rangle_\omega = \kappa\bar{n}\left(1 + \frac{2Q\kappa\delta}{\delta^2 + \omega^2}\right) \tag{2.28}$$

where $Q = \frac{\sigma^2 - \bar{n}}{\bar{n}}$.

Far above threshold $\left(\frac{\bar{n}}{n_s} \gg 1\right)$ we find the following results.

At zero frequency

$$(\delta i^2)_{\omega=0} = \kappa\bar{n}\,(1 + 2Q) \tag{2.29}$$

The factor 1 represents the shot noise contribution. A negative Q factor will cause a reduction in the shot noise level. Far above threshold we find the following results. For the case where the decay rate of the lower level is much faster than that of the upper level ($\gamma_b \gg \gamma_a$)

$$(\delta i^2)_{\omega=0} = 0 \tag{2.30}$$

In this limit spontaneous emission is minimised since reabsorption is improbable and the suppression of pump fluctuations is sufficient to give zero intensity fluctuations in the output light at $\omega = 0$.

In the case $\gamma_a = \gamma_b$

$$(\delta i^2)_{\omega=0} = \frac{\kappa\bar{n}}{2} \tag{2.31}$$

Spontaneous emission still contributes to intensity fluctuations even far above threshold since the atoms are continually returned to the upper level by reabsorption. In this case suppression of pump fluctuations is only sufficient to reduce the shot noise by a factor of two.

3. Squeezed Pumps

In a recent series of papers MARTE *et al.* [4] have considered pumping lasers with squeezed light. They have demonstrated that the output of such a laser will oscillate with one of two macroscopically distinct phases. If the atoms are pumped with a squeezed vacuum (no coherent amplitude) the laser line width may be reduced. If a coherent amplitude is added to the pump light, it is possible to simulate the pumping of a laser with a subpoissonian source [2]. In this case it was found that the output light may have amplitude fluctuations reduced below the quantum fluctuations of the vacuum.

The analysis of Marte *et al.* modelled the active medium by an ensemble of N two level atoms. This is a standard model used in laser theory [6,10] but requires the atoms to be pumped by an inverted bath in order to achieve atomic inversion. In this paper we wish to demonstrate how the features of this model may be realized in practice. That is, we shall describe a pumping mechanism which both achieves atomic inversion and squeezes the fluctuations on the lasing transition.

We consider an ensemble of three level atoms inside an optical cavity. The atomic energy level scheme is shown in Fig. 2. The three level atom interacts with two cavity modes a_1 and a_2. The difference frequency between the two modes is resonant with the active laser transition between levels 0 and 1. The two cavity modes are pumped externally with squeezed light fields.

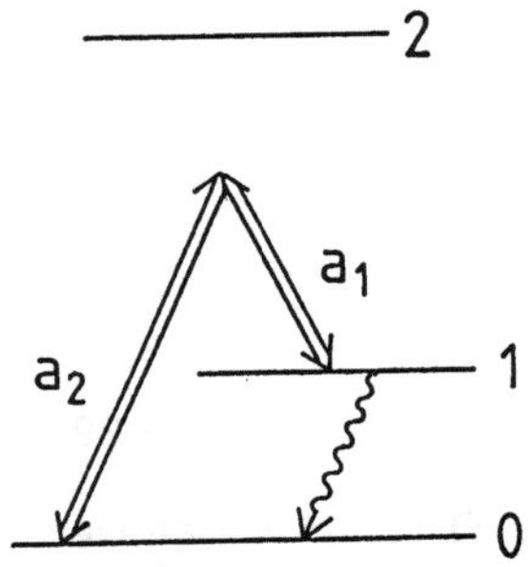

Figure 2 Three level atomic scheme for a squeezed pump laser.

The upper level two of the atom is assumed highly detuned and hence may be eliminated adiabatically giving an effective two level approximation for the interaction of the atom with the two field modes

$$H_{AF} = \hbar g\,(\sigma_+ a_2 a_1^\dagger + \sigma_- a_2^\dagger a_1) \tag{3.1}$$

where σ_+, σ_- are the atomic operators for the 1, 0 transition of the atom and g is the dipole coupling strength.

The inputs to the two cavity modes are taken to be two independently squeezed baths with zero mean field.

The time evolution of the density operator of the atom field system is given by

$$W(t) = (L + \Lambda)\,W(t) \tag{3.2}$$

where

$$LW = -\frac{i}{\hbar}[H,W] \tag{3.3}$$

and the cavity modes coupled to the squeezed baths obey the equations [11]

$$\begin{aligned} \Lambda_i W = {} & \gamma_i(N_i + 1)\left\{[a_i, W a_i^\dagger] + [a_i W, a_i^\dagger]\right\} \\ & + \gamma_i N_i \left\{[a_i^\dagger, W a_i] + [a_i^\dagger W, a_i]\right\} \\ & - \gamma_i M_i^* \left\{[a_i, W a_i] + [a_i W, a_i]\right\} \\ & - \gamma_i M_i \left\{[a_i^\dagger, W a_i^\dagger] + [a_i^\dagger W, a_i]\right\} \end{aligned} \tag{3.4}$$

where the squeezing parameters M_i obey the restrictions

$$|M_i|^2 \leq N_i(N_i + 1)\ . \tag{3.5}$$

We are interested in the limit in which the photon decay rates γ_i are much larger than the dipole coupling constant g of the atom, ($g \ll \gamma_1, \gamma_2$). We shall derive a master equation for the reduced density operator of the atom

$$\rho(t) = Tr_F\, W(t) \tag{3.6}$$

in this limit.

We choose an initial state with no correlations between the atom and the field modes

$$W(0) = F\rho(0)\ . \tag{3.7}$$

The initial atomic density operator is unspecified but the field density operator is specified by the requirement

$$\Lambda F = 0,\ TrF = 1\ . \tag{3.8}$$

The unique solution F of (3.8) describes the two oscillators in equilibrium with their heat baths, with the following correlations

$$Tr_F(a_i F) = 0,\ Tr_F(a_i^\dagger a_i F) = N_i$$

$$Tr_F(a_i^2 F) = M_i \tag{3.9}$$

and zero correlations between the modes $Tr_F(a_1^\dagger a_2 F) = 0$. The solution for $\rho(t)$ may be obtained from (3.2)-(3.6) as

$$\rho(t) = Tr_F\, e^{(\Lambda+L)t} F\rho(0)$$

$$\equiv U(t)\,\rho(0) \tag{3.10}$$

where we have introduced the time evolution operator U(t). Differentiating with respect to time we obtain

$$\dot{\rho}(t) = \dot{U}(t)\,\rho(0) = \dot{U}(t)\,U(t)^{-1}\,\rho(t)$$

$$\equiv \mathcal{L}(t)\,\rho(t)\ . \tag{3.11}$$

The generator of infinitesimal time translations $\mathcal{L}(t)$ in (3.11) can be calculated quite easily as a power series in the small parameter $\frac{g}{\gamma_i}$. Technically the series in question arises when $\dot{U}(t)$ and $U(t)$ are expanded in powers of the coherent interaction Liouvillian L and the quotient series $\dot{U}U^{-1}$ is evaluated. $\dot{U}(t)$ is given by

$$\dot{U}(t) = Tr_F(\Lambda + L)\, e^{(L+\Lambda)t} F$$
$$= Tr_F\, L\, e^{(L+\Lambda)t} F\ . \tag{3.12}$$

It follows that the expansion for $\mathcal{L}(t)$ has no zero order term. The first order term also vanishes

$$\dot{U}_1(t) = Tr_F\, L e^{\Lambda t} F$$
$$= Tr_F\, LF = 0 \tag{3.13}$$

which follows since we have $Tr_F(a_1^\dagger a_2 F) = 0$. Since $U_0(t) = Tr_F(e^{\Lambda t} F) = 1$ the second order contribution to $\mathcal{L}(t)$ reads

$$\mathcal{L}(t) = Tr_F \int_0^t dt'\, L\, e^{\Lambda t'}\, L\, e^{\Lambda(t-t')}\, F$$

$$= Tr_F \int_0^t dt'\, L\, e^{\Lambda t'}\, LF\ . \tag{3.14}$$

It is clear that the explicit time dependence of the generator $\mathcal{L}(t)$ arises on the time scale $\frac{1}{\gamma_1 + \gamma_2}$ through field correlations of the form

$$Tr_F(a_2^\dagger a_1 \, e^{\Lambda t} X) = e^{-(\gamma_1+\gamma_2)t} \, Tr_F \, a_2^\dagger a_1 \, X$$

$$Tr_F(a_1^\dagger a_2 \, e^{\Lambda t} X) = e^{-(\gamma_1+\gamma_2)t} \, Tr_F \, a_1^\dagger a_2 \, X \tag{3.15}$$

with $X = a_2^\dagger a_1 F$, $a_1^\dagger a_2 F$ or the hermitian conjugates $F a_1^\dagger a_2$, $F a_2^\dagger a_1$. Since the time $\frac{1}{\gamma_1 + \gamma_2}$ is assumed to be much smaller than the time scale on which the atomic density operator $\rho(t)$ changes, we may put the upper limit in the time integral in (3.14) to infinity, thus completing the adiabatic elimination of the cavity modes. The corresponding asymptotic generator reads

$$\mathcal{L} = \int_0^\infty dt \, Tr_F \, L \, e^{\Lambda t} \, LF \quad . \tag{3.16}$$

Writing out the commutator prescriptions, we arrive at the equation

$$\begin{aligned} \mathcal{L}\rho = -\frac{g^2}{(\gamma_1 + \gamma_2)} \Big\{ & <\Gamma\Gamma^+>(\sigma_+\sigma_-\rho - 2\sigma_-\rho\sigma_+ + \rho\sigma_+\sigma_-) \\ & + <\Gamma^+\Gamma>(\sigma_-\sigma_+\rho - 2\sigma_+\rho\sigma_- + \rho\sigma_-\sigma_+) \\ & - <\Gamma\Gamma>(2\sigma_+\rho\sigma_+) - <\Gamma^+\Gamma^+>(2\sigma_-\rho\sigma_-) \Big\} \end{aligned} \tag{3.17}$$

where $\Gamma = a_1 + a_2$.

We now consider the correlations present in the input fields to the cavity. For independently squeezed reservoirs, we have from (3.4)

$$\begin{aligned} <\Gamma\,\Gamma^+> &= N_1(1 + N_2) \\ <\Gamma^+\,\Gamma> &= N_2(1 + N_1) \\ <\Gamma^2> &= M_1^*\,M_2 \\ <\Gamma^{+2}> &= M_1 M_2^* \end{aligned} \tag{3.18}$$

where the squeezing parameters M_i obey the restrictions given by (3.5).

The master equation (3.17) can now be written in the form

$$\begin{aligned} \dot\rho = \; & \bar\gamma \, \bar N \left\{[\sigma_-,\rho\sigma_+] + [\sigma_-\rho,\sigma_+]\right. \\ & + \bar\gamma \, (\bar N + 1)\left\{[\sigma_+,\rho\sigma_-] + [\sigma_+\rho,\sigma_-]\right\} \\ & - \bar\gamma \; 2\bar M \sigma_- \rho\sigma_- - \bar\gamma \; 2\bar M^* \sigma_+ \rho\sigma_+ \end{aligned} \tag{3.19}$$

where

$$\bar\gamma = \frac{g^2}{\gamma_1 + \gamma_2}\left(N_1 - N_2\right)$$

$$\bar N = \frac{N_1 (N_2 + 1)}{N_1 - N_2}$$

and

$$M = \frac{M_2 M_1^*}{N_1 - N_2}$$

where $|\bar{M}|^2 \lessapprox \bar{N}(\bar{N}+1)$ (3.20)

Equation (3.19) is of the same form as the master equation derived by Marte and Walls. It is possible to achieve a positive inversion of the atoms for $N_2 > N_1$. We note that a master equation of the same form as (3.19) results if the pumping reservoirs share quantum correlations such as the output from a frequency upconverter. However, in this case $|\bar{M}| \lessapprox \bar{N}$ and it is not possible to achieve both inversion and squeezing of the atoms.

Spontaneous emission from the atoms may be included in the master equation (3.19). In this case, the equality in (3.20) can never be obeyed and $|M|^2$ is always less than $\bar{N}(\bar{N}+1)$.

The properties of the laser which follow from a pumping of the form described by the master equation (3.19) have been discussed in detail by MARTE *et al.* [4] The effect of the squeezed pump is to break the phase symmetry of the usual laser. There are two stable solutions for the complex field amplitude differing by π in phase. These stable solutions are in phase with the low noise quadrature of the squeezed bath. The effect of the symmetry breaking is to reduce the phase diffusion and hence narrow the laser linewidth.

Pumping the laser with a subpoissonian excitation may be modelled by adding a coherent component to the pump in phase with the quadrature with reduced fluctuations. In this case, the fluctuations in the amplitude quadrature of the output laser field are reduced. The spectrum of fluctuations in the amplitude quadrature of the output field is shown in Fig.(3) where the vacuum level is unity.

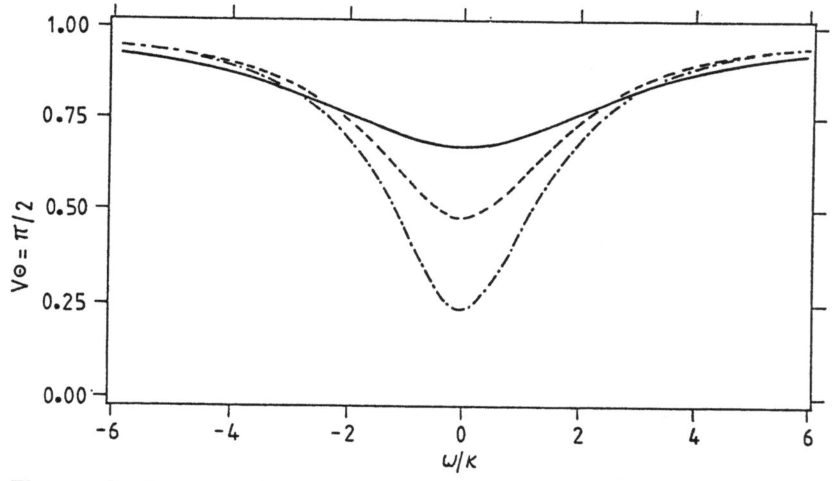

Figure 3 Spectrum of fluctuations in the amplitude quadrature with subpoissonian pump for fixed output intensity. Solid line: 40% squeezing; dashed line: 58% squeezing; dash dotted line 81% squeezing.

References

1. Y.M. Golubev and I.V. Sokolov: Sov.Phys. J.E.T.P. **60**, 234 (1984).
2. Y. Yamamoto, S. Machida and O. Nilsson: Phys.Rev **A34** 4025 (1986).
3. S. Machida, Y. Yamamoto and Y. Itaya: Phys.Rev.Lett. **58**, 1000 (1987).
   S. Machida and Y. Yamamoto: Phys.Rev.Lett. **60**, 792 (1988).
4. M.A. Marte and D.F. Walls: Phys.Rev. **A37**, 1235 (1988).
   M.A. Marte, H. Ritsch and D.F. Walls: Phys.Rev.Lett. **61**, 1093 (1988), Phys.Rev **A38** 3577 (1988).
5. P. Filipowicz, J. Javanainen and P. Meystre, Phys.Rev. **A34**, 3077 (1986).
6. M.O. Scully and W.E. Lamb: Phys.Rev. **159**, 208 (1967).
7. M. Sargent, M.O. Scully and W.E. Lamb: Laser Physics, p.297, (Addison Wesley 1974).
8. R. Görtz and D.F. Walls: Zeitschrift für Physik **B25**, 423 (1976).
9. M.O. Scully (private communication).
10. H. Haken: *Laser Theory* Vol.2 (Springer, Berlin 1984).
11. C.W. Gardiner and M.J. Collett, Phys.Rev. **A31**, 3761 (1985).

Squeezing, Bistability and Instability in the Optical Parametric Oscillator

E. Giacobino[1], *C. Fabre*[1], *A. Heidmann*[1], *S. Reynaud*[1], *and L. Lugiato*[2]

[1]Laboratoire de Spectorscopie Hertzienne de l'Ecole Normale Supérieure, Université Pierre et Marie Curie, F-75252 Paris, France

[2]Dipartimento di Fisica del Politecnico, Corso Duca degli Abruzzi 24, I-10129 Torino, Italy

Optical parametric interaction has been one of the most successful processes to date to generate squeezed and non classical light[1-3]. It involves the generation of two signal fields at frequencies ω_1 and ω_2 by a nonlinear crystal irradiated with a pump field at frequency ω_0 (energy conservation requires that $\omega_1+\omega_2=\omega_0$). The crystal can be placed in a resonant optical cavity, operating either below[4-6] or above[7-8] threshold.

A recent paper[9] has shown that, above the oscillation threshold, bistability and instabilities should occur when both signal and pump fields are detuned from resonance. The presence of these new features suggests the possibility of squeezing in regions of the parameter space which have not been explored before in the optical parametric oscillator (OPO) . Since squeezing is obtained near the bistability threshold of passive optical cavities[5,10-11], it is expected that quantum noise is reduced in the vicinity of the thresholds associated to bistability and instability in the OPO.

In this paper, we study the quantum fluctuations of the fields emitted or reflected by a degenerate parametric oscillator. Using the semiclassical linear input output theory[12] already used for the OPO[7-8] and for bistability[13], we are able to obtain analytical expressions for the fluctuation spectra in the case of the bistable OPO.

1. The model

We assume that the only noise sources are the field fluctuations incident on the coupling mirror. We consider that the incident pump field is in a coherent state and that the incident signal field is in the vacuum state.

Inside the cavity, the evolution of the field is described by the effective hamiltonian H[9]:

$$H = \hbar\, [\gamma_0\Delta_0\, \alpha_0{}^*\alpha_0 + \gamma_1\Delta_1\, \alpha_1{}^*\alpha_1 + i(g/2)\, (\alpha_1{}^*\alpha_1{}^*\alpha_0 - \alpha_0{}^*\alpha_1\alpha_1)] \qquad (1)$$

where g is the coupling parameter for parametric generation in the nonlinear crystal, α_0 and α_1 are the pump and signal fields, Δ_0 and Δ_1 are the detunings normalized to the damping parameters γ_0 and γ_1. The intracavity field variables α_i and $\alpha_i{}^*$ are scaled as the annihilation and creation quantum operators a_i and $a_i{}^+$.

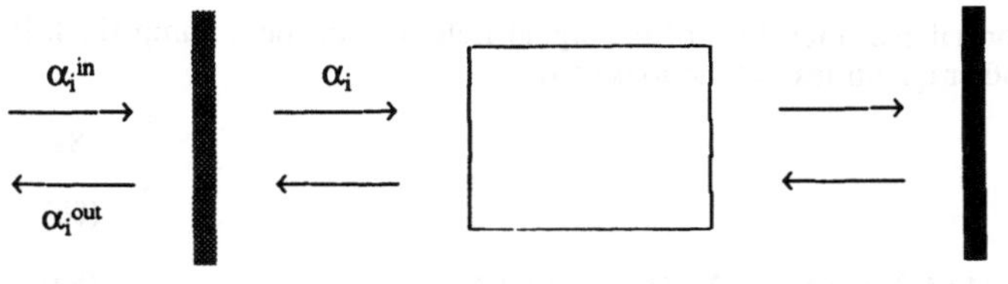

Figure 1: The model of the Degenerate Optical Parametric Oscillator studied in this paper. A high finesse cavity with one coupling mirror contains a noise-free nonlinear medium which parametrically couples a pump mode and a signal mode. The various field variables are the intracavity fields α_i, the input fields α_i^{in} and the output fields α_i^{out} (i=0 for the pump mode, i=1 for the signal mode).

Editors: J.D. Harvey and D.F. Walls © Springer-Verlag Berlin, Heidelberg 1989

The equations describing the variation of the intracavity fields α_i in a round trip time τ are[9]:

$$\tau\, d\alpha_1/dt = -\,\gamma_1(1+i\Delta_1)\,\alpha_1 + g\,\alpha_1^*\alpha_0 + t_1\,\alpha_1^{in} \tag{2a}$$

$$\tau\, d\alpha_0/dt = -\,\gamma_0(1+i\Delta_0)\,\alpha_0 - g\,\alpha_1\alpha_1/2 + t_0\,\alpha_0^{in} \tag{2b}$$

These equations involve the driving terms $t_1\,\alpha_1^{in}$ and $t_0\,\alpha_0^{in}$. For the signal field α_1, the driving term is only due to the semiclassical random field $\delta\alpha_1^{in}$ which represents the vacuum fluctuations coming onto the coupling mirror[12]. For the pump field α_0^{in}, the driving term has a non zero mean value and fluctuations equal to the vacuum fluctuations.

The transmission amplitude coefficients t_i and the damping coefficients γ_i are related by:

$$t_i^2 = 2\,\gamma_i \tag{3}$$

The output fields will be obtained from the classical equation for reflection and transmission at the mirror:

$$\alpha_i^{out} = t_i\,\alpha_i - \alpha_i^{in} \tag{4}$$

2. Working points

The stationnary solutions of equations (2) give the working points of the OPO. Two kinds of solutions are found[9]:

i) the trivial solution, in which the signal field has a zero mean value:

$$\alpha_1 = 0 \tag{5a}$$

$$\alpha_0 = t_0\,\alpha_0^{in} / [\gamma_0(1+i\Delta_0)] \tag{5b}$$

ii) the non trivial solution, which corresponds to above threshold operation of the OPO:

$$g\,\alpha_0 = \gamma_1\,(1+i\Delta_1)\,\alpha_1/\alpha_1^* \tag{6a}$$

$$t_0\,g\,\alpha_0^{in} = \gamma_0\,(1+i\Delta_0)\,g\,\alpha_0 + g^2\,\alpha_1\alpha_1/2 \tag{6b}$$

We choose a phase reference such that α_1 is real and we define the normalized quantities A_1, A_P, I_1 and I_P by:

$$A_1 = g\,\alpha_1 / (2\,\gamma_0\,\gamma_1)^{1/2} \tag{7a}$$

$$A_P = t_0\,g\,\alpha_0^{in} / (\gamma_0\,\gamma_1) \tag{7b}$$

$$I_1 = A_1^2 \tag{7c}$$

$$I_P = A_P^2 \tag{7d}$$

The quantities I_1 and I_P are the normalized intensities of the signal field and incident pump field. It will be convenient to express the other quantities as functions of I_1:

$$A_1 = I_1^{1/2} \tag{8a}$$

$$A_P = (1+i\Delta_0)\,(1+i\Delta_1) + I_1 \tag{8b}$$

$$I_P = (1-\Delta_0\Delta_1+I_1)^2 + (\Delta_0+\Delta_1)^2 = (1+\Delta_0^2)\,(1+\Delta_1^2) + 2I_1\,(1-\Delta_0\Delta_1) + I_1^2 \tag{8c}$$

Equation (8c) gives the relation between the incident and the intracavity intensities. As shown in ref.[9], there are two regimes (Fig.2):

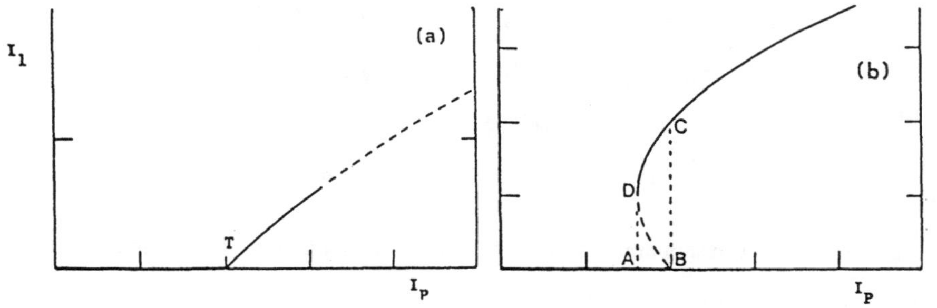

Figure 2: Normalized signal intensity I_1 as a function of the normalized pump intensity I_P, for (a) Δ_0=-1, Δ_1=3, (b) Δ_0=1, Δ_1=3 (the dashed lines indicate the unstable branches).

1) $\Delta_0\Delta_1<1$: there only one value of I_1 for each value of the input intensity I_P. The oscillation threshold is given by:

$$I_P > (1+\Delta_0^2)\,(1+\Delta_1^2) \tag{9a}$$

2) $\Delta_0\Delta_1>1$: bistability appears for values of the input intensity I_P such that:

$$(\Delta_0+\Delta_1)^2 < I_P < (1+\Delta_0^2)\,(1+\Delta_1^2) \tag{9b}$$

The signal intensity at the upper turning point D is given by:

$$I_1 = \Delta_0\Delta_1 - 1 \tag{9c}$$

3. Linear stability analysis

After a standard procedure, the equations for the fluctuations are obtained by linearizing eqs (2) around a given working point. In order to write these equations in a matrix form, we denote $[\delta\alpha]$ the column matrix containing the fluctuations of the intracavity field variables:

$$[\delta\alpha] = \begin{vmatrix} \delta\alpha_1 \\ \delta\alpha_1^* \\ \delta\alpha_0 \\ \delta\alpha_0^* \end{vmatrix} \tag{10}$$

The same notations will be used for the input field variables $[\delta\alpha^{in}]$ and the output ones $[\delta\alpha^{out}]$. We also define the diagonal matrices T containing the transmission coefficients:

$$T = \begin{vmatrix} t_1 & 0 & 0 & 0 \\ 0 & t_1 & 0 & 0 \\ 0 & 0 & t_0 & 0 \\ 0 & 0 & 0 & t_0 \end{vmatrix} \tag{11}$$

With these notations, the equations for the intracavity field fluctuations can be written as:

$$\tau\, d[\delta\alpha]/dt = -A\,[\delta\alpha] + T\,[\delta\alpha^{in}] \tag{12}$$

with:

$$A = \begin{vmatrix} \gamma_1(1+i\Delta_1) & -g\alpha_0 & -g\alpha_1^* & 0 \\ -g\alpha_0^* & \gamma_1(1-i\Delta_1) & 0 & -g\alpha_1 \\ g\alpha_1 & 0 & \gamma_0(1+i\Delta_0) & 0 \\ 0 & g\alpha_1^* & 0 & \gamma_0(1-i\Delta_0) \end{vmatrix} \quad (13)$$

Inserting the relations corresponding to the non trivial solution and introducing the new parameter:

$$\gamma = \gamma_1 / \gamma_0 \quad (14)$$

one gets:

$$A = \gamma_1 \underline{A} \quad (15a)$$

with:

$$\underline{A} = \begin{vmatrix} (1+i\Delta_1) & -(1+i\Delta_1) & -(2I_1/\gamma)^{1/2} & 0 \\ -(1-i\Delta_1) & (1-i\Delta_1) & 0 & -(2I_1/\gamma)^{1/2} \\ (2I_1/\gamma)^{1/2} & 0 & (1+i\Delta_0)/\gamma & 0 \\ 0 & (2I_1/\gamma)^{1/2} & 0 & (1-i\Delta_0)/\gamma \end{vmatrix} \quad (15b)$$

The stability condition is that all the eigenvalues of $\underline{A}$ have a positive real part. This stability analysis is worked out in detail in ref.[9]. For $\Delta_0\Delta_1>1$, the stability analysis shows that the negative slope branch of the bistability curve is unstable while the two positive slope branches are stable. For $\Delta_0\Delta_1<1$, the stability condition can be written:

$$(1+\Delta_0^2)\,[1+\Delta_0^2+4\gamma(1+\gamma)] < -2I_1(1+\gamma)^2\,[1+\Delta_0^2+2\gamma(1+\Delta_0\Delta_1)] \quad (16)$$

So, there is an instability for:

$$1+\Delta_0^2+2\gamma(1+\Delta_0\Delta_1) <0 \quad (17a)$$

$$I_1 > \frac{(1+\Delta_0^2)\,[1+\Delta_0^2+4\gamma(1+\gamma)]}{-2(1+\gamma)^2\,[1+\Delta_0^2+2\gamma(1+\Delta_0\Delta_1)]} \quad (17b)$$

Conditions (17) require that $\Delta_0\Delta_1<-1$: bistability and instability do not occur in the same region of the parameter space (Fig.2).

4. Linear input output transformation for the fluctuations

Equation (12) allows to compute the intracavity field fluctuations in terms of the incident ones. In the frequency domain, it leads to an algebraic form :

$$[\delta\alpha(\omega)] = (i\omega+A)^{-1}\, T\, [\delta\alpha^{in}(\omega)] \quad (18)$$

In this equation, the frequency ω is a dimensionless parameter equal to the product of the actual circular frequency by the cavity round trip time τ.

The output fields will be obtained from equation (4):

$$[\delta\alpha^{out}(\omega)] = T\,[\delta\alpha(\omega)] - [\delta\alpha^{in}(\omega)] \qquad (19)$$

This leads to the following expression for the linear input output transformation:

$$[\delta\alpha^{out}(\omega)] = [T\,(i\omega+A)^{-1}\,T - 1]\,[\delta\alpha^{in}(\omega)] \qquad (20)$$

We now introduce the two covariance matrices $V^{in}(\omega)$ and $V^{out}(\omega)$ associated with the input and output fluctuations[12]:

$$V^{in}(\omega) = <[\delta\alpha^{in}(\omega)]\,[\delta\alpha^{in}(\omega)]^{+}> \qquad (21a)$$

$$V^{out}(\omega) = <[\delta\alpha^{out}(\omega)]\,[\delta\alpha^{out}(\omega)]^{+}> \qquad (21b)$$

They are related through the linear transformation:

$$V^{out}(\omega) = [T\,(i\omega+A)^{-1}\,T - 1]\,V^{in}(\omega)\,[T\,(i\omega+A)^{-1}\,T - 1]^{+} \qquad (22)$$

5. Case of input vacuum fluctuations

In the particular case where the input pump field is a coherent state and the signal one the vacuum state, one can write:

$$V^{in}(\omega) = I/2 \qquad (23)$$

where I is the 4x4 identity matrix. Equation (22) becomes:

$$V^{out}(\omega) = (1/2)\,[T\,(i\omega+A)^{-1}\,T - 1]\,[T\,(i\omega+A)^{-1}\,T - 1]^{+} \qquad (24a)$$

$$= I/2 + T\,(i\omega+A)^{-1}\,D\,(-i\omega+A^{+})^{-1}\,T \qquad (24b)$$

with:

$$D = (1/2)\,[T^2 - (i\omega+A) - (-i\omega+A^{+})] \qquad (25)$$

From the relation (3) between the transmission and damping coefficients and the expression (15) of A, one obtains[12]:

$$D = \gamma_1\,\underline{D} \qquad (26a)$$

with:

$$\underline{D} = \begin{vmatrix} 0 & 1+i\Delta_1 & 0 & 0 \\ 1-i\Delta_1 & 0 & 0 & 0 \\ 0 & 0 & 0 & 0 \\ 0 & 0 & 0 & 0 \end{vmatrix} \qquad (26b)$$

Finally, we obtain the variance of the output field fluctuations as:

$$V^{out}(\omega) = I/2 + (1/\gamma_1)\,T\,(ix+\underline{A})^{-1}\,\underline{D}\,(-ix+\underline{A}^{+})^{-1}\,T \qquad (27)$$

with:

$$x = \omega / \gamma_1 \qquad (28)$$

The expressions of the squeezing spectra can be computed from equation (27) by inverting the 4x4 matrix $\underline{A}$.

6. Squeezing spectra

We give the results of the computation for the fluctuations of one particular quadrature $\delta q_i(\omega,\theta)$ (having a phase θ and a frequency ω) of one of the two output fields (i=0 for the pump field, i=1 for the signal field):

$$\delta q_i(\omega,\theta) = W_i(\theta)\ [\delta\alpha^{out}(\omega)] \tag{29a}$$

with:

$$W_1(\theta) = \begin{vmatrix} e^{-i\theta} & e^{i\theta} & 0 & 0 \end{vmatrix} \tag{29b}$$

$$W_0(\theta) = \begin{vmatrix} 0 & 0 & e^{-i\theta} & e^{i\theta} \end{vmatrix} \tag{29c}$$

We denote $S_i(\omega,\theta)$ the variance of $\delta q_i(\omega,\theta)$ which is also the squeezing factor for this quadrature:

$$S_i(\omega,\theta) = W_i(\theta)\ V^{out}(\omega)\ W_i(\theta)^+ \tag{30}$$

From eqs (29) and (30), this squeezing factor can be expressed as:

$$S_i(\omega,\theta) = 1 + [a_i + b_i \cos(2\theta) + c_i \sin(2\theta)]/d \tag{31}$$

The coefficients a_i, b_i, c_i and d, which depend on ω, are calculated from eq. (27). They have the following values:

$$a_1 = 8(1+\Delta_1^2)\ [\gamma^4x^4+2\gamma^2x^2(1-\Delta_0^2)+(1+\Delta_0^2)^2] + 16I_1\ (1+\Delta_1^2)\ (\gamma^2x^2+1+\Delta_0^2) \tag{32a}$$

$$b_1 = a_1 + 4x^2\ [\gamma^4x^4+2\gamma^2x^2(1-\Delta_0^2)+(1+\Delta_0^2)^2] + 16I_1(I_1-\gamma x^2)\ (\gamma^2x^2+1-\Delta_0^2-2\Delta_0\Delta_1) \tag{32b}$$

$$c_1 = 4\Delta_1x^2\ [\gamma^4x^4+2\gamma^2x^2(1-\Delta_0^2)+(1+\Delta_0^2)^2] + 16\Delta_0I_1(1+\Delta_1^2)\ (-\gamma^2x^2+1+\Delta_0^2)$$
$$+ 16I_1(I_1-\gamma x^2)\ [\Delta_1(\gamma^2x^2+1-\Delta_0^2)+2\Delta_0] \tag{32c}$$

$$a_0 = 16I_1(1+\Delta_1^2)\ (\gamma^2x^2+1+\Delta_0^2+2I_1) \tag{32d}$$

$$b_0 = 16I_1(1+\Delta_1^2)\ (\gamma^2x^2+1-\Delta_0^2+2I_1) + 8I_1x^2\ (\gamma^2x^2+1-\Delta_0^2+2\Delta_0\Delta_1) + 32I_1^2\ (I_1-\gamma x^2) \tag{32e}$$

$$c_0 = -32\Delta_0I_1\ (1+\Delta_1^2)\ (1+I_1) + 8I_1x^2\ [\Delta_1(\gamma^2x^2+1-\Delta_0^2)-2\Delta_0] + 32\Delta_1I_1^2\ (I_1-\gamma x^2) \tag{32f}$$

$$d = 4x^2\ [1+\Delta_0^2+(1+\gamma)(2I_1-\gamma x^2)]^2 + [4I_1(1-\Delta_1\Delta_0)+(2I_1-\gamma x^2)^2-x^2(1+\Delta_0^2+4\gamma)]^2 \tag{32g}$$

7. Optimum squeezing

When the phase θ is varied, the minimum and maximum values of $S_i(\omega,\theta)$ are:

$$S_i^{min}(\omega) = [\ d+a_i - (b_i^2+c_i^2)^{1/2}\ ]/d \tag{33a}$$

$$S_i^{max}(\omega) = [\ d+a_i + (b_i^2+c_i^2)^{1/2}\ ]/d \tag{33b}$$

Let us first study the squeezing at zero frequency and at the oscillation threshold as a function of the detunings Δ_0 and Δ_1 of pump and signal fields. The oscillation threshold is the usual one when there is no bistability (Fig.2a, point T) or the upper turning point of the bistability curve (Fig.2b, point D). It can be shown that:

$$S_1^{min}(0) = \frac{2I_1}{1 + \Delta_0^2 + 2I_1} \tag{34a}$$

$$S_0^{min}(0) = \frac{1 + \Delta_0^2}{1 + \Delta_0^2 + 2I_1} \tag{34b}$$

where I_1 is the signal intensity at the threshold (I_1=0 at the oscillation threshold T, I_1 given by eq.(9c) at the bistability turning point D).

In this case, the fluctuations are independent of the ratio γ of the cavity widths for signal and pump fields. It turns out that the sum $[S_1^{min}(0)+S_0^{min}(0)]$ is equal to 1. This striking property appears in Fig.3 where it is clear that the signal and pump fields are best squeezed in complementary regions of the parameter space. The bistable region gives a good squeezing for the pump, whereas the "monostable" region gives perfect squeezing for the signal.

However, the signal field has interesting squeezing properties in the bistable region for some non zero frequencies. The noise spectrum is shown in Fig.4a as a function of the pump intensity on the upper branch of the bistability curve. A large noise reduction is obtained for non zero frequencies. It

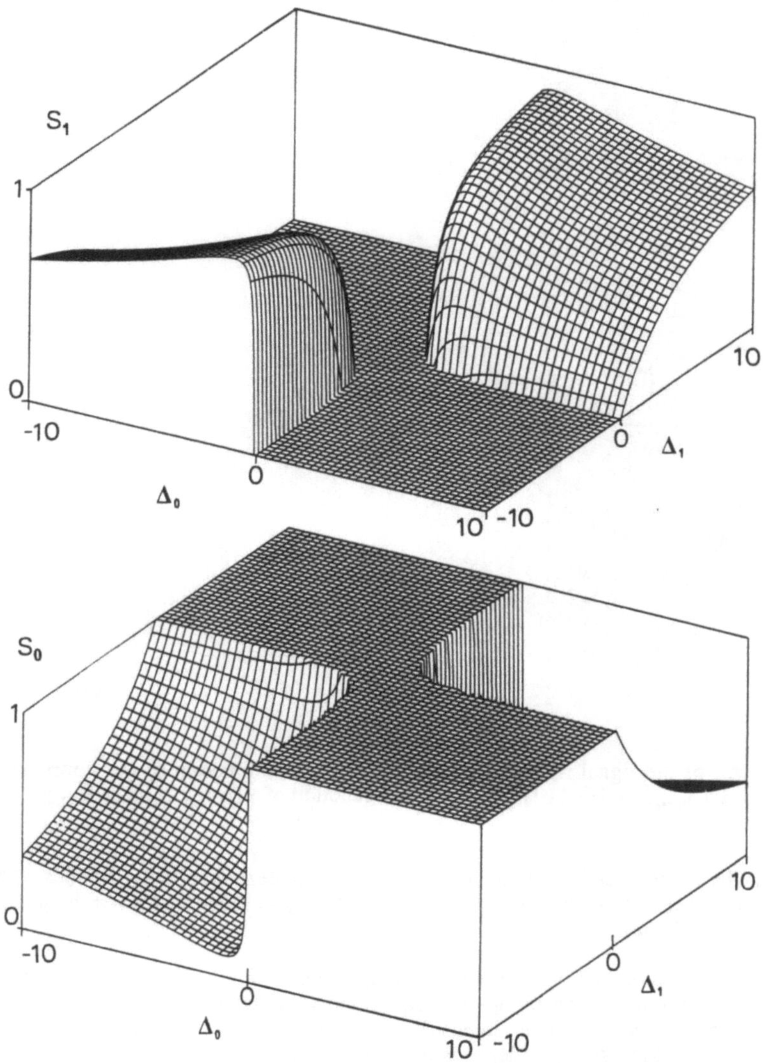

Figure 3: Optimum squeezing at zero frequency $S_i^{min}(0)$, in the vicinity of the threshold as a function of the detunings Δ_0 and Δ_1: (a) for the signal field (i=1); (b) for the pump field (i=0).

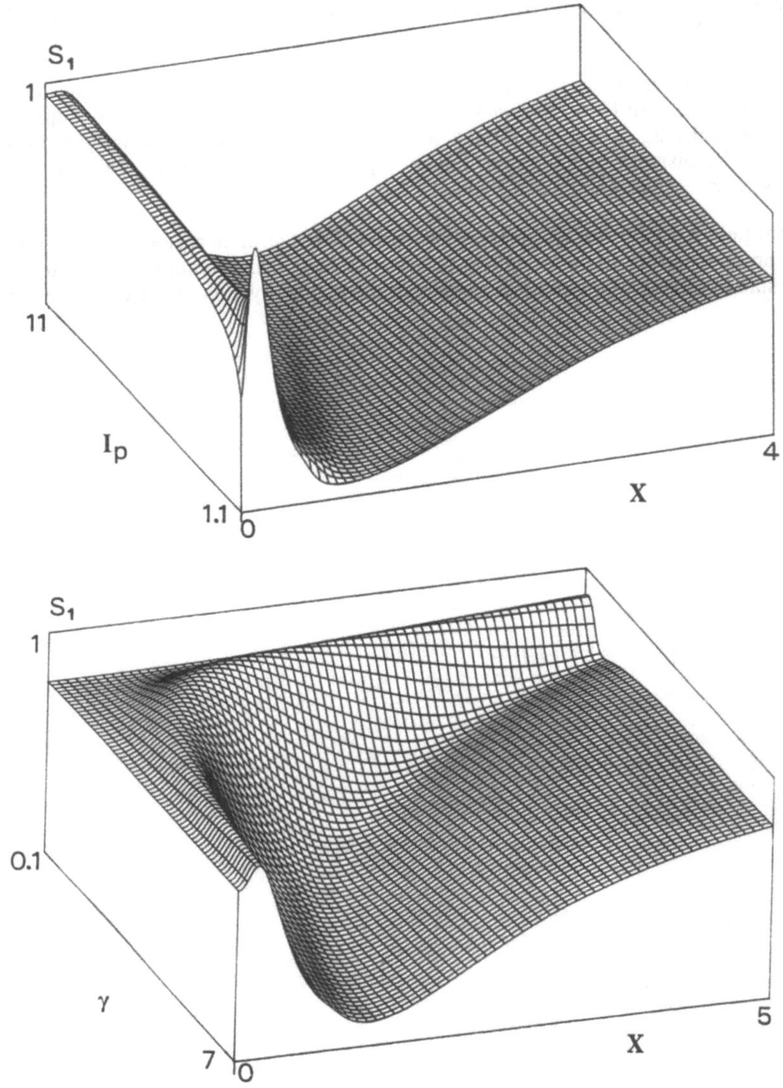

Figure 4: Optimum spectra of the signal field for $\Delta_0 = \Delta_1 = 1.2$: (a) as a function of the normalized pump intensity I_P ($\gamma = 10$, $1.1\ I_{threshold} < I_P < 11\ I_{threshold}$); (b) as a function of γ ($I_p = 2\ I_{threshold}$, $0 < \gamma < 7$).

does not vary much with the pump intensity. On the other hand, this squeezing strongly depends on γ as shown in Fig.4b. Good squeezing is obtained only when the cavity is much better for the pump field than for the signal field (γ large).

The pump field is also squeezed in a broad frequency range (Fig.5). Fig.5a shows the optimum squeezing as a function of the pump intensity I_P. The squeezing is larger close to threshold, but substancial noise reduction remains at higher intensities. Similarly to the preceeding case, the behaviour of the squeezing spectrum strongly depends on γ. To obtain good noise reduction on the pump field, small values of γ are more favorable (much better cavity for the signal field than for the pump field).

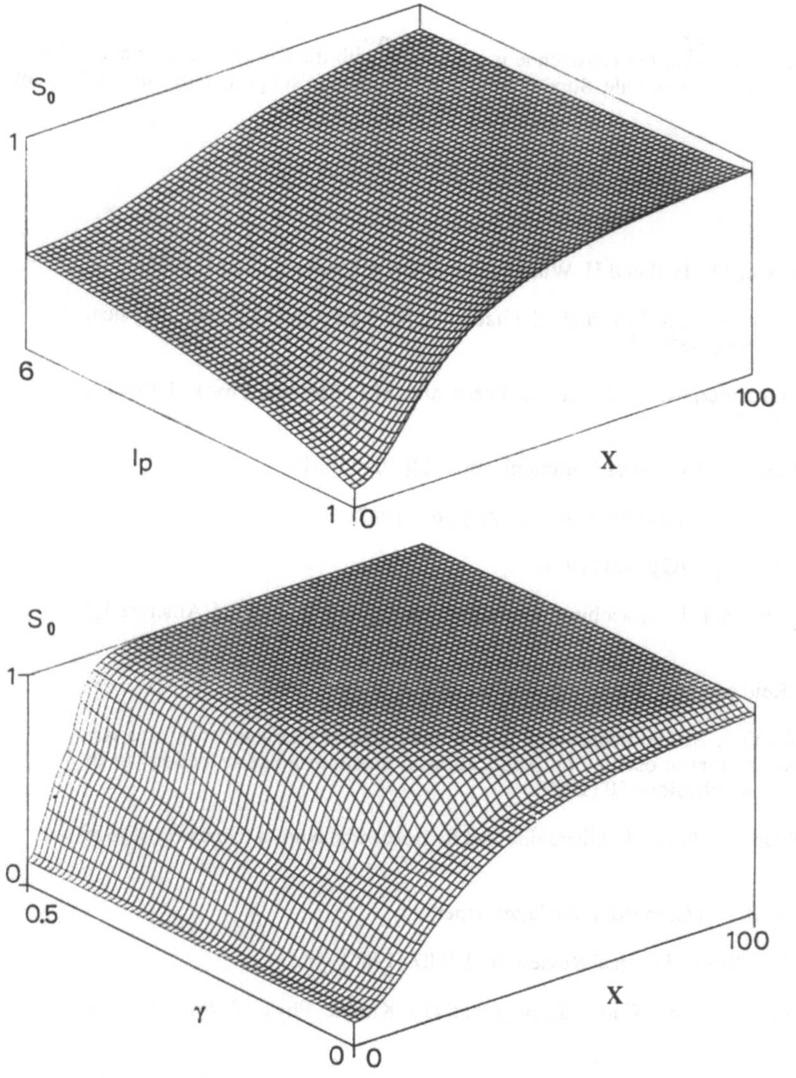

Figure 5: Optimum squeezing spectra of the pump field for $\Delta_0 = 1$, $\Delta_1 = 10$: (a) as a function of the normalized pump intensity I_P ($\gamma = 10^{-3}$, $I_{threshold} < I_P < 6\ I_{threshold}$); (b) as a function of γ ($I_p = 1.2\ I_{threshold}$, $0 < \gamma < 0.5$).

Conclusion

Up to now, most of the work concerning quantum noise reduction in the OPO has been restricted to the resonant case (Δ_0=Δ_1=0). We have shown that new features are predicted in the detuned case. In this paper, we have focused on the bistable region. Other regions of the parameter space are very interesting too, especially in the vicinity of the instability thresholds. This case will be studied elsewhere.

Acknowledgements

The Laboratoire de Spectroscopie Hertzienne is associated with the CNRS, the Université Pierre et Marie Curie and the Ecole Normale Supérieure. This research is supported by the EEC grant ST2J-0278-C

References

1 L.A. Wu, H.J. Kimble, J.L. Hall and H. Wu, Physical Review Letters 57 2520 (1986)

2 A. Heidmann, R. Horowicz, S. Reynaud, E. Giacobino, C. Fabre and G. Camy, Physical Review Letters 59 2555 (1987)

3 R.E. Slusher, P. Grangier, A. La Porta, B. Yurke and M.J. Potasek, Physical Review Letters 59 2566 (1987)

4 G.J. Milburn and D.F. Walls, Optics Communications 39 401 (1981)

5 L. Lugiato and G. Strini, Optics Communications 41 67 (1982)

6 B. Yurke, Physical Review A29 408 (1984)

7 S. Reynaud, C. Fabre and E. Giacobino, Journal of the Optical Society of America B4 1520 (1987)

8 A.S. Lane, M.D. Reid and D.F. Walls, Physical Review A38 788 (1988).

C. Fabre, E. Giacobino, A. Heidmann and S. Reynaud, "Noise characteristics of a non degenerate optical parametric oscillator - Application to quantum noise reduction" to be published in Journal de Physique 50 (1989)

9 L. Lugiato, C. Oldano, C. Fabre, E. Giacobino and R. Horowicz, Nuovo Cimento 10D 959 (1988).

Previous related work is referenced in the latter paper.

10 M.J. Collett and D.F. Walls, Physical Review A32 2887 (1985)

11 M.G. Raizen, L.A. Orozco, M. Xiao, R.L. Boyd and H.J. Kimble, Physical Review Letters 59 198 (1987)

12 S. Reynaud and A. Heidmann, "A semiclassical linear input output transformation for quantum fluctuations" to be published in Optics Communications (1989)

13 S. Reynaud and E. Giacobino, Journal de Physique 49 C2-477 (1988)

Non-linear Beam Splitters in the Production of Squeezed Coherent Light

M.J. Collett and D.F. Walls

Physics Department, University of Auckland, Auckland, New Zealand

The generation of highly squeezed light which is not simply a squeezed vacuum but includes a strong coherent component is clearly a desirable goal. It would, for instance, be a necessary prerequisite of any practical application of squeezed states in the area of optical communications. There is, however, a problem in that direct generation of squeezed light with a coherent component is highly sensitive to such non-idealities as laser noise, small detunings or phase instabilities [1]. The result is that such light turns out to be far more difficult to produce in practice than a similarly squeezed vacuum.

There is available, however, an alternative approach to the production of squeezed coherent light. We may consider first producing a squeezed vacuum and then somehow *adding on* the coherent excitation. The simplest way of implementing this idea is the familiar beam splitter scheme, using a weak coupling beam splitter to mix a small fraction of an intense coherent beam into our squeezed vacuum [2]. The arrangement is depicted in Fig.1. Ignoring phase shifts, we can write

$$a_{out} = \sqrt{1-\eta^2}\; a_{in} + \eta\, b_{in} \;,$$

$$b_{out} = \sqrt{1-\eta^2}\; b_{in} - \eta\, a_{in} \;.$$

If now b_{in} is highly squeezed, a_{in} is a strong coherent beam, and the coupling η is small, the output b_{out} can be a squeezed coherent beam. For instance, if b_{in} is 90% squeezed, and $\eta^2 = 0.1$, then b_{out} is approximately 80% squeezed with a coherent intensity

$$|\langle b_{out}\rangle^2 = 0.1|\langle a_{in}\rangle| \;.$$

If the aim is simply to produce a squeezed coherent beam at any cost, then this scheme does indeed work. It is, however, open to a serious practical objection, namely that it is extremely wasteful of coherent intensity. 90% of the incident coherent light comes out as a_{out}, which is not of interest. And the more squeezed we want b_{out} to be, the worse this problem becomes. The fundamental problem is that since the beam splitter merely

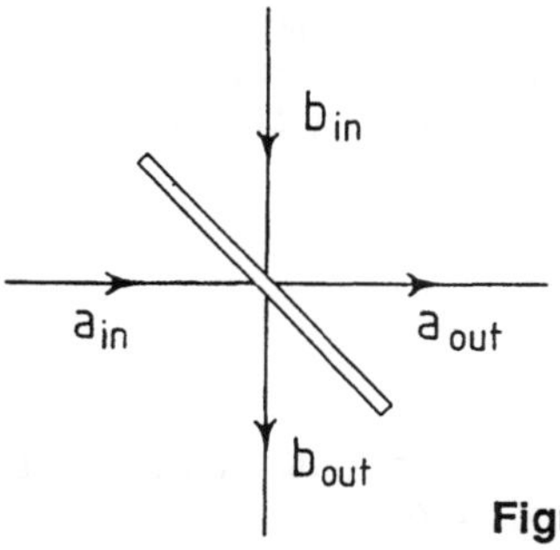

Figure 1 *Simple linear beamsplitter*

Springer Proceedings in Physics, Vol. 41 **Quantum Optics V**
Editors: J.D. Harvey and D.F. Walls © Springer-Verlag Berlin, Heidelberg 1989

performs a simple linear combination of the inputs, the signal-to-noise ratio of the output can be no better than that of the input. The output coherent intensity must thus always be reduced by at least the same factor as the noise.

Is it possible to beat this limit? Can we improve on the simple beam splitter? The answer is - yes, we can, with a **non-linear** beam splitter. Non-linearity can allow the response to small changes in the amplitude (i.e. the noise) to be different from the response to the overall coherent amplitude. This holds out the possibility of putting into one output the coherent part of one input and the noise of the other. To see how this possibility may be realized, we shall consider the non-linear beam splitters which can be constructed using familiar non-linear systems in (double-ended) cavities. The arrangement and notation are indicated schematically in Fig.2.

As a first example, we may consider the degenerate parametric amplifier. This is, of course, familiar as a **generator** of squeezing, but now we wish to consider it in a quite different role. The Hamiltonian describing the non-linear interaction is

$$H_{int} = i\hbar\chi\,(a_1^{+2}\,a_2 - a_2^{+}\,a_1^{2})\ ,$$

where a_2 is the mode operator for the pump and a_1 that for the sub-harmonic. It is convenient to introduce the scaled mode amplitudes

$$\varepsilon_i = \chi\langle a_i\rangle\ .$$

The amplitude and phase quadrature operators are then

$$X_{i+} = \frac{1}{2|\varepsilon_i|}\,(\varepsilon_i^{*}a_i + \varepsilon_i\,a_i^{+})\ ,$$

$$X_{i-} = \frac{1}{2i|\varepsilon_i|}\,(\varepsilon_i^{*}a_i - \varepsilon_i\,a_i^{+})\ .$$

Our principle interest is in the fluctuation operators

$$\Delta\hat{A} = \hat{A} - \langle\hat{A}\rangle\ .$$

If the degenerate paramp is operated above threshold, the classical amplitude of the pump mode in the cavity is independent of the input pump amplitude. If small changes in the pump amplitude (i.e. pump fluctuations) have no effect on the internal pump mode field, they certainly cannot have any effect on the output from the far side of the cavity. Above threshold then, we expect no transmission of pump fluctuations; they should all be reflected. On the other hand, there is certainly significant coherent excitation of the pump mode, which **will** contribute to the transmitted field. In short, if $a_{2,in}^{(L)}$ is a coherent pump and $a_{2,in}^{(R)}$ is a squeezed vacuum, $a_{2,out}^{(R)}$ should be a squeezed coherent field. And this is indeed just the result we find. Performing a linearised fluctuation analysis around the classical steady state [3] gives for the internal pump mode in the frequency domain

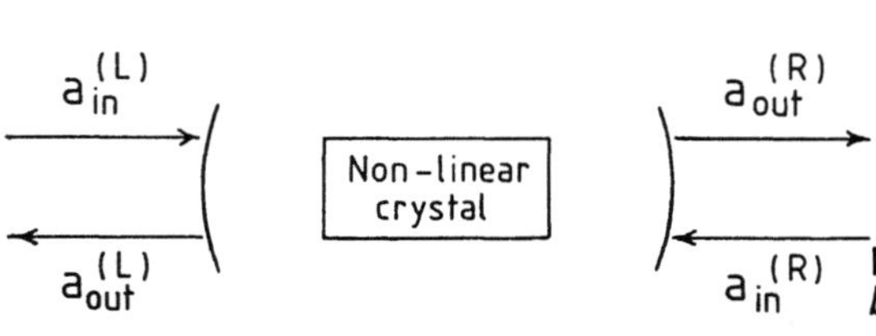

Figure 2 *Non-linear beamsplitter modelled by an intracavity interaction*

$$\Delta X_{2+}(\omega) = \frac{|\varepsilon_1| \sqrt{2\gamma_1}\, \Delta X_{1+,in} + i\omega\left(\sqrt{2\gamma_2^{(L)}}\, \Delta X_{2+,in}^{(L)} + \sqrt{2\gamma_2^{(R)}}\, \Delta X_{2+,in}^{(R)}\right)}{|\varepsilon_1|^2 - i\omega\left(\gamma_2^{(L)} + \gamma_2^{(R)}\right) - \omega^2}$$

where $\gamma_2^{(L)}$ and $\gamma_2^{(R)}$ describe the pump amplitude loss through each end of the cavity and γ_1 the total subharmonic amplitude loss. On resonance, input noise from the pump makes no contribution to amplitude noise in the pump frequency cavity mode. The output of interest is then [4]

$$\begin{aligned}\Delta X_{2+out}^{(R)}(\omega) &= \sqrt{2\gamma_2^{(R)}}\, \Delta X_{2+}(\omega) + \Delta X_{2+,in}^{(R)} \\ &= \left\{2|\varepsilon_1| \sqrt{\gamma_1\gamma_2^{(R)}}\, \Delta X_{1+,in} + 2i\omega \sqrt{\gamma_2^{(L)}\gamma_2^{(R)}}\, \Delta X_{2+,in}^{(L)} \right. \\ &\quad \left. + \left[|\varepsilon_1|^2 - \omega^2 + i\omega(\gamma_2^{(R)} - \gamma_2^{(L)})\right] \Delta X_{2+,in}^{(R)} \right\} \\ &\quad \times \left(|\varepsilon_1|^2 - \omega^2 - i\omega(\gamma_2^{(L)} + \gamma_2^{(R)})\right)^{-1} .\end{aligned}$$

On resonance, this is independent of $\Delta X_{2+,in}^{(L)}$. If the right-hand input is highly squeezed in the appropriate phase, so that $\Delta X_{2+,in}^{(R)}$ may be assumed to be small, the output noise is limited only by the contribution from the subharmonic.

So this system does appear to produce the desired output. The question is whether it does so any more efficiently than the simple beam splitter. The best that the beam splitter can achieve is (with b_{in} perfectly squeezed) on output signal-to-noise ratio equal to the input one

$$\frac{SNR_{out}}{SNR_{in}} = 1 .$$

For the paramp we find

$$\frac{\text{Output coherent intensity}}{\text{Input coherent intensity}} = \frac{2\gamma_2^{(R)}\gamma_1^2 / |\chi|^2}{|\langle a_2^{(L)},in\rangle|^2} ,$$

and, assuming perfect squeezing of $X_{2+,in}^{(R)}$ and vacuum input at the subharmonic

$$\frac{\text{Output noise}}{\text{Input noise}} = \frac{4\gamma_1\gamma_2^{(R)}}{|\varepsilon_1|^2} .$$

Combining these, and using the classical steady state solutions, gives

$$\begin{aligned}\frac{SNR_{out}}{SNR_{in}} &= \frac{\gamma_1\left(\sqrt{2\gamma_2}\, |\chi\langle a_{2,in}\rangle| - \gamma_1\gamma_2\right)}{|\chi\langle a_{2,in}\rangle|^2} \\ &\leq 1 .\end{aligned}$$

So we can only improve on the simple beam splitter if we relax our assumption of vacuum input at the subharmonic and allow this also to be squeezed in the appropriate phase. Such an arrangement is of course possible in principle, but the practical difficulties of matching the phases of all three inputs would presumably be at least as much of a problem as the difficulties noted earlier with a direct generation approach.

Since fluctuations feeding in from the second mode are the problem with the paramp, we may perhaps do better with a single mode non-linear system. Our second example then, will be dispersive bistability via a Kerr medium [5]. In this case, the interaction Hamiltonian is

$$H_{int} = \hbar\chi a^{\dagger 2} a^2$$

and the useful intensity parameter is

$$\varepsilon = 2\chi|\langle a\rangle|^2.$$

The detuning is denoted by δ. A linearised treatment of the fluctuations, just as for the paramp, gives (inside the cavity)

$$\Delta X_\pm(\omega) = \frac{\sqrt{2\gamma^{(R)}}\,\Delta X^{(R)}_{\pm,in} - \sqrt{2\gamma^{(L)}}\,\Delta X^{(L)}_{\pm,in}}{\gamma^{(R)} + \gamma^{(L)} - i(\omega + \delta + 2\varepsilon \pm \varepsilon)}$$

Evidently the internal fluctuations always depend on both inputs, and there is therefore no chance of suppressing the transmitted fluctuations. However, there remains the possibility of interference effects reducing the **reflected** fluctuations. For the output we find

$$\Delta X^{(L)}_{\pm,out}(\omega) = \frac{-2\sqrt{\gamma^{(L)}\gamma^{(R)}}\,\Delta X^{(R)}_{\pm,in} + \left[\gamma^{(L)} - \gamma^{(R)} + i(\omega + \delta + 2\varepsilon \pm \varepsilon)\right]\Delta X^{(L)}_{\pm,in}}{\gamma^{(R)} + \gamma^{(L)} - i(\omega + \delta + 2\varepsilon \pm \varepsilon)}$$

On resonance, this is independent of $\Delta X^{(R)}_{\pm,in}$ provided $\gamma^{(R)} = \gamma^{(L)}$, and either $\delta + |\varepsilon| = 0$ (suppressing reflected **phase** fluctuations) or $\delta + 3|\varepsilon| = 0$ (suppressing reflected **amplitude** fluctuations).

Once again we must compare the reduction in fluctuations with the reduction in amplitude. Using the classical solution for the internal mode amplitude and the appropriate boundary conditions, we find

$$\langle a^{(L)}_{out}\rangle = \frac{-2\gamma^{(L)}\langle a^{(L)}_{in}\rangle}{(\gamma^{(R)} + \gamma^{(L)})^2 + (\delta + \varepsilon)^2} + \langle a^{(L)}_{in}\rangle$$

or

$$|\langle a^{(L)}_{out}\rangle|^2 = \frac{(\delta + \varepsilon)^2}{\gamma^2 + (\delta + \varepsilon)^2}\,|\langle a^{(L)}_{in}\rangle|^2$$

assuming $\gamma^{(R)} = \gamma^{(L)} = \frac{1}{2}\gamma$, and that, as before, there is no coherent input from the right-hand side. The first possibility for suppression of reflected fluctuations was $\delta + \varepsilon = 0$: this also results in no reflected amplitude. The second possibility was $\delta + 3\varepsilon = 0$: in this case the reflected coherent intensity is

$$|\langle a^{(L)}_{out}\rangle|^2 = \frac{4\delta^2}{\gamma^2 + 4\delta^2}\,|\langle a^{(L)}_{in}\rangle|^2 .$$

For $\gamma \leq \delta$ most of the input intensity is thus retained in the output of interest, while the resonant amplitude noise can be provided entirely by the transmitted squeezed input.

We can understand a little more clearly what is actually happening here by considering the optical bistability curve. This has two points, one on each branch, at which the slope of the curve describing the dependence of intracavity on input intensity (i.e. the small signal response) is the same as that of an empty cavity and its resonant component is accordingly completely transmitted. On the upper branch, the mean field (the large signal response) is also the same as for an empty cavity, and the coherent input is transmitted in the same way as its attendant noise. On the lower branch, however, the mean field is much smaller than the empty cavity case - most of the incident coherent light is not being allowed into the cavity at all, but is simply being reflected off the front mirror. The further one is into the bistable regime (i.e. the larger δ/γ) the smaller the lower branch intensity and hence the more efficient the reflection of the coherent beam.

This initial analysis has thus indicated that under ideal operating conditions, a dispersively bistable cavity system is an efficient way of combining squeezed and coherent light. The sensitivity of these results to non-ideal conditions (small parameter changes, laser noise etc.) remains to be investigated.

References

1. P.D. Drummond and M.D. Reid, Phys.Rev. **A37**, 1806 (1988): S.F. Pereira, Min Xiao, H.J. Kimble and J.L. Hall, Phys.Rev.**A38**, 4931 (1988): T.A.B. Kennedy T.B. Anderson and D.F. Walls, preprint (1989).

2. L. Mandel, Phys.Rev.Lett. **49**, 136 (1982): H. Fearn and R. Loudon, Opt.Comm. **64**, 485 (1987).

3. P.D. Drummond, K.J. McNeil and D.F. Walls, Optica Acta **27**, 321 (1980); Optica Acta **28**, 211 (1980); M.J. Collett and D.F. Walls, Phys.Rev. **A32**, 2887 (1985).

4. M.J. Collett and C.W. Gardiner, Phys.Rev. **A30**, 1386 (1984): C.W. Gardiner and M.J. Collett, Phys.Rev.**A31**, 3761 (1985).

5. P.D. Drummond and D.F. Walls, J.Phys.**A13**, 725 (1980).

Photon Statistics of Broadband Squeezed Vacuum*

P. Kumar and J. Huang

Department of Electrical Engineering and Computer Science,
The Technological Institute, Northwestern University,
Evanston, IL 60208, USA

1. Introduction

Within the last few years, sources of squeezed light have become available in many laboratories [1]. The attention of many researchers in the field of quantum optics have now diverted from the generation schemes to the applications and the investigation of some other novel quantum properties of squeezed light. Applications in interferometry [2] and precision measurements [3] have already been demonstrated whereas those in optical communications [4] and spectroscopy [5] are yet to be realized.

Since homodyne detection best utilizes the potential of quadrature squeezed light, it has been the measurement of choice in the generation experiments and the applications that have followed. In homodyne detection of squeezed light, the measurement noise falls below the local-oscillator shot noise when the latter's phase is appropriately chosen. There are other measurements, however, that can be made on quantum states of light, such as squeezed states, which would manifest other nonclassical features of these states. One such measurement is direct detection or photon counting. Direct detection is also the optimum measurement for number states or near-number states of light [6]. Experiments which have generated amplitude squeezed light – states with the squeezed quadrature parallel to the mean field – have demonstrated their generation by measuring noise in direct detection [7]. For amplitude squeezed light, the direct-detection noise falls below the shot-noise level produced by the coherent state of light of the same power.

In addition to the noise in direct detection (variance of the photon count), it is also of interest to measure the full photon-counting distribution of squeezed light. This is because the probability of counting a given number of photons (for example, an odd number of photons in squeezed vacuum) can be exactly zero for squeezed light [8]. Below in Sec. 2, we show that this zero probability is another manifestation of the nonclassical nature of squeezed light.

Squeezed states also exhibit interesting photon-counting distributions which show oscillatory behavior under appropriate conditions [9]. Although these distributions for single-mode degenerate squeezed light are well studied [10], not much

* Supported in part by the National Science Foundation under Grant No. EET-8844003.

Springer Proceedings in Physics, Vol. 41 **Quantum Optics V**
Editors: J.D. Harvey and D.F. Walls

attention has been paid to those of multimode squeezed light of the kind generated by a below-threshold parametric oscillator [11] – the most successful light squeezer to date. Such a source generates squeezing over a finite bandwidth and it is of interest to know how the counting distributions depend upon the photon-counting interval, which can be less than or greater than the inverse-bandwidth time. For classical thermal light, for example, the photon-counting distribution changes from a Bose-Einstein to a Poisson when the counting interval becomes longer than the inverse-bandwidth time [12]. Photon-counting distributions of coherent light, on the other hand, are insensitive to the counting interval and stay Poisson for all counting times.

In a recent paper, using a characteristic function technique [8], we have calculated the photon-counting distribution of squeezed-vacuum light of bandwidth B. We have shown that, similar to the broadband thermal-light case, the counting distributions are sensitive to the counting interval T. In this paper, in Secs. 4 through 8, we present extensive formulae for the photon-counting distributions under both $BT \gg 1$ and $BT \ll 1$ conditions. It is shown that, similar to the single-mode result, the probability for counting an odd number of photons stays identically zero under the former condition. The latter condition is, however, equivalent to having a photodetector with quantum efficiency that is much less than unity. Furthermore in Sec. 8, we also examine conditions under which the oscillatory behavior of the counting distributions might be experimentally observable.

2. No Odd Counts: A Manifestation of the Nonclassical Nature of Squeezed Light

Squeezed light manifests its nonclassical behavior in many ways. The fact that its quadrature-noise variance falls below that of coherent light is one such example. The probability of counting an odd number of photons in a squeezed vacuum can be identically zero [8]. Here we present a mathematical argument which shows that this zero probability is another manifestation of the nonclassical nature of squeezed light.

The density operator $\hat{\rho}$ for an arbitrary state of light can always be expanded in the coherent-state representation as

$$\hat{\rho} = \int \frac{d^2\alpha}{\pi} P(\alpha, \alpha^*) |\alpha >< \alpha| \,, \tag{2.1}$$

provided one allows for the possibility of a highly singular $P(\alpha, \alpha^*)$ [13]. Customarily, a state of light has been called nonclassical if it does not possess a well-behaved, positive-definite P function. That is, the density operator is not expressible as a statistical mixture of almost-classical coherent states with probability density $P(\alpha, \alpha^*)$.

Our argument proceeds by assuming that the density operator associated with the state $|\psi>$ of squeezed vacuum possesses a P representation (2.1) with a well-behaved positive-definite P function. The probability P_n for counting n photons

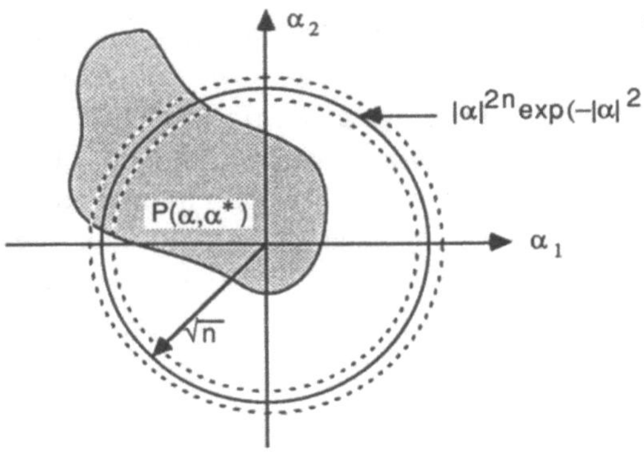

Figure 1: α-plane representation of the double integral in (2.2).

can then be expressed as

$$\begin{aligned} P_n &= |<n|\psi>|^2 = \mathrm{Tr}(\hat{\rho}|n><n|) \\ &= \int \frac{d^2\alpha}{\pi} P(\alpha,\alpha^*)|<n|\alpha>|^2 \\ &= \frac{1}{n!}\int \frac{d^2\alpha}{\pi} P(\alpha,\alpha^*)|\alpha|^{2n}\exp(-|\alpha|^2)\,, \end{aligned} \tag{2.2}$$

which is an integral in the complex α plane as depicted in Fig. 1. The integrand is a product of two well-behaved positive-definite functions $P(\alpha,\alpha^*)$ and $|\alpha|^{2n}\exp(-|\alpha|^2)$; the former is only so because of our assumption. The latter is circularly symmetric, peaks at $|\alpha| = \sqrt{n}$, and decays to 0 at $|\alpha| = 0$ and $|\alpha| = \infty$. Thus P_n can be identically 0 only if $P(\alpha,\alpha^*)$ becomes negative for some values of α contrary to our assumption. Moreover, $P(\alpha,\alpha^*)$ for squeezed vacuum must be highly singular as well because only $P_{n=2k+1} = 0$ and $P_{n=2k} \neq 0$. A similar argument has been given in the past to show that $P(\alpha,\alpha^*)$ must attain negative values in order for the quadrature-noise variance of squeezed light to fall below that of coherent light [14].

3. Mode Expansion of Finite Bandwidth Fields

The positive frequency part of the electric-field operator $\hat{E}^{(+)}(t)$ can be expanded in terms of the relevant field modes over the photon-counting interval $[t-T/2, t+T/2]$ as

$$\hat{E}^{(+)}(t) = \frac{1}{\sqrt{T}}\sum_{\ell=-\infty}^{\infty} \hat{a}_\ell \exp\{-i(\ell 2\pi/T + \omega_0)t\}\,\mathrm{rect}(\frac{t}{T}). \tag{3.1}$$

Here ω_0 is the center radian-frequency of the squeezed light; $\{\hat{a}_\ell\}$ are the modal annihilation operators and the choice of mode functions is dictated by the photon-counting interval [15]; $\mathrm{rect}(t) = 1$ from $t = -1/2$ to $1/2$ and zero elsewhere, and we have suppressed the spatial dependence by assuming a single transverse mode.

For a given bandwidth B, the above expansion can be truncated at $\ell = N \equiv \mathrm{int}(BT/2)$ so that the Fourier transform of $\hat{E}^{(+)}(t)$ can be approximated by

$$\hat{E}^{(+)}(\omega) = \sum_{\ell=-N}^{N} \hat{a}_\ell \operatorname{sinc}\{\omega - (\omega_0 + \ell 2\pi/T)\}, \tag{3.2}$$

where $\operatorname{sinc}(x) \equiv \sin(\pi x)/(\pi x)$, and $2N+1$ is the total number of modes required to describe the squeezed field of bandwidth B over the time interval T. The above approximation is quite accurate for $BT \gg 1$. However, when $BT \ll 1$, although only one mode $\hat{a}_0$ is required to describe the field, it must be mixed with a vacuum mode according to

$$\hat{a}_0 = \sqrt{1-BT}\,\hat{c}_0 + \sqrt{BT}\,\hat{b}_0\,, \tag{3.3}$$

where $\hat{c}_0$ is in the vacuum state and $\hat{b}_0$ is in the state generated by the squeezer [8]. Finally, when $BT \simeq 1$, the detailed functional dependence of the squeezing bandwidth becomes important and one needs to consider the multimode extension of (3.3). Equation (3.3) is equivalent to having a photodetector with quantum efficiency $q = BT$ for detecting the squeezed light. The effect of less-than-unity quantum efficiency will be discussed in Sec. 8.

4. Joint Photon Distribution

Using a characteristic function approach, we have recently calculated the joint probability $P(n_{-N},\ldots,n_0,\ldots,n_N)$ for counting $n_{-N},\ldots,n_0,\ldots,n_N$ photons in modes $-N,\ldots,0,\ldots,N$, respectively [8]. For squeezed vacuum, it is given by

$$\begin{aligned} P(n_{-N},\ldots,n_0,\ldots,n_N) &\equiv\ < n|\hat{\rho}|n > \\ &= f_0(n_0)\prod_{i=1}^{N} f_i(n_i,n_{-i}), \end{aligned} \tag{4.1}$$

$$f_0(n_0) = \int \exp[\operatorname{Re}(\eta_0^2\alpha_0 - |\eta_0|^2\beta_0)]L_{n_0}(|\eta_0|^2)\frac{d^2\eta_0}{\pi}, \tag{4.2}$$

$$\begin{aligned} f_i(n_i,n_{-i}) &= \int \frac{d^2\eta_i d^2\eta_{-i}}{\pi^2}\exp[2\operatorname{Re}(\eta_i\eta_{-i}\alpha_i) - \beta_i(|\eta_i|^2+|\eta_{-i}|^2)] \\ &\quad \cdot L_{n_i}(|\eta_i|^2)L_{n_{-i}}(|\eta_{-i}|^2), \end{aligned} \tag{4.3}$$

where $|n> \equiv \prod_{i=-N}^{N}|n_i>_i$ is the n-mode number state, $\{L_{n_i}\}$ are the Laguerre polynomials, $\alpha_i \equiv \mu_i^*\nu_{-i}^*$, $\beta_i \equiv |\mu_i|^2$, $\mu_{\pm i} \equiv \mu(\omega_0 \pm \omega_i)$, $\nu_{\pm i} \equiv \nu(\omega_0 \pm \omega_i)$, and $\omega_i = 2\pi i/T$. The squeezed-state generating Bogoliubov transformation

$$\hat{a}_i = \mu_i\hat{b}_i + \nu_i\hat{b}_{-i}^\dagger, \tag{4.4}$$

for $i = -N$ to N, relates the output mode $\hat{a}_i$ of the squeezer to the input mode $\hat{b}_i$ via the squeeze parameters μ_i and ν_i. Since we are mainly interested in the Wu et al. [11] experiment, $\hat{b}_i$'s are all in vacuum states. Explicit expressions for $\mu(\omega_0 \pm \omega_i)$ and $\nu(\omega_0 \pm \omega_i)$ in the case of a below-threshold parametric oscillator [11] can be found in the literature [16]. Furthermore, because both $\hat{a}_i$ and $\hat{b}_i$ must obey the bosonic commutators, the following relations exist between μ_i and ν_i [17]: $|\mu_i| = |\mu_{-i}|$, $|\nu_i| = |\nu_{-i}|$, $|\mu_i|^2 = 1 + |\nu_i|^2$, and $\mu_i\nu_{-i} = \mu_{-i}\nu_i$.

The manifestly factored nature of (4.1) implies that photons are emitted independently into modes 0, $\pm 1, \ldots$, and $\pm N$ by the squeezer. The function $f_i(n_i, n_{-i})$ represents the joint probability of counting n_i photons in mode i and n_{-i} photons in mode $-i$. Moreover, below we show that

$$f_i(n_i, n_{-i} \neq n_i) = 0, \tag{4.5}$$

further implying that photons are emitted simultaneously into the paired modes $\pm i$. To prove (4.5), let us consider the joint generating function

$$Q_j(\lambda_1, \lambda_2) = \sum_{n_i=0}^{\infty} \sum_{n_{-i}=0}^{\infty} f_i(n_i, n_{-i})(1-\lambda_1)^{n_i}(1-\lambda_2)^{n_{-i}}, \tag{4.6}$$

from which the joint probability $f_i(n_i = j, n_{-i} = k) \equiv f(j,k)$ can be calculated via

$$f(j,k) = \frac{(-1)^{j+k}}{j!\,k!} \frac{\partial^{j+k}}{\partial\lambda_1^j \partial\lambda_2^k} Q_j(\lambda_1, \lambda_2)|_{\lambda_1=1,\lambda_2=1}\,. \tag{4.7}$$

Substituting (4.3) in (4.6), the two double integrals in the former can be carried out to yield

$$Q_j(\lambda_1, \lambda_2) = [-\lambda_1\lambda_2\beta_i + \beta_i\lambda_1 + \beta_i\lambda_2 + (1-\lambda_1)(1-\lambda_2)]^{-1}\,. \tag{4.8}$$

By taking the various derivatives, it is easily shown that

$$f(j,k) = \frac{|\nu_i|^{2j}}{(|\nu_i|^2+1)^{j+1}} \delta_{jk}\,. \tag{4.9}$$

5. Photon Statistics of the Degenerate Mode

Let us first consider the case when $BT = 1$. This condition can be met, in principle, by matched filtering the squeezed light prior to photodetection. In this case only the $i = 0$ mode is required to describe the field. The photon distribution is then given by (4.2). The form of (4.2) is not very convenient for calculational purposes. Further progress is made by calculating the degenerate-mode generating function

$$Q_d(\lambda) = \sum_{n=0}^{\infty} f_0(n_0 = n)(1-\lambda)^n. \tag{5.1}$$

Substituting (4.2) in (5.1), the double integral can be evaluated and we obtain

$$\begin{aligned} Q_d(\lambda) &= (-\lambda^2|\nu_0|^2 + 2\lambda|\nu_0|^2 + 1)^{-1/2} \\ &= (1+|\nu_0|^2)^{-1/2}\left[1 - \frac{|\nu_0|^2}{|\nu_0|^2+1}(1-\lambda)^2\right]^{-1/2} \\ &= \sum_{n=0}^{\infty} \frac{(2n-1)!!}{2^n\,n!} \frac{|\nu_0|^{2n}}{(|\nu_0|^2+1)^{n+1/2}}(1-\lambda)^{2n}. \end{aligned} \tag{5.2}$$

The various probabilities can be straightforwardly read from the above expansion and we obtain the well known photon distribution for a single-mode degenerate squeezed-vacuum state [9,10]:

$$\begin{aligned} P_{n=2k+1} &= 0, \\ P_{n=2k} &= \frac{(2k-1)!!}{2^k\, k!} \frac{|\nu_0|^{2k}}{(1+|\nu_0|^2)^{k+1/2}} \\ &= \frac{(2k-1)!!}{2^k\, k!} \frac{\bar{n}^k}{(\bar{n}+1)^{k+1/2}}, \end{aligned} \tag{5.3}$$

where $\bar{n} = |\nu_0|^2$ is the average number of photons counted during T. From the above equation, the probability of counting an odd number of photons is clearly zero; implying that photons are emitted in pairs by the squeezer.

6. Photon Statistics of a Coupled Nondegenerate Mode

A coupled nondegenerate mode can be optically filtered, for example, by the apparatus sketched in Fig. 2. The filter-cavity bandwidth B_c is so chosen ($= 1/T$) as to pass the degenerate mode and to reflect the remaining nondegenerate modes. Furthermore, by appropriately matched filtering the reflected nondegenerate modes, a single coupled nondegenerate mode-pair can be photodetected with a wideband detector. Such a detector will count the total number of photons $z_\ell = n_\ell + n_{-\ell}$ in the paired set of modes. From (4.3), the probability distribution $P_\ell(z_\ell)$ is given by

$$\begin{aligned} P_\ell(z_\ell) &= \int \frac{d^2\eta_\ell}{\pi}\frac{d^2\eta_{-\ell}}{\pi} \exp\{2\mathrm{Re}(\eta_\ell\eta_{-\ell}\alpha_\ell) - \beta_\ell(|\eta_\ell|^2 + |\eta_{-\ell}|^2)\} \\ &\quad \cdot L^1_{z_\ell}(|\eta_\ell|^2 + |\eta_{-\ell}|^2). \end{aligned} \tag{6.1}$$

Once again, further progress is made by calculating the generating function

$$Q_n(\lambda) = \sum_{n=0}^{\infty} P_\ell(z_\ell = n)(1-\lambda)^n \tag{6.2}$$

for the ℓth coupled nondegenerate mode. It is easily calculated to be

$$Q_n(\lambda) = (-\lambda|\nu_\ell|^2 + 2\lambda|\nu_\ell|^2 + 1)^{-1}$$

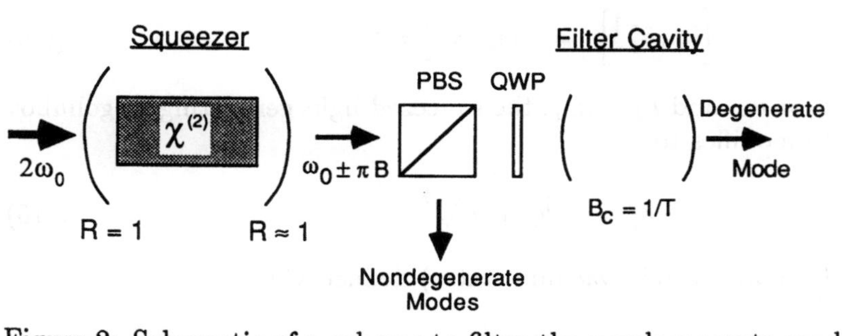

Figure 2: Schematic of a scheme to filter the nondegenerate modes.

$$= (1+|\nu_\ell|^2)^{-1}\left[1-\frac{|\nu_\ell|^2}{|\nu_\ell|^2+1}(1-\lambda)^2\right]^{-1}$$

$$= \sum_{n=0}^{\infty}\frac{|\nu_\ell|^{2n}}{(|\nu_\ell|^2+1)^{n+1}}(1-\lambda)^{2n}. \qquad (6.3)$$

Reading coefficients in the above expansion, we obtain the following photon distribution for a coupled nondegenerate mode:

$$\begin{aligned} P_{n=2k+1} &= 0, \\ P_{n=2k} &= \left(|\nu_\ell|^2\right)^k / \left(1+|\nu_\ell|^2\right)^{k+1} \\ &= (\bar{n}/2)^k / (\bar{n}/2+1)^{k+1}, \end{aligned} \qquad (6.4)$$

where $\bar{n}$ is the average number of counted photons. The above probability distribution is similar to the Bose-Einstein distribution except that the probability of odd counts is zero. We call the above distribution a *modified* Bose-Einstein distribution.

The nondegenerate photon distribution (6.4) is quite different from the degenerate photon distribution (5.3). The difference is due to the fact that nondegenerate squeezing of the ℓth mode arises from two independent identically squeezed double-sideband modes with annihilation operators

$$\hat{a}_\ell^+ = (\hat{a}_\ell + \hat{a}_{-\ell})/\sqrt{2}, \qquad (6.5)$$

$$\hat{a}_\ell^- = (\hat{a}_\ell - \hat{a}_{-\ell})/i\sqrt{2}. \qquad (6.6)$$

The positive-frequency part of the electric-field operator can be expressed in terms of these double-sideband modes as follows [cf. (3.1)]:

$$\begin{aligned} \hat{E}^{(+)}(t) &= \left[\sum_{\ell=1}^{\infty}\left\{\hat{a}_\ell \exp(-i\frac{2\pi\ell}{T}t) + \hat{a}_{-\ell}\exp(i\frac{2\pi\ell}{T}t)\right\} + \hat{a}_0\right]\exp(-i\omega_0 t) \\ &= \left[\sum_{\ell=1}^{\infty}\left\{\hat{a}_\ell^+\sqrt{2}\cos(\frac{2\pi\ell}{T}t) + \hat{a}_\ell^-\sqrt{2}\sin(\frac{2\pi\ell}{T}t)\right\} + \hat{a}_0\right]\exp(-i\omega_0 t). \end{aligned} \qquad (6.7)$$

The temporal mode function associated with $\hat{a}_\ell^+$ is $\sqrt{2}\cos(2\pi\ell t/T)$ whereas that with $\hat{a}_\ell^-$ is $\sqrt{2}\sin(2\pi\ell t/T)$. The $\hat{a}_\ell^+$ and $\hat{a}_\ell^-$ obey

$$\left[\hat{a}_\ell^\pm, \hat{a}_\ell^{\pm\dagger}\right] = 1, \qquad (6.8)$$

$$\left[\hat{a}_\ell^+, \hat{a}_\ell^{-\dagger}\right] = \left[\hat{a}_\ell^+, \hat{a}_\ell^-\right] = 0. \qquad (6.9)$$

If we assume that $\nu_\ell = \nu_{-\ell}$ and $\mu_\ell = \mu_{-\ell}$, the squeezed-light generating Bogoliubov transformation (4.4) modifies to

$$\hat{a}_\ell^\pm = \mu_\ell \hat{b}_\ell^\pm + \nu_\ell \hat{b}_\ell^{\pm\dagger} \qquad (6.10)$$

with the new input modes (also in vacuum states) defined via

$$\hat{b}_\ell^+ = (\hat{b}_\ell + \hat{b}_{-\ell})/\sqrt{2}, \tag{6.11}$$

$$\hat{b}_\ell^- = (\hat{b}_\ell - \hat{b}_{-\ell})/i\sqrt{2}. \tag{6.12}$$

The characteristic function and the joint probability-distribution function for the ℓth coupled-nondegenerate mode factorize when expressed in the new variables. The latter is given by

$$f_\ell(n_\ell, n_{-\ell}) = f_{\ell^+}(n_\ell^+) \cdot f_{\ell^-}(n_\ell^-) \tag{6.13}$$

$$f_{\ell^\pm}(n_\ell^\pm) = \int \exp[\mathrm{Re}(\eta_\ell^{\pm 2}\alpha_\ell - |\eta_\ell^\pm|^2\beta_\ell)] L_{n_\ell^\pm}(|\eta_\ell^\pm|^2) \frac{d^2\eta_\ell^\pm}{\pi}, \tag{6.14}$$

where $\alpha_\ell = \mu_\ell^* \nu_\ell^*$, $\beta_\ell = |\mu_\ell|^2$, and n_ℓ^+ are the emitted photons in the + mode whereas n_ℓ^- are those in the − mode. The factored form of the joint probability-distribution function tells us that the coupled nondegenerate mode consists of two identically squeezed ± modes. Identical squeezing results from our assumption that $\nu_\ell = \nu_{-\ell}$ and $\mu_\ell = \mu_{-\ell}$.

Due to the similarity between (6.14) and (4.2), the photon distributions of the ℓth + and − modes are identical to the degenerate squeezed-mode photon distribution (5.3) and are given by

$$P_{n_\ell^\pm = 2k+1} = 0,$$

$$P_{n_\ell^\pm = 2k} = \frac{(2k-1)!!}{(2k)!!} \frac{|\nu_\ell|^{2k}}{(|\nu_\ell|^2 + 1)^{k+1/2}}. \tag{6.15}$$

Furthermore, the factored form of the joint probability-distribution function (6.13) implies that n_ℓ^+ and n_ℓ^- are independent random variables. The distribution for the total photon count $z_\ell = n_\ell^+ + n_\ell^-$ can then be calculated via

$$P_{z_\ell} = P_{n_\ell^+} * P_{n_\ell^-}, \tag{6.16}$$

where $*$ stands for convolution. Substituting (6.15) in (6.16), we get the coupled nondegenerate mode photon distribution:

$$P_{z_\ell = n} = \sum_{m=0}^{n} P_{n_\ell^+ = m} P_{n_\ell^- = n-m}$$

$$= \begin{cases} 0 & \text{for} \quad n = 2k+1 \\ |\nu_\ell|^{2k}/(|\nu_\ell|^2+1)^k & \text{for} \quad n = 2k\,. \end{cases} \tag{6.17}$$

Finally, the joint probability-distribution function $P(n_{-N}, \ldots, n_0, \ldots, n_N)$ for multimode squeezed light [cf. (4.1)] completely factorizes in the double-sideband-mode photon variables n_0, $n_1^\pm$, …, $n_\ell^\pm$, …, $n_N^\pm$. This suggests that the double-sideband modes are the natural modes for the description of multimode squeezed light.

7. Multimode Photon Statistics

For long counting times so that $BT \gg 1$, a wideband photodetector will count all the photons in the $2N+1$ modes. Defining a total count variable $n = \sum_{i=0}^{N} z_i$, equation (4.1) leads to the following photon distribution for multimode squeezed-vacuum light:

$$P_n = \int \exp\{\sum_{i=-N}^{N} \mathrm{Re}(\eta_i \eta_{-i} \alpha_i - \beta_i |\eta_i|^2)\} L_n^{2N}(\sum_{i=-N}^{N} |\eta_i|^2) \prod_{i=-N}^{N} \frac{d^2\eta_i}{\pi}. \tag{7.1}$$

The multimode generating function is given by

$$\begin{aligned} Q_m(\lambda) &= \sum_{n=0}^{\infty} P_n (1-\lambda)^n \\ &= (-\lambda^2|\nu_0|^2 + 2\lambda|\nu_0|^2 + 1)^{-1/2} \prod_{\ell=1}^{N} (-\lambda^2|\nu_\ell|^2 + 2\lambda|\nu_\ell|^2 + 1)^{-1}. \end{aligned} \tag{7.2}$$

In the special case when $|\nu_\ell| = |\nu|$ for all ℓ, we obtain

$$\begin{aligned} Q_m(\lambda) &= (-\lambda^2|\nu|^2 + 2\lambda|\nu|^2 + 1)^{-(N+1/2)} \\ &= (1+|\nu|^2)^{-(N+1/2)} \left[1 - \frac{|\nu|^2}{|\nu|^2+1}(1-\lambda)^2\right]^{-(N+1/2)} \\ &= \sum_{n=0}^{\infty} \frac{[2(N+n)-1]!!}{2^n\, n!\,(2N-1)!!} \frac{|\nu|^{2n}}{(|\nu|^2+1)^{N+n+1/2}} (1-\lambda)^{2n}. \end{aligned} \tag{7.3}$$

Thus the probability distribution for multimode squeezed-vacuum light is:

$$\begin{aligned} P_{n=2k+1} &= 0, \\ P_{n=2k} &= \frac{[2(N+k)-1]!!}{2^k\, k!(2N-1)!!} \frac{|\nu|^{2k}}{(1+|\nu|^2)^{N+k+1/2}}. \end{aligned} \tag{7.4}$$

We conclude, therefore, that the probability of counting an odd number of photons in squeezed vacuum is identically *zero* even when the counting interval is much *longer* than the inverse-bandwidth time.

The photon distribution simplifies a great deal in the limit $N \to \infty$, yielding

$$\begin{aligned} P_{n=2k+1} &= 0, \\ P_{n=2k} &\overset{N\to\infty}{=} \frac{\exp(-\bar{n}/2)[\bar{n}/2]^k}{k!}, \end{aligned} \tag{7.5}$$

where $\bar{n} = (2N+1)|\nu|^2$ is, once again, the average number of photons counted during T. We call the above distribution a *modified* Poisson distribution because the probability of even counts follows a Poisson formula whereas that of odd counts is zero. In Fig. 3, we have plotted this distribution for various values of $\bar{n}$.

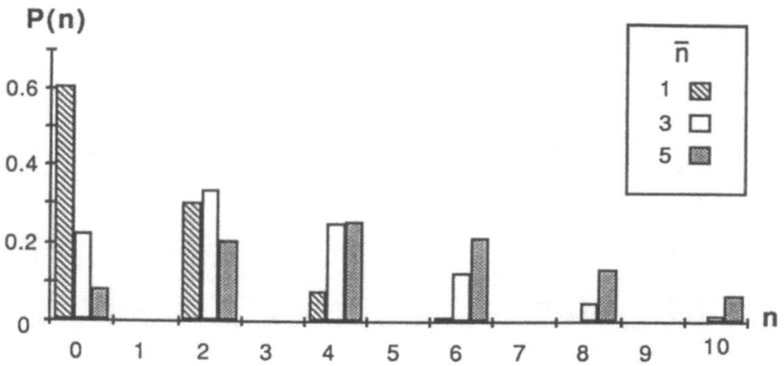

Figure 3: Photon distribution of broadband squeezed vacuum in the limit $BT \gg 1$.

8. Effect of Nonunity Quantum Efficiency

Nonunity quantum efficiency q of the photodetector is easily taken into account by coupling each mode to a loss reservoir represented by

$$\hat{d}_i = q^{\frac{1}{2}}\hat{a}_i + (1-q)^{\frac{1}{2}}\hat{c}_i, \tag{8.1}$$

where $\{\hat{c}_i\}$ are in vacuum states and we have assumed frequency independent q. As pointed out in Sec. 3, the $BT \ll 1$ case can be treated in the same way by choosing $q = BT$. Substituting (4.4) in (8.1), and following the characteristic function technique that was used to arrive at (7.2), we obtain the following generating function in the presence of loss or nonunity quantum efficiency:

$$Q_q(\lambda) = (-q^2\lambda^2|\nu_0|^2 + 2q\lambda|\nu_0|^2+1)^{-1/2}\prod_{\ell=1}^{N}(-q^2\lambda^2|\nu_\ell|^2 + 2q\lambda|\nu_\ell|^2+1)^{-1}. \tag{8.2}$$

The details will be presented elsewhere; however, we note that $q \neq 1$ modifies P_n in a drastic way because the pair-correlated photons are lost independently. In the limit $q \ll 1$, (5.3), (6.4), and (7.5) modify to

$$P_k = \frac{(2k-1)!!}{2^k\, k!}\frac{(2\bar{n})^k}{(2\bar{n}+1)^{k+1/2}}, \tag{8.3}$$

$$P_k = \bar{n}^k/(\bar{n}+1)^{k+1}, \tag{8.4}$$

$$P_k = \exp(-\bar{n})\,\bar{n}^k/k!, \tag{8.5}$$

respectively. That is, the modified Bose-Einstein and Poisson distributions reduce to ordinary Bose-Einstein and Poisson distributions in the presence of large losses (detection or any other).

For the purpose of illustration in Fig. 4, we have plotted the photon distribution of single-mode squeezed vacuum for various values of q. We have chosen $|\nu_0|^2 = 1$, a value realizable by the current experiments which would give about 8dB of squeezing

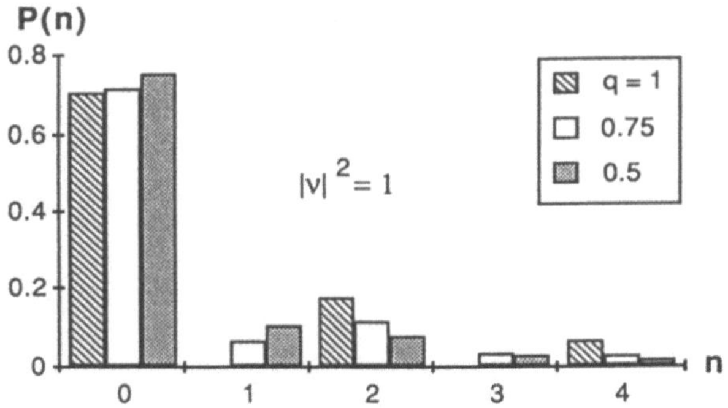

Figure 4: Effect of nonunity quantum efficiency on the photon statistics of single-mode squeezed light.

in the absence of detection losses [11]. We see that the oscillation in the photon number is observable even with $q = 0.75$, a value obtainable with state of the art photodetectors [18].

References

[1] For references see the squeezed states feature issue: J. Opt. Soc. Am. B **4**, (1987).

[2] M. Xiao, L.-A. Wu, and H. J. Kimble, Phys. Rev. Lett. **59**, 278 (1987); P. Grangier, R. E. Slusher, B. Yurke, and A. LaPorta, Phys. Rev. Lett. **59**, 2153 (1987).

[3] M. Xiao, L.-A. Wu, and H. J. Kimble, Opt. Lett. **13**, 476 (1988).

[4] H. P. Yuen and J. H. Shapiro, IEEE Trans. Info. Th. **IT-25**, 179 (1978).

[5] B. Yurke and E. A. Whittaker, Opt. Lett. **12**, 236 (1987).

[6] H. P. Yuen, Opt. Lett. **12**, 789 (1987); H. P. Yuen, Phys. Rev. Lett. **56**, 2176 (1986); G. Björk and Y. Yamamoto, Phys. Rev. A **37**, 125 (1988).

[7] A. Heidmann, R. J. Horowicz, S. Reynaud, E. Giacobino, C. Fabre, and G. Camy, Phys. Rev. Lett. **59**, 2555 (1987); S. Machida, Y. Yamamoto, and Y. Itaya, Phys. Rev. Lett. **58**, 1000 (1987).

[8] J. Huang and P. Kumar, submitted to Phys. Rev. A.

[9] R. S. Bondurant, B. S. Thesis, MIT, 1978, unpublished; W. Schleich and J. A. Wheeler, Nature **326**, 574 (1987).

[10] H. P. Yuen, Phys. Rev. A **13**, 2226 (1976); W. Schleich and J. A. Wheeler, J. Opt. Soc. Am. B **4**, 1715 (1987); A. Vourdas and R. M. Weiner, Phys. Rev. A **36**, 5866 (1987); G. S. Agarwal and G. Adam, Phys. Rev. A **38**, 750 (1988).

[11] L.-A. Wu, H. J. Kimble, J. L. Hall, and H. Wu, Phys. Rev. Lett. **57**, 2520 (1986).

[12] R. M. Gagliardi and S. Karp, *Optical Communications* (Wiley, New York, 1976), p. 97-98.

[13] R. J. Glauber, Phys. Rev. **131**, 2766 (1963); E. C. G. Sudarshan, Phys. Rev. Lett. **10**, 277 (1963).

[14] D. F. Walls, Nature **306**, 141 (1983).

[15] C. W. Helstrom, *Quantum Detection and Estimation Theory* (Academic, New york, 1978), p. 202-204.

[16] C. W. Gardiner and C. M. Savage, Opt. Commun. **50**, 173 (1985).

[17] B. Yurke, Phys. Rev. A **32**, 300 (1985).

[18] K. Kikuchi, T. Okoshi, and A. Hirose, Electron. Lett. **21**, 801 (1985); M. G. Stapelbroek, J. Opt. Soc. Am. A **4**, P117 (1987).

Nearly Transparent Optical Networks

P. Tombesi

Dipartimento di Fisica, Università di Roma "La Sapienza",
P. le A. Moro 2, I-00185 Roma, Italy

In a recent letter Yuen[1] showed that a photon-number amplifier and a parametric amplifier may be used in a local area network to get a transparent optical line. In order to carry information down a line without degrading the signal-to-noise ratio he suggested that such a device should in principle be used.

In the present paper the properties of the radiation at the end of a line with losses, where N phase sensitive linear amplifiers have been inserted, will be analysed and discussed. The argument is similar to the one used by Loudon[2] in studing the noise properties of a chain of phase-sensitive optical amplifiers, however, in this paper the equivalence between the squeezing of the loss and the amplification of the signal in an optical waveguide will be shown.

Let us consider an isolated single-mode optical waveguide, carrying a monochromatic signal at frequency ω along the +z direction; our first task is to obtain a *spatial* differential equation for the propagation of the field throughout the medium. One way to get such a spatial differential equation for the propagating mode is to replace the time variable t with z/v where v is the phase velocity in the medium[3]. The validity of such a procedure, however, could be questionable[4] expecially for a dispersive medium. Thus, restricting our attention to a single mode of the electromagnetic field, we assume there is no dispersion in our actual medium. In order to include the absorption phenomenologically we use a simplified version of the Caves and Crouch[5] procedure. We replace the actual medium by a sequence of slabs of ideal nonlinear medium separated by beamsplitters. By ideal medium it is meant a medium which has no absorption, no dispersion and uniform refraction index. Reflection at the beamsplitters models a linear loss mechanism that in the actual medium is due to the absorption[5]. The beamsplitter scattering matrix relates the propagating signal entering the actual medium at $z+\Delta z$ to the signal leaving it at z. If we denote by γ the absorption coefficient - loss per unit length - of the actual medium, we get $\gamma\,\Delta z$ is the loss within the slab Δz at the signal frequency. The transformation performed by the beamsplitter on the signal mode, described by the boson annihilation operator a(z), is therefore

$$a(z+\Delta z) = (1-\gamma\Delta z)^{1/2}\, a(z) + (\gamma\Delta z)^{1/2}\, c_i. \tag{1}$$

The annihilation operator c_i accounts for fluctuations associated with the medium and are due to absorption. Indeed, to conserve the energy, or to preserve the unitarity of the beamsplitter scattering matrix, a free field must enter into a second input port of the beamsplitter. The operator c_i satisfies the canonical commutation relation $[c_i, c_j^+] = \delta_{ij}$ and the subscript i indicates the position of the beamsplitter at z. It is useful to define the continuum operator c(z) by the relation

$$(\Delta z)^{1/2}\, c_i = \int_z^{z+\Delta z} d\zeta\; c(\zeta) \quad . \tag{2}$$

It is easy to check that the operators $c(\zeta)$ obey the continuum commutation relation

$$[c(\zeta), c^+(\zeta')] = \delta(\zeta-\zeta') \quad . \tag{3}$$

We now consider the limit $\Delta z \to 0$ in Eq.(1) and we get

Springer Proceedings in Physics, Vol. 41 **Quantum Optics V**
Editors: J.D. Harvey and D.F. Walls

$$a(z+\Delta z) - a(z) = -\frac{1}{2}\gamma\Delta z\, a(z) + \gamma^{1/2}(\Delta z)^{1/2} c_i + O((\Delta z)^2) \qquad (4)$$

thus, by making use of Eq.(2) we obtain

$$\frac{da(z)}{dz} = -\frac{1}{2}\gamma a(z) + \gamma^{1/2} c(z) \qquad (5)$$

where we used

$$\lim_{\Delta z \to 0} \frac{1}{\Delta z} \int_z^{z+\Delta z} d\zeta\, c(\zeta) = c(z) \ . \qquad (6)$$

The first term in Eq.(5) describes the attenuation due to absorption and the second term represents the Langevin-force operator associated with absorption fluctuations. Due to Eq.(3), all Langevin operators commute with one another at different positions. By considering an ideal nonlinear medium with second-order susceptibility tensor $\chi^{(2)}$ within the slabs, the drift term only would be different, as was shown in ref.(5) and, for the degenerate case, we get

$$\frac{da(z)}{dz} = -\frac{1}{2}\gamma a(z) - g\, a^{+}(z) + \gamma^{1/2} c(z) \qquad (7)$$

where g, the nonlinear gain coefficient of the medium, is proportional to the second-order susceptibility tensor/5/ and to an external pump assumed very intense to be considered classical. The use of a $\chi^{(2)}$ medium keeps away/6/ the Brillouin light scattering/7/ of the strong pump into the sidebands of the pump beam, from the signal frequency which is at half the pump frequency.

Our next step is to introduce the correlation function for c(z), averaging on a "bath" state. We can assume for simplicity $\langle c_i \rangle = \langle c_i^{+} \rangle = 0$ and the expectation $\langle \cdots \rangle$ is performed on the bath state. To get the usual dissipation we may assume

$$\langle c_i^{+} c_i \rangle = n \ ; \ \langle c_i c_i^{+} \rangle = n + 1 \ ; \qquad \langle c_i c_i \rangle = \langle c_i^{+} c_i^{+} \rangle = 0 \ . \qquad (8)$$

Assumptions (8) correspond to averaging on an ensemble with "ideal thermal white-noise"/8/, where n represents the average number of photons at a given frequency. In terms of continuum operators Eq.(8) becomes

$$\langle c^{+}(z)\, c(z') \rangle = n\ \delta(z-z') \ , \quad \langle c(z)\, c^{+}(z') \rangle = (n+1)\ \delta(z-z') \qquad (9)$$

and the quantum operator c(z) corresponds to an input field, entering the beamsplitter, with random uniformly distributed phase fluctuations. After the landmark experiment of Slusher *et al.*,/9/ we know that the theoretically predicted/10/ states, with anisotropic distribution of phase fluctuations, can be produced and a great degree of squeezing/11/ can also be achieved./12/ Then, if c could be in a squeezed vacuum state, beside the correlation functions for the Langevin force operators Eqs.(8), there will also be/8/

$$\langle c(z)\, c(z') \rangle = m\, e^{2i\theta}\, \delta(z-z') \ , \quad \langle c^{+}(z)\, c^{+}(z') \rangle = m\, e^{-2i\theta}\ \delta(z-z') \qquad (10)$$

where m is a real positive number and θ a phase angle. These relations exhibit the phase-sensitive nature of the Langevin force. The values of m and n, however, cannot be independent, as a consequence of the Heisenberg principle, and consequently

$$m^2 \le n\,(n+1) \ . \qquad (11)$$

The characteristics of the vacuum state entering the port c are, however, determined by the scattering loss which is phase independent, and hence cannot be squeezed. Nevertheless, we will show that the squeezing of the scattering loss can be achieved by amplifying the signal down the line. In order to show this let us consider now a transmission system associated with a homodyne detection. In this case the information is transmitted only in one of the quadrature of the electromagnetic field, defined as $A(\theta) = \frac{a\, e^{-\theta} + a^{+}\, e^{i\theta}}{2}$ and the signal-to-noise ratio (SNR) reads

$$SNR = \frac{<A(\theta)>^2}{<\Delta A(\theta)^2>} \tag{12}$$

where $<\Delta A(\theta)^2> = <A(\theta)^2> - <A(\theta)^2>$.

The loss can be modelled[13,14] by introducing an ideal beam-splitter with reflection coefficient Γ_o ($0 \leq \Gamma_o \leq 1$). If we consider only the case of a single frequency ω, the input signal a_{in} is transformed by the beam-splitter into the output signal

$$a_{out} = \sqrt{1-\Gamma_o}\, a_{in} + \sqrt{\Gamma_o}\, c \tag{13}$$

where c is an annihilation operator accounting for absorption fluctuations, entering the unused port of the beam-splitter, which are associated with the medium. The operator c satisfies the canonical commutation relationship $[c,c^+] = 1$. The radiation entering this port of the beam-splitter has the same statistical properties, of c_i in Eq.(8). The third of Eq.(8) reflects the phase independence of the scattering loss.

The loss has the effect of degrading SNR and at the exit port of the beam splitter, for $\theta = 0$ Eq.(12) becomes

$$SNR_{out} = \frac{<A_{out}(0)>^2}{<\Delta A_{out}(0)^2>} = \frac{<A_{in}(0)>^2}{<\Delta A_{in}(0)^2> + \frac{\Gamma_o}{1-\Gamma_o}<\Delta C(0)^2>} \tag{14}$$

where $<\Delta C(0)^2> = <C(0)^2> - <C(0)>^2 = \frac{2n+1}{4}$ for the quadrature phase of the noise.

SNR_{out} is always smaller than SNR_{in} due to the presence of the noise contribution $<\Delta C(0)^2>$.

An ideal lossless parametric amplifier with gain coefficient G_o performs the modal transformation[14,15]

$$b + b^{+} = G_o^{\frac{1}{2}} (a_{in} + a_{in}^{+}) \; ; \qquad b - b^{+} = G_o^{-\frac{1}{2}} (a_{in} - a_{in}^{+}) \tag{15}$$

amplifying the quadrature $A_{in}(0) = \frac{a_{in}+a_{in}^{+}}{2}$ and squeezing the quadrature $A_{in}\left(\frac{\pi}{2}\right) = \frac{a_{in}-a_{in}^{+}}{2i}$. Thus a phase sensitive amplifier preserves the homodyne-detection SNR's for both quadratures. Another parametric amplifier can be used to realize the inverse transformation, i.e. squeezing $B(0) = \frac{b+b^{+}}{2}$ while amplifying $B\left(\frac{\pi}{2}\right) = \frac{b-b^{+}}{2i}$. Let us now suppose that a parametric amplifier is inserted before the loss, and the "reverse" amplifier (squeezer) after it. Due to the nature of the transformation (13), in which the noise term introduced by the loss is present additively, at the exit of the squeezer the signal field will be unchanged, while the noise term will be squeezed. The new modal transformation can be written

$$a_{out} = \sqrt{1-\Gamma_o}\, a_{in} + \sqrt{\Gamma_o}\, d(G_o) \tag{16}$$

where

$$d(G_o) + d(G_o)^{+} = G_o^{-\frac{1}{2}} (c + c^{+}) \; ; \qquad d(G_o) - d(G_o)^{+} = G_o^{\frac{1}{2}} (c - c^{+}) \tag{17}$$

and the new noise terms have the statistical properties

$$\langle d(G_0)\rangle = 0 \;; \qquad \langle d(G_0)d(G_0)\rangle = \frac{2n+1}{4}(G_0^{-1} - G_0) \tag{18a}$$

$$\langle d(G_0)^+ d(G_0)\rangle = \langle d(G_0)d(G_0)^+\rangle - 1 = \frac{2n+1}{4}(G_0+G_0^{-1}) - \frac{1}{2} \tag{18b}$$

clearly showing the phase dependent nature of the noise. In the particular case of a zero temperature thermal bath ($n = 0$) Eqs.(18) are the expectation values of an ideal squeezed vacuum state. The signal-to-noise ratio at the end of the squeezer is then

$$SNR_{out} = \frac{\langle A_{out}(0)\rangle^2}{\langle \Delta A_{out}(0)^2\rangle} = \frac{\langle A_{in}(0)\rangle^2}{\langle \Delta A_{in}(0)^2\rangle + \dfrac{\Gamma_0}{1-\Gamma_0}\langle \Delta D(G_0,0)^2\rangle} \tag{19}$$

where

$$\langle \Delta D(G_0,0)^2\rangle = \langle D(G_0,0)^2\rangle - \langle D(G_0,0)\rangle^2 = \frac{2n+1}{4} G_0^{-1}. \tag{20}$$

The noise contribution to SNR can be vanishingly small, provided the amplifier gain is sufficiently high. It appears that nearly ideal narrow-band parametric amplifiers of arbitrary gain can be built with the help of a cavity[12]; thus, the loss can be less effective in degrading SNR. It has also to be noted that the squeezer has no effect on SNR, and it is just used to get the loss squeezed, and the signal unchanged (apart the scaling factor $(1-\Gamma_0)$).

Recently Kimble *et al.*[17] found that there is essentially no excess noise in parametric amplification as they observed an actual minimum uncertainity squeezed state or two-photon coherent state[10]. Furthermore, Loudon[2] has shown that the SNR degradation down the line does not grow with the number of amplifiers as far as the quantum noise is concerned.

Let us now consider the scheme of Fig.1 representing a transmission line in which N phase sensitive amplifiers are present along the line. Each step of free propagation between two amplifiers has loss Γ_0 and G_0 is the amplification factor of each parametric amplifier. We can introduce along the waveguide, after the i-th loss, N identity operators $I = G_0^{i-1} \times (G_0^{-1})^{i-1}$. If we group the amplifier-loss-squeezer to have the squeezer realizing the reverse transformation of the amplifier, we see that our line is equivalent to a line of squeezed losses. Each i-th step has loss squeezed by a factor $(G_0^{-1})^{i-1}$ and at the end of the line there is a parametric amplifier with gain G_0^N (see Fig. 1). The last amplifier does not change, however, SNR and hence, as far as this particular quantity is concerned, our line is equivalent to a waveguide having squeezed losses.

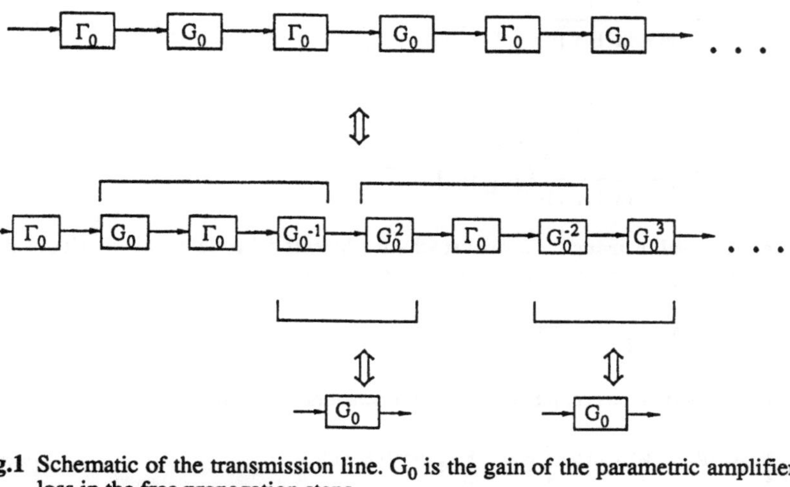

Fig.1 Schematic of the transmission line. G_0 is the gain of the parametric amplifiers, while Γ_0 is the loss in the free propagation steps.

Each step of amplifier-loss-squeezer realizes the transformation

$$a_i = \sqrt{1-\Gamma_0}\, a_{i-1} + \sqrt{\Gamma_0}\, d(G_0^{i-1}) \tag{21}$$

where i = 1,2,...,N. The first step (i = 1) has unsqueezed loss. At the end of the line, the field is

$$b_N + b_N^+ = (G_0^{1/2})^N (a_N + a_N^+) \;; \qquad b_N - b_N^+ = (G_0^{-1/2})^N (a_N - a_N^+) \tag{22}$$

where

$$a_N = (1-\Gamma_0)^{\frac{N}{2}} a_0 + \Gamma_0^{\frac{1}{2}} \sum_{i=1}^{N} (1-\Gamma_0)^{\frac{i-1}{2}} d(G_0^{N-i}). \tag{23}$$

The SNR can be calculated directly on a_N because, as noted before, the last amplification does not change it. Using the independence between the different noise sources ($<c_ic_j> = 0$, $<c_ic_j^+> = 0$, $i\neq j$), we obtain

$$SNR_N = \frac{<A_N(0)>^2}{<\Delta A_N(0)^2>} =$$

$$= \frac{(1-\Gamma_0)^N <A_0(0)>^2}{(1-\Gamma_0)^N <\Delta A_0(0)^2> + \Gamma_0 \sum_{i=1}^{N} (1-\Gamma_0)^{i-1} <\Delta D(G_0^{N-i},0)^2>} \tag{24}$$

being

$$<\Delta D(G_0^{N-i},0)^2> = \frac{2n+1}{4} (G_0^{-1})^{N-i}. \tag{25}$$

Substituting in (24) and performing the sum, we obtain

$$SNR_N = \frac{<A_0(0)>^2}{<\Delta A_0(0)^2> + \frac{2n+1}{4} \frac{1-(1-\Gamma)^{1/N}}{(1-\Gamma)^{1/N}} \frac{1-(1-\Gamma)^{-1}G^{-1}}{1-(1-\Gamma)^{-1/N}G^{-1/N}}} \tag{26}$$

where Γ and G, the total losses and gains respectively, are such that $(1-\Gamma) = (1-\Gamma_0)^N$ and $G = G_0^N$. It is now interesting to perform the continuum limit $N \to \infty$. Taking into account

$$\lim_{N\to\infty} \frac{1-(1-\Gamma)^{1/N}}{(1-\Gamma)^{1/N} - G^{-1/N}} = \frac{-\ln(1-\Gamma)}{\ln G + \ln(1-\Gamma)} \tag{27}$$

we obtain

$$SNR_\infty = \frac{<A_0(0)>^2}{<\Delta A_0(0)^2> + \frac{2n+1}{4} \frac{-\ln(1-\Gamma)}{\ln G + \ln(1-\Gamma)} \left[1-(1-\Gamma)^{-1}G^{-1}\right]} \tag{28}$$

This result can also be obtained by using the spatial stochastic differential equation approach of Caves and Crouch(5). Indeed from Eq.(7) and its hermitian conjugate we get:

$$\frac{d}{dz} A(z;\tfrac{\pi}{2}) = \left(g - \tfrac{\gamma}{2}\right) A(z;\tfrac{\pi}{2}) + \gamma^{1/2}\, C(z;\tfrac{\pi}{2}) \tag{29}$$

with the obvious generalization to the continuous case of the quadrature operators.
The solution reads

$$A(z;\tfrac{\pi}{2}) = A(z_0;\tfrac{\pi}{2})\, e^{(g-\gamma/2)(z-z_0)} + \gamma^{1/2} \int_{z_0}^{z} C(z';\tfrac{\pi}{2})\, e^{(g-\gamma/2)(z-z')}\, dz' \tag{30}$$

by plugging this equation into Eq.(12) and performing the expectations by using Eq.(9) with n=0 we get

$$\mathrm{SNR} = \frac{\langle A(z_o;\frac{\pi}{2})\rangle^2}{\langle \Delta A(z_o;\frac{\pi}{2})^2\rangle + \frac{\gamma}{4(\gamma-2g)}[e^{(\gamma-2g)(z-z_o)}-1]} \tag{31}$$

Once we make the following positions

$$A(z_o;\tfrac{\pi}{2}) \equiv A_o(0); \qquad (1-\Gamma) = e^{-\gamma L}; \qquad G = e^{2gL}, \tag{32}$$

where L represents the length of the waveguide, Eq.(31) is equivalent to Eq.(28).
We are led to conclude that in order to carry the signal down a line without degrading its signal-to-noise ratio a waveguide with a $\chi^{(2)}$ medium should be employed. Kumar/6/ has recently explored the possibility of squeezed-light generation in optical waveguides with the dominant nonlinearity of second order. We suggest, however, that this device should be used as a parametric amplifier to get a nearly transparent optical network.

Acknowledgements

This work has been done in collaboration with A.Mecozzi. The financial support of the Gruppo Nazionale di Struttura della Materia and of the Centro Interuniversitario di Fisica della Materia is greatly appreciated.

References

1. H.P.Yuen, Opt. Lett. **12**, 789 (1987).
2. R.Loudon, IEEE J.Quantum Electron. **QE-21**, 766 (1985).
3. Y.R.Shen, Phys.Rev. **155**, 921 (1967).
4. J.Tucker and D.F.Walls, Phys.Rev. **178**, 2036 (1969).
5. C.M.Caves and D.D.Crouch, J. Opt. Soc. Am. **B4**, 1535 (1987).
6. P.Kumar in *Squeezed and Nonclassical Light*, Proceedings of the NATO ARW held at Cortina January 25-29, 1988. Eds. P.Tombesi and E.R.Pike (Plenum, to be published).
7. R.M.Shelby, M.D.Levenson, S.H.Perlmutter, R.G.DeVoe and D.F.Walls, Phys.Rev.Lett. **57**, 691 (1986).
8. C.W.Gardiner and M.J.Collett, Phys.Rev. **A31**, 3761 (1985).
9. R.E.Slusher, L.W.Hollberg, B.Yurke, J.C.Mertz and J.F.Valley, Phys.Rev.Lett. **55**, 2409 (1985).
10. H.P.Yuen, Phys.Rev. A **13**, 2226 (1976).
11. D.F.Walls, Nature (London) **306**, 141 (1983); R.Loudon and P.L.Knight, J.Mod.Opt. **34**, 709 (1987).
12. L.Wu, H.J.Kimble, J.L.Hall and H.Wu, Phys.Rev.Lett. **57**, 2520 (1986).
13. H.P.Yuen and J.H.Shapiro, in *Coherence and Quantum Optics IV*. Eds. L.Mandel and E.Wolf (Plenum, N.Y. 1978) p. 719.
14. H.P.Yuen, in *Quantum Optics, Experimental Gravitation and Measurement Theory* . Eds. P.Meystre and M.O.Scully (Plenum, N.Y. 1983).
15. H.Takahasi, Adv. Commun. Sys. **1**, 227 (1965).
16. P.S.Henry, IEEE J.Quantum Electron, **QE-21**, 1862 (1985).
17. H.J.Kimble, in *Digest of International Quantum Electronics Conference* (Optical Society of America, Washington D.C. 1987) p.34.

Input-Output Fields and Their Squeezing for Arbitrary Linear Response

R. Graham

Fachbereich Physik, Universität-Gesamthochschule Essen, D-4300 Essen, Fed. Rep. of Germany

1. Introduction

As an essential element of a quantum theory of squeezing a formalism has been developed in papers by Yurke [1], Yurke and Denker [2] and Collett and Gardiner [3,4], which is known in the field as 'input-output' formalism [3,4]. This is a kind of S-matrix formalism which relates input fields, which may be any kind of quantum noise, and/or coherent fields entering some quantum optical device to output fields emitted by such devices. The great advantage of this formalism, like of any S-matrix formalism, is the fact that it deals directly with the experimentally accessible quantities. In fact, the theoretical conversion from the internally generated quantum fields (the central objects before the advent of the input-output formalism) to the observed output fields proved to be essential in order to obtain quantitative agreement between the theoretical analysis of quantum optical models [3,5] and experiments in the laboratory [6,7]. In view of the importance of this formalism the present article is devoted to its further development. We wish to construct here the relation between input and output fields for systems with arbitrary linear response, including, in particular, arbitrarily retarded linear response. In this case questions of causality, of central importance in any S-matrix formalism, have to be analyzed with more generality and greater depth than appears to have been done until now in the literature. Retardation in the input output formalism has been dealt with in ref. [6] within the context of a special model. The treatment we shall present here is model--independent and quite general, and built on the requirement of causality. It gives back results obtained in ref. [6] and thereby substantuates a claim of model-independence made there.

The formalism is developed here in the context of the quantum Langevin equation. For recent references cf. [9-11]. As an application we treat squeezing for arbitrary linear response in the presence of arbitrary dispersive linear losses [13].

2. Quantum Langevin Equation

We consider a 'small' system (the quantum optical device of interest) linearly coupled to a large reservoir at temperature $T = \beta^{-1}$. The temperature may vanish, in which case the reservoir is in the ground state (e.g. the vacuum of the electromagnetic field outside the device). For simplicity of notation we represent the small system by a single pair of canonically conjugate quantum variables q,p and write the Hamiltonian in the form

$$H = \frac{p^2}{2} + V(q) - \Gamma q + \frac{1}{2}\lambda q^2 + H_{res} \tag{1}$$

where V(q) is the potential in the absence of the reservoir coupling, Γ is a reservoir operator and H_{res} represents the free reservoir dynamics. The λ-term turns out to be necessary [9] and λ remains to be determined. The free reservoir

Springer Proceedings in Physics, Vol. 41 **Quantum Optics V**
Editors: J.D. Harvey and D.F. Walls

dynamics $\Gamma(t) = \Gamma^{(0)}(t)$ is assumed to be stationary in the sense that the expectation value of its commutator at different times depends only on the time-difference. We define the response function of the reservoir by

$$\frac{i}{\hbar}\langle[\Gamma^{(o)}(t), \Gamma^{(o)}(\tau)]\rangle\, \Theta(t-\tau) = 2i\, \chi''_{\Gamma}(t-\tau)\, \Theta(t-\tau) = \chi_{\Gamma}(t-\tau)$$

$$= -\frac{d}{dt}\gamma(t-\tau) \tag{2}$$

In case we are dealing with a thermal reservoir we may invoke the fluctuation dissipation theorem, which implies

$$\frac{1}{2}\langle\, \Gamma^{(o)}(t)\, \Gamma^{(o)}(\tau) + \Gamma^{(o)}(\tau)\, \Gamma^{(o)}(t)\, \rangle = \int \frac{d\omega}{2\pi}\, e^{-i\omega(t-\tau)}\, S_{\Gamma}(\omega) \tag{3}$$

$$S_{\Gamma}(\omega) = \hbar\, \chi''_{\Gamma}(\omega)\, \coth \frac{1}{2}\hbar\beta\omega \tag{4}$$

where

$$\chi''_{\Gamma}(\omega) = \int dt\, e^{i\omega t}\, \chi''_{\Gamma}(t) = \frac{\omega}{2}\, (\gamma(\omega) + \gamma(-\omega)) \equiv \omega\, \gamma'(\omega) \tag{5}$$

$$\gamma(\pm\omega) = \gamma(\pm\omega + i0)$$

$$\gamma(z) = \int_0^{\infty} dt\, e^{izt}\, \gamma(t) \tag{6}$$

Thermodynamic stability of the reservoir guarantees that $\gamma'(\omega) > 0$. The loss function $\gamma(z)$ is analytic for $\mathrm{Im}\, z > 0$ and has no zeros there [9]. The continuation of $\gamma(z)$ from $\mathrm{Im}\, z > 0$ to $\mathrm{Im}\, z < 0$, in general, has poles for $\mathrm{Im}\, z < 0$, but we shall later need the assumption that it has no zeros for $\mathrm{Im}\, z < 0$. The idea now is to keep only the systematic part of the back-action of the small system on the reservoir and to treat it in linear response only, which gives

$$\Gamma(t) = \Gamma^{(o)}(t) + \int_{-\infty}^{t} d\tau\, \chi_{\Gamma}(t-\tau)\, q(\tau) \tag{7}$$

Inserting eq. (7) in the Heisenberg equation of motion

$$\ddot{q} = -V'(q) - \lambda q + \Gamma(t) \tag{8}$$

using $\chi_{\Gamma}(t) = -\dot{\gamma}(t)$, and performing a partial integration yields the quantum Langevin equation

$$\ddot{q} = -V'(q) - \int_{-\infty}^{t} d\tau\, \gamma(t-\tau)\, \dot{q}(\tau) + \Gamma^{(o)}(t) \tag{9}$$

where we have chosen the parameter λ as

$$\lambda = 2\int_0^{\infty} \frac{d\omega}{2\pi}\, \gamma'(\omega) \tag{10}$$

As seen in (9) this choice guarantees that V(q) is the potential in the absence of dissipation. We note that thermodynamic stability implies that λ is always non-vanishing and positive [9]. The generalization of (9) to an arbitrary number of variables is merely a matter of introducing indices where appropriate. It may be worth recalling the independence, of the derivation presented here, from any particular reservoir model. In fact H_{res} was never specified. The essential ingredient is the treatment of the back-action in linear response.

Frequently the loss function $\gamma(t)$ is a sum of independent terms $\gamma(t) = \sum \gamma_\lambda(t)$ corresponding to independent reservoir operators $\Gamma(t) = \sum \Gamma_\lambda(t)$ with

$$\frac{i}{\hbar} \langle [\Gamma_\lambda^{(o)}(t), \Gamma_{\lambda'}^{(o)}(\tau)] \rangle \, \Theta(t-\tau) = - \delta_{\lambda\lambda'} \, \dot{\gamma}_\lambda(t-\tau) = \delta_{\lambda\lambda'} \, \chi_\lambda(t-\tau) \tag{2a}$$

Eq. (7) is then generalized and reads

$$\Gamma_\lambda(t) = \Gamma_\lambda^{(o)}(t) + \int_{-\infty}^{t} d\tau \, \chi_\lambda(t-\tau) \, q(\tau) \tag{7a}$$

3. Input-Output Formalism

Eq. (7a) is the basic relation on which an input-output formulation can be based. As noted by Collett and Gardiner [3,4] it is useful to define an advanced reservoir field $\Gamma^{adv}(t)$ by

$$\Gamma_\lambda(t) = \Gamma_\lambda^{(adv)}(t) + \int_{t}^{\infty} d\tau \, \chi_\lambda^{adv}(t-\tau) \, q(\tau) \tag{11}$$

where $\Gamma_\lambda^{(adv)}(t)$ has the same properties as $\Gamma_\lambda^{(O)}(t)$ (eqs. (2)-(6)) and $\chi_\lambda^{adv}(t-\tau) = \chi_\lambda(\tau-t)$ is the advanced response function. We remark that an advanced quantum Langevin equation is obtained when eq. (11) is used to express Γ in eq. (8). While $\Gamma^{(O)}(t)$ is a quantum mechanical input to the system provided by the reservoir, the advanced field $\Gamma^{(adv)}(t)$ may be considered as a corresponding output [3,4]. From eqs. (7), (11) we have that

$$\Gamma_\lambda^{(adv)}(t) - \Gamma_\lambda^{(o)}(t) = 2i \int_{-\infty}^{+\infty} d\tau \, \chi_\lambda''(t-\tau) \, q(\tau)$$

$$\Gamma_\lambda^{(adv)}(\omega) - \Gamma_\lambda^{(o)}(\omega) = 2i\omega \, \gamma_\lambda'(\omega) \, q(\omega) \tag{12}$$

However, $\Gamma^{(o)}$ and $\Gamma^{(adv)}$ are not yet appropriate physical input-output fields. The reason is that fundamental causality relations are not satisfied.

In order to make manifest the lack of causality let us specialize to linear systems, where $V(q) = (1/2) \, \omega_o^2 q^2$. Then eq. (9) can be solved

$$q(t) = \sum_\lambda \int_{-\infty}^{t} d\tau \, \chi_o(t-\tau) \, \Gamma_\lambda^{(o)}(\tau) = \sum_\lambda \int_{t}^{\infty} d\tau \, \chi_o(\tau-t) \, \Gamma_\lambda^{(adv)}(\tau) \tag{13}$$

with the linear response function

$$\chi_o(t-\tau) = \int d\omega\, e^{-i\omega\tau}\, \chi_o(\omega)$$

$$\chi_o(\omega) = (\omega_o^2 - \omega^2 - i\omega\, \gamma(\omega))^{-1} \tag{14}$$

The linear response function is analytic in the upper half of the complex ω-plane. This is sufficient for a causal relation of the expectation value $\langle q(t)\rangle$ to $\langle \Gamma_\lambda^{(o)}(\tau)\rangle$. However, a causal relationship of $\langle \Gamma_\lambda^{(adv)}(t)\rangle$ to $\langle q(\tau)\rangle$ would require that also $\chi_o^{-1}(-\omega)$ be analytic for Im $\omega > 0$ and quantum mechanical causality would only be achieved if the commutators $\langle [q(t), \Gamma_\lambda^{(o)}(t')]\rangle$ and $\langle [\Gamma_\lambda^{(adv)}(t), q(t')]\rangle$ vanish for $t'>t$. Using eqs. (13), (14) and (2) these commutators are easily evaluated as

$$\langle [q(t), \Gamma_\lambda^{(o)}(t')]\rangle = 2\hbar \int \frac{d\omega}{2\pi}\, e^{-i\omega(t-t')}\, \chi_o(\omega)\, \chi_\lambda''(\omega)$$

$$\langle [\Gamma_\lambda^{(adv)}(t), q(t')]\rangle = 2\hbar \int \frac{d\omega}{2\pi}\, e^{-i\omega(t-t')}\, \chi_o(\omega)\, \chi_\lambda''(-\omega) \tag{15}$$

The commutators would vanish for $t'>t$ only if $\chi_\lambda''(\omega) = \frac{\omega}{2}(\gamma_\lambda(\omega) + \gamma_\lambda(-\omega))$ $= -\chi_\lambda''(-\omega)$ were analytic in the upper half of the complex ω-plane. For constant $\gamma_\lambda(\omega) = \gamma_\lambda$ this would be the case, but, in general, $\gamma_\lambda(-\omega)$ has singularities if analytically continued to ω with positive imaginary part. Therefore, the above commutators are non-vanishing for $t'>t$. This lack of causality can also be observed in the relation between $\Gamma_\lambda{}^{(adv)}$ and $\Gamma_\lambda^{(o)}$ which is easily obtained from eqs. (12), (13). Taking Fourier transforms we find

$$\Gamma_\lambda^{(adv)}(\omega) = \sum_{\lambda'} (\delta_{\lambda\lambda'} + 2i\omega\, \gamma_\lambda'(\omega)\, \chi_o(\omega))\, \Gamma_{\lambda'}{}^{(o)}(\omega) \tag{16}$$

where we can easily spot the absence of causality because $\gamma_\lambda(-\omega+io)$, in general, fails to be analytic in the upper half of the complex ω-plane. As already noted we have to conclude from these considerations that, unlike the case γ = const., the fields $\Gamma_\lambda^{(o)}(t)$, $\Gamma_\lambda^{(adv)}(t)$ are <u>not</u> appropriate input-output fields. In order to construct causally related input-output fields $E_\lambda^{(in)}(t)$, $E_\lambda^{(out)}(t)$ we make the ansatz

$$\Gamma_\lambda^{(o)}(t) = \int_{-\infty}^{t} d\tau\, \kappa_\lambda(t-\tau)\, E_\lambda^{(in)}(\tau)$$

$$\Gamma_\lambda^{(adv)}(t) = \int_{t}^{\infty} d\tau\, \kappa_\lambda(\tau-t)\, E_\lambda{}^{(out)}(\tau) \tag{17}$$

The kernels $\kappa_\lambda(t)$ remain to be determined. In order to comply with eq. (17) the

Fourier transform $\kappa_\lambda(\omega)$ must be analytic for Im $\omega>0$, and it must satisfy the symmetry relation $\kappa_\lambda(\omega) = \kappa_\lambda^*(-\omega)$. Let us try to construct the input-output fields in such a way that they can be expanded in normal modes

$$E_\lambda^{(in)}(t) = \int_0^\infty \frac{d\omega}{2\pi} \left[\frac{\hbar\omega}{2}\right]^{1/2} (b_\lambda^{(in)}(\omega)e^{-i\omega t+i\varphi_\lambda(\omega)} + b_\lambda^{(in)^+}(\omega)e^{i\omega t-i\varphi_\lambda(\omega)}) \qquad (18)$$

with

$$\langle[b_\lambda^{(in)}(\omega),\ b_{\lambda'}^{(in)^+}(\omega')]\rangle = 2\pi\ \delta(\omega-\omega')\ \delta_{\lambda\lambda'} \qquad (19)$$

hence

$$\langle[E_\lambda^{(in)}(\omega),\ E_{\lambda'}^{(in)}(\omega')]\rangle = \frac{\hbar\omega}{2}\ 2\pi\ \delta(\omega+\omega')\ \delta_{\lambda\lambda'} \qquad (20)$$

and similarly for the out-field. The phases $\varphi_\lambda(\omega)$ need not to be specified further. They take account of the fact that the fields E_λ, in general, are propagating. Eq. (20) is consistent with eqs. (17), (2) if

$$|\kappa_\lambda(\omega)|^2 = \frac{4\chi_\lambda''(\omega)}{\omega} = 4\gamma_\lambda'(\omega) \qquad (21)$$

Consistently, both sides of eq. (21) are positive. The condition (21) also serves to satisfy eqs. (18)-(20) for $E_\lambda^{(out)}$. While eq. (21) fixes the absolute value $\kappa_\lambda(\omega)$ the phase of $\kappa_\lambda(\omega)$ remains to be determined. This has to be done in such a way that $\kappa_\lambda(\omega)$ is analytic for Im $\omega > 0$. Thus we infer from eq. (21) that the phase-factor must cancel the singularities of $\sqrt{\gamma(-\omega)}$ for Im $\omega > 0$ without introducing new singularities. It must also respect the $\omega \rightarrow -\omega$ symmetry. This seems to leave no alternative to the choice

$$\frac{\kappa_\lambda(\omega)}{\kappa_\lambda(-\omega)} = \frac{\gamma_\lambda(\omega)}{\gamma_\lambda(-\omega)} \qquad (22)$$

from which

$$\kappa_\lambda(\omega) = \gamma_\lambda(\omega)\ M_\lambda(\omega)$$

$$M_\lambda(\omega) = \sqrt{2}\ \left(\frac{1}{\gamma_\lambda(\omega)} + \frac{1}{\gamma_\lambda(-\omega)}\right)^{1/2} \qquad (23)$$

follows. We note that $\kappa_\lambda(\omega+io)$ exists even if $\chi_o(\omega)$ looses its meaning. The singularities of $\gamma_\lambda(-\omega)$ for Im $\omega > 0$ are now seen to cancel. However, in order not to introduce new singularities we have to assume that there are no zeros in the denominator, $|\gamma_\lambda(-\omega)| > 0$ for $\infty >$ Im $\omega > 0$, otherwise it would seem to be impossible to construct proper input - output fields via eq. (17). The assumption is satisfied by all the phenomenological expressions used for $\gamma(\omega)$, we are aware off, like the Drude formula, $\gamma_\lambda(\omega) = \gamma_\lambda/(1-i\omega\tau_\lambda)$, but the general status of this additional condition remains to be clarified at the time of this writing. We are now in a position to check causality by evaluating

$$\langle[q(t), E_\lambda^{(in)}(t')]\rangle = \int \frac{d\omega}{2\pi} e^{-i\omega(t-t')} \frac{\hbar\omega}{2} \chi_o(\omega)\, \kappa_\lambda(\omega)$$
$$= \langle[E_\lambda^{(out)}(t), q(t')]\rangle \tag{24}$$

As both χ_o and κ_λ are analytic for Im $\omega > 0$, the commutators are seen to vanish for $t' > t$. The direct relation between $E^{(out)}(\omega)$ and $E^{(in)}(\omega)$ easily follows from eqs. (16), (17) and reads

$$E_\lambda^{(out)}(\omega) = \sum_{\lambda'} S_{\lambda\lambda'}(\omega)\, E_{\lambda'}^{(in)}(\omega) \tag{25}$$

with

$$S_{\lambda\lambda'} = \delta_{\lambda\lambda'} + \frac{(\gamma_\lambda(\omega)-\gamma_\lambda(-\omega))M_\lambda(\omega)\delta_{\lambda\lambda'}+i\omega(\gamma_\lambda(\omega)+\gamma_\lambda(-\omega))\chi_o(\omega)\gamma_{\lambda'}(\omega)M_{\lambda'}(\omega)}{\gamma_\lambda(-\omega)\, M_\lambda(\omega)} \tag{26}$$

Again the singularities of $\gamma(-\omega)$ for Im $\omega > 0$ are seen to cancel. Hence, provided there are no zeros of $\gamma(-\omega)$ for Im $\omega > 0$, the function $S(\omega)$, the analogue of the S-matrix, is seen to be analytic for Im $\omega > 0$ and $E^{(out)}$ is causally related to $E^{(in)}$. The matrix $S_{\lambda\lambda'}$ as given by eq. (26) can be checked to be unitary. The preceding analysis is readily generalized to the case of n variables q_i and n fields Γ_i. All the response functions we have introduced then become n x n matrices, and the S-matrix becomes a matrix in the product space labelled by the two indices i, λ. Eqs. (17) are now replaced by

$$\Gamma_\lambda^{(o)}(t) = \int_{-\infty}^{t} d\tau\, \underset{\sim}{\kappa}_\lambda(t-\tau)\cdot E_\lambda^{(in)}(\tau)$$

$$\Gamma_\lambda^{(adv)}(t) = \int_{-\infty}^{t} d\tau\, \underset{\sim}{\kappa}_\lambda^{T}(\tau-t)\cdot E_\lambda^{(out)}(\tau) \tag{27}$$

where the Γ and E are understood as vectors with n components, and $\underset{\sim}{\kappa}^T$ is the transposed of $\underset{\sim}{\kappa}$. The reality, symmetry and positivity properties of $\underset{\sim}{\gamma}_\lambda(\omega)$ are now expressed by $\underset{\sim}{\gamma}_\lambda(\omega) = \underset{\sim}{\gamma}_\lambda^{*}(-\omega)$ and the fact that $(\underset{\sim}{\gamma}_\lambda(\omega) + \underset{\sim}{\gamma}_\lambda^{T}(-\omega))$ is a symmetric, hermitian, real and non-negative matrix. The relation (21) now takes the form

$$\underset{\sim}{\kappa}_\lambda(\omega)\cdot\underset{\sim}{\kappa}_\lambda^{T}(-\omega) = 2(\underset{\sim}{\gamma}_\lambda(\omega) + \underset{\sim}{\gamma}_\lambda^{T}(-\omega)) \tag{28}$$

where $\underset{\sim}{\kappa}_\lambda(-\omega) = \underset{\sim}{\kappa}_\lambda^{*}(\omega)$. We may take advantage of the fact that $(\underset{\sim}{\gamma}_\lambda^{-1}(\omega) + \underset{\sim}{\gamma}_\lambda^{T-1}(-\omega))$ must also be hermitian and non-negative and hence has a unique hermitian non-negative square root

$$\underset{\sim\lambda}{M}(\omega) = \sqrt{2}\,(\underset{\sim\lambda}{\gamma}^{-1}(\omega+io) + \underset{\sim\lambda}{\gamma}^{T-1}(-\omega+io))^{1/2} \tag{29}$$

Then eq. (23) takes the form

$$\underset{\sim\lambda}{\kappa}(\omega+io) = \underset{\sim\lambda}{\gamma}(\omega+io)\cdot\underset{\sim\lambda}{M}(\omega+io) \tag{30}$$

and the S-matrix reads

$$\underset{\sim\lambda\lambda'}{S} = \underset{\sim\lambda}{M}^{-1}(\omega)\cdot\underset{\sim\lambda}{\gamma}^{T-1}(-\omega)\cdot(\underset{\sim}{1}\delta_{\lambda\lambda'}+i\omega(\underset{\sim\lambda}{\gamma}(\omega)+\underset{\sim\lambda}{\gamma}^{T}(-\omega))\cdot\underset{\sim}{\chi}_o(\omega))\cdot\underset{\sim\lambda'}{\gamma}(\omega)\cdot\underset{\sim\lambda'}{M}(\omega) \tag{31}$$

4. Squeezing for Arbitrary Linear Response

As an illustration of the formalism developed in the preceding section we discuss briefly squeezing (cf. [1] - [8]) by coupling of pairs of frequencies from the present point of view. A more extensive treatment will be given in [13]. In this case the positive frequency part $E^{(+)}(\omega) = -i\omega q^{(+)}(\omega)/\sqrt{\epsilon(\omega)}$ ($\epsilon(\omega)$ = dielectric constant) of the internal electric field is generated from input (including vacuum fluctuations) at two frequencies which we denote by ω and $\omega_p-\omega$ where ω_p is the frequency of an external pump field. Assuming operation in the linear regime and the contribution from input at $n\omega_p-\omega$ for $n>1$ and $-m\omega_p+\omega$ for $m\geq1$ to be negligible we have generally

$$q^{(+)}(\omega) = \tilde{\chi}_1(\omega)\,\Gamma^{(+)}(\omega) + \tilde{\chi}_2(\omega)\,\Gamma^{(-)}(\omega_p-\omega) \tag{32}$$

where ω lies in the interval $0 < \omega < \omega_p$ and $\Gamma^{(+)}$, $\Gamma^{(-)}$ denote the positive and negative frequency parts of $\Gamma^{(o)}(\omega)$, respectively. $\tilde{\chi}_1(\omega)$ and $\tilde{\chi}_2(\omega)$ have to be determined from the equations of motion of the system and depend on the specific device under consideration. We shall not discuss this part of the problem here ([6], [12], [13]) and assume $\tilde{\chi}_1(\omega)$, $\tilde{\chi}_2(\omega)$ as given. However, we give an example at the end. If the external pump is switched off the two response functions are given by $\tilde{\chi}_1 = (\omega_o^2 - \omega^2 - i\omega\gamma(\omega))^{-1}$, $\tilde{\chi}_2 = 0$. Let us now consider the case where the output field to be measured is coupled out via a mirror with loss-function $\gamma_1(\omega)$, while the total loss-function is $\gamma(\omega)$. Then the output field for $0 < \omega < \omega_p$ is given as

$$E^{(out)(+)}(\omega) = \frac{1}{M_1(\omega)\gamma_1(-\omega)}\,[\Gamma_1^{(+)}(\omega) + 2i\omega\gamma_1'(\omega)\,(\tilde{\chi}_1(\omega)\Gamma^{(+)}(\omega) + \tilde{\chi}_2(\omega)\Gamma^{(-)}(\omega_p-\omega))] \tag{33}$$

where M_1 is given by eq. (23).

The fact that $E^{(out)(+)}$ satisfies free-field commutation relations places general conditions on $\tilde{\chi}_1$ and $\tilde{\chi}_2$. These are

$$\omega\gamma'(\omega)\,|\tilde{\chi}_1(\omega)|^2 - \mathrm{Im}\,\tilde{\chi}_1(\omega) = (\omega_p-\omega)\,\gamma'(\omega_p-\omega)\,|\tilde{\chi}_2(\omega)|^2 \tag{34}$$

and

$$\tilde{\chi}_2(\omega)(1+2i(\omega_p-\omega)\gamma'(\omega_p-\omega)\tilde{\chi}_1(\omega_p-\omega)) = \tilde{\chi}_2(\omega_p-\omega)(1+2i\omega\gamma'(\omega)\tilde{\chi}_1(\omega)) \qquad (35)$$

These relations are independent of γ_1, containing only γ', and are therefore the same for all output fields. By taking absolute values on both sides of (35) and using (34) to eliminate $\tilde{\chi}_1(\omega)$ it follows that $|\tilde{\chi}_2(\omega)| = |\tilde{\chi}_2(\omega_p-\omega)|$.

The power spectrum can now be calculated from

$$< E^{(out)(-)}(\omega)\ E_1^{(out)(+)}(\omega') > = S(\omega)\ 2\pi\ \delta(\omega-\omega') \qquad (36)$$

We note the correlation functions

$$< \Gamma^{(-)}(\omega)\ \Gamma^{(+)}(\omega') > = 2\pi\ \delta(\omega-\omega')\ 2\hbar\omega\ \gamma'(\omega)\ n(\omega)$$

$$<[\Gamma^{(+)}(\omega),\ \Gamma^{(-)}(\omega')]> = 2\pi\ \delta(\omega-\omega')\ 2\hbar\omega\ \gamma'(\omega) \qquad (37)$$

with $n(\omega) = (\exp \beta\hbar\omega-1)^{-1}$.

For the power spectrum we find, using eq. (34) to eliminate $\tilde{\chi}_1(\omega)$

$$S(\omega) = \frac{1}{2}\ \hbar\omega\ n(\omega) + (1 + n(\omega) + n(\omega_p-\omega))\ 2\hbar\omega^2\ (\omega_p-\omega)$$

$$\cdot\ \gamma_1'(\omega)\ \gamma'(\omega_p-\omega)\ |\tilde{\chi}_2(\omega)|^2 \qquad (38)$$

If the external pump is switched off $S(\omega)$ reduces to the thermal spectrum $S_o(\omega) = n(\omega)\hbar\omega/2$ as, of course, it must. In the limit $n(\omega) \to 0$ appropriate for optical fields, only the contribution due to spontaneous emission survives.

In order to see squeezing the output-field is analyzed by homodyning with a strong field at the frequency $\omega_p/2$ between the coupled frequencies [5]. This generates the quadrature signal, for $-\omega_p/2 < \omega < \omega_p/2$,

$$X(\omega) = [E^{(out)(+)}(\omega + \frac{\omega_p}{2})\ e^{-i\varphi} + E^{(out)(-)}(-\omega + \frac{\omega_p}{2})\ e^{i\varphi}] \qquad (39)$$

whose normally ordered spectrum is measured in a balanced homodyne detection experiment [6,7]

$$< : X(\omega)\ X(\omega') : > = 2\pi\ \delta(\omega+\omega')\ S_x(\omega) \qquad (40)$$

In a squeezed state $S_x(\omega)$ becomes negative for an appropriate choice of φ but remains always larger than $-\hbar(\omega+\omega_p/2)/2$ since $\langle X^2\rangle$ is positive. The percentage of squeezing in power is therefore given by $-2S_x(\omega)/\hbar(\omega + \omega_p/2)$. With the definition

$$< E^{(out)(-)}(\omega)\ E^{(out)(-)}(\omega') > = S_{--}(\omega)\ 2\pi\ \delta(\omega+\omega') \qquad (41)$$

we may write

$$S_x(\omega) = 2\mathrm{Re}\left\{S_{--}(-\omega + \omega_p/2)\ e^{2i\varphi}\right\}$$

$$+ S(-\omega + \omega_p/2) + S(\omega + \omega_p/2) \tag{42}$$

The correlation function $S_{--}(-\omega + \omega_p/2)$ is given by a somewhat lengthy expression, which can be reduced considerably, by using eq. (35). We obtain

$$S_{--}(-\omega + \omega_p/2) = \frac{-i\hbar(\omega_p^2-4\omega^2)\gamma_1'(\omega + \frac{\omega_p}{2})\gamma_1'(-\omega + \frac{\omega_p}{2})\tilde{\chi}_2^*(-\omega + \frac{\omega_p}{2})}{M_1(-\omega + \frac{\omega_p}{2})M_1(\omega + \frac{\omega_p}{2})\gamma_1(-\omega + \frac{\omega_p}{2})\gamma_1(\omega + \frac{\omega_p}{2})} \cdot$$

$$\cdot\ (1+n(\omega + \omega_p/2) + n(-\omega + \omega_p/2)) \cdot$$

$$\cdot\ (1-2i(\omega + \omega_p/2)\gamma'(\omega + \omega_p/2)\tilde{\chi}_1^*(\omega + \omega_p/2))$$

$$= S_{--}(\omega + \omega_p/2) \tag{43}$$

Only the absolute value of $S_{--}(-\omega + \omega_p/2)$ is, of course, of real importance, because its phase can be compensated, at a given frequency ω, by the choice of φ in eq. (37). Forming the absolute value, and again using eq. (34) we may completely eliminate $\tilde{\chi}_1$ and obtain

$$|S_{--}(-\omega + \omega_p/2)|^2 = \hbar^2(\omega_p^2/4-\omega^2)^2\ \gamma_1'(-\omega + \omega_p/2)\ \gamma_1'(\omega + \omega_p/2)\ |\tilde{\chi}_2(\omega + \omega_p/2)|^2$$

$$\cdot\ (1+4(\omega_p^2/4-\omega^2)\ \gamma'(\omega + \omega_p/2)\ \gamma'(\omega_p/2 -\omega)\ |\tilde{\chi}_2(\omega + \omega_p/2)|^2)$$

$$\cdot\ (1+n(\omega + \omega_p/2)+n(-\omega + \omega_p/2))^2$$

$$= |S_{--}(\omega + \omega_p/2)|^2 \tag{44}$$

Here we also used the symmetry $|\tilde{\chi}_2(-\omega + \omega_p/2)| = |\tilde{\chi}_2(\omega + \omega_p/2)|$ noted after eq. (35). The squeezing spectrum obtained is therefore

$$S_{x\,\min}(\omega) = S(-\omega + \omega_p/2) + S(\omega + \omega_p/2) - 2|S_{--}(\omega + \omega_p/2)| \tag{45}$$

and may be written as

$$S_{x\,\min}(\omega) = \frac{\hbar}{2}\ (-\omega + \omega_p/2)\ n(-\omega + \omega_p/2) + \frac{\hbar}{2}\ (\omega + \omega_p/2)\ n(\omega + \omega_p/2)$$

$$+\ 2\hbar(\omega_p^2/4 - \omega^2)\ (1+n(-\omega + \omega_p/2) + n(\omega + \omega_p/2))|\tilde{\chi}_2(\omega + \omega_p/2)|^2$$

$$\cdot\Big\{(\omega_p/2 - \omega)\ \gamma_1'(\omega_p/2 - \omega)\ \gamma'(\omega_p/2 + \omega)$$

$$+(\omega_p/2 + \omega)\ \gamma_1'(\omega_p/2 + \omega)\ \gamma'(\omega_p/2 - \omega)$$

$$-(\gamma_1'(\omega_p/2 - \omega)\gamma_1'(\omega_p/2 + \omega))^{1/2}$$

$$\cdot\left[|\tilde{\chi}_2(\omega + \omega_p/2)|^{-2} + (\omega_p^2 - 4\omega^2)\gamma'(\omega_p/2 + \omega)\gamma'(\omega_p/2 - \omega)\right]^{1/2}\Big\} \quad (46)$$

$S_{x\ min}(\omega)$ has thus been expressed entirely in terms of the absolute value of the response function $\tilde{\chi}_2(\omega)$. We note that this result holds for a general linear response of the form (32) and for a general linear loss function, also independent of any Markoff approximation. It is therefore applicable to any squeezing mechanism in which waves at two frequencies ω, $\omega_p - \omega$ are coupled in a phase-sensitive way. As an example we consider subharmonic generation. With $E^{(+)}(\omega) = -i\omega q^{(+)}(\omega) / \sqrt{\epsilon(\omega)}$ we have the equation of motion [14]

$$\chi_o^{-1}(\omega)\ q^{(+)}(\omega) = -\tilde{\chi}^{nl}(\omega)\ q^{(-)}(\omega_p - \omega) + \Gamma^{(+)}(\omega) \quad (47)$$

and the hermitian conjugate for $q^{(-)}(\omega)$ with

$$\tilde{\chi}^{nl}(\omega) = \frac{4\pi\omega(\omega_p - \omega)}{\sqrt{\epsilon(\omega)\epsilon(\omega_p - \omega)}}\ E_o^{(+)}(\omega_p)\ \chi^{nl}(\omega, \omega_p - \omega) \quad (48)$$

where $E_o^{(+)}(\omega_p)$ is the electric field strength of the pump field and $\chi^{nl}(\omega, \omega_p - \omega)$ is the nonlinear susceptibility of the crystal. The susceptibilities $\tilde{\chi}_1(\omega)$ and $\tilde{\chi}_2(\omega)$ are readily determined and read

$$\tilde{\chi}_1(\omega) = [1 - \tilde{\chi}^{nl}(\omega)\tilde{\chi}^{nl*}(\omega_p - \omega)\chi_o(\omega)\chi_o^*(\omega_p - \omega)]^{-1}\chi_o(\omega)$$

$$\tilde{\chi}_2\ (\omega) = -\ \chi_o^*(\omega_p - \omega)\tilde{\chi}_1(\omega)\tilde{\chi}^{nl}(\omega) \quad (49)$$

It is possible to check that they satisfy the relations (34), (35) if and only if $\chi^{nl}(\omega, \omega_p - \omega) = \chi^{nl}(\omega_p - \omega, \omega)$. This relation is well known to be satisfied [14] if losses in the nonlinear susceptibilities χ^{nl} (not in the linear response functions χ_o, $\tilde{\chi}_1$, $\tilde{\chi}_2$) are negligible. If losses in $X^{nl}(\omega, \omega_p - \omega)$ occur, eqs. (47) have to be generalized by including additional noise terms [15], the linear response is no longer given by eq. (32), and eqs. (34), (35) are no longer valid.

References

1. B. Yurke: Phys. Rev. A29, 408 (1984)
2. B. Yurke, J.S. Denker: Phys. Rev. A29, 1419 (1984)
3. M.J. Collett, C.W. Gardiner: Phys. Rev. A30, 1386 (1984)

4. C.W. Gardiner, M.J. Collett: Phys. Rev. A31, 3761 (1985)
5. M.J. Collett, D.F. Walls: Phys. Rev. A32, 2887 (1985)
6. R.E. Slusher, L.W. Holberg, B. Yurke, J.C. Mertz, J.F. Valley: Phys. Rev. Lett. 55, 2409 (1985)
7. L.A. Wu, H.J. Kimble, J.L. Hall, H. Wu: Phys. Rev. Lett. 57, 2520 (1986)
8. C.W. Gardiner, A.S. Parkins, M.J. Collett: J. Opt. Soc. Am. B4, 1683 (1987)
9. G. W. Ford, J.T. Lewis, R.F. O'Connell: Phys. Rev. A37, 4419 (1988)
10. C.W. Gardiner: IBM J. Res. and Dev. 32, 127 (1988)
11. R. Graham: Europhys. Lett. 6, 477 (1988)
12. R. Graham: in 'Squeezed and Non-Classical Light', ed. E.R. Pike, P. Tombesi, Plenum, New York, to appear
13. R. Graham, A. Kramer: in preparation
14. N. Bloembergen: 'Nonlinear Optics', Benjamin, New York, 1965
15. R. Graham, H. Haken: Z. Physik 210, 276 (1968)

Time Dependent Quantum Correlations and Squeezing

P.D. Drummond

Physics Department, University of Queensland, St. Lucia 4067, Queensland, Australia

A unified theory of the detection of quantum correlations and squeezing is presented, using normally ordered photo-detector theory. The results hold for time-dependent local-oscillators, and are applicable to examples in nonlinear quantum optics.

1. INTRODUCTION

Recent developments and success in generating radiation with nonclassical statistics has opened the way to new types of experiments in fundamental quantum mechanics. In particular, the prediction of squeezing [1] in four-wave-mixing and parametric oscillation has been recently verified in experiment [2]. More recent theoretical work has led to the calculation of nonclassical correlations at causally distinct locations [3]. In principle, this would allow a test of quantum-mechanics in a directly comparable experiment to that proposed by EINSTEIN, PODOLSKY and ROSEN [4]. While related work on the BELL [4] inequalities has resulted in experimental tests, there is no experimental test yet on the type of correlation proposed in the E.P.R. paper. In addition, there have been recent suggestions that quantum field theories can also be tested in squeezing experiments [5].

These new types of experiments in laser physics utilize the efficient detection of radiation fields using a local oscillator, which permits the detection of phase-dependent information. We note that the detection of a quantum field phase in this sense is not possible in conventional particle physics. For this reason, quadrature-phase experiments have little or no resemblance to traditional scattering experiments. Nevertheless, they reveal quantum-mechanical information that is complementary to that obtained in more conventional ways. It is clearly important to discover just what new types of measurement are now possible, using quadrature phase techniques. The possible applications are also of some interest.

The obvious applications of quadrature phase detection are in the areas of measurement and communications. In fact one of the motivations behind the experimental developments in squeezing was a suggestion by CAVES [6] that squeezed radiation could be used to efficiently detect gravity waves. More recent is the suggestion that fundamental quantum measurement theory is testable using quadrature phase techniques, which could lead to new types of precision measurement. Nonlinear quantum field theories can also be tested with quadrature phase measurements. This is the case in soliton propagation, which is relevant to optical communications systems.

The usual local oscillator measurements use local oscillator fields with a constant amplitude [7]. However, the early work of SHAPIRO [7], and recent experimental developments [8] indicate that more general time-dependent local-oscillator measurements could be useful. In particular, mode-locked lasers naturally lend themselves to time-dependent local oscillator experiments. The advantage of mode-locked lasers is the large peak intensities that are achieved. This allows nonlinear interactions of the type that lead to squeezing without using narrow-band intra-cavity methods. The present paper analyses the physics of measurements with time-varying local oscillators, utilizing normally-ordered [9] correlation methods to treat both the cases of squeezing and quantum correlations.

2. PHOTOMULTIPLIER THEORY

We wish to analyse experiments with groups of photo-detectors, each illuminated by combinations of a signal and a local oscillator. For this reason, we start by reviewing the correlations of photon-counts in photo-multipliers. We note that GLAUBER [9] has shown that the probability of N-photon coincidence measurements in a time-interval Δt with N photo-detector atoms is proportional to the normally-ordered, time-ordered correlation function of 2N electric field operators:

$$P^N(t_1,\mathbf{x}_1, \dots t_N, \mathbf{x}_N) \propto (\Delta t)^N \langle \hat{E}^{(-)}(t_1,\mathbf{x}_1) \dots \hat{E}^{(-)}(t_N,\mathbf{x}_N)\, \hat{E}^{(+)}(t_N,\mathbf{x}_N) \dots \hat{E}^{(+)}(t_1,\mathbf{x}_1)\rangle$$

$$\text{where: } t_1 < t_2 \dots < t_N \,. \tag{2.1}$$

In general, the GLAUBER analysis uses perturbation theory, and therefore applies to weakly coupled photo-detectors. We will assume, however, that it can be extended to nearly perfectly efficient photo-detectors, of the type used in quadrature-phase measurements. In addition, real detectors are necessarily of finite volume and area. The relative efficiency at location $\mathbf{x}$ will be written as $\eta_n(\mathbf{x})$ per unit volume for the n–th detector. The actual coincidence count-rate for N counts at times $t_1, \dots, t_N$ depends on the combination of detectors that register the counts. The list of detectors that register counts is therefore written in order of arrival times as $\{\nu\}$, where $\{\nu\} = \{\nu(1),\dots,\nu(N)\}$. The total counting probability is then obtained on integrating (2.1) over the relevant active volumes and efficiency factors:

$$P^{(N)}_{\{\nu\}}(t_1, \dots t_N) \propto \iiint P(t_1,\mathbf{x}_{\nu(1)} \dots t_N,\mathbf{x}_{\nu(N)}) \prod_{n=1}^{N} \eta_n(\mathbf{x}_{\nu(n)}) d^3\mathbf{x}_{\nu(n)} . \tag{2.2}$$

Next, it is useful to expand in the orthogonal modes at each detector. As long as these do not spatially overlap, the electromagnetic fields can be expanded with distinct localized operators for the fields in the vicinity of each detector. That is, near the n-th detector with a narrow-band field of mean frequency ω_n, the postive frequency component of the electric field operator is written:

$$\hat{\mathbf{E}}^{(+)}(t,\mathbf{x}) = \left(\frac{h\omega_n}{4\pi\varepsilon}\right)^{1/2} \sum_j \hat{\Phi}_{nj}(t,\mathbf{x})\, \mathbf{u}_{nj}(\mathbf{x}) \,. \tag{2.3}$$

Here $\hat{\Phi}_{nj}(t,x)$ is defined as a sum over longitudinal modes within the j-th transverse mode $\mathbf{u}_{nj}$. In a paraxial approximation, and a quantization length of l :

$$\hat{\Phi}_{nj}(t,x) = \left(\frac{c}{l}\right)^{1/2} \sum_k \hat{a}_{knj}(t)\, e^{ikx} . \tag{2.4}$$

In addition, in the paraxial approximation, the mode normalization is:

$$\int |\mathbf{u}_{nj}(\mathbf{x})|^2 \, d^2\mathbf{r} = 1$$

$$\text{where: } \mathbf{r} = (y,z) . \tag{2.5}$$

The annihilation operator $\hat{a}_{knj}(t)$ describes the k-th longitudinal mode at the n-th detector, in the j-th transverse mode. The physical interpretation of $\hat{\Phi}_{nj}(t,x)$ is that it corresponds to the input amplitude of the photon flux in the j-th transverse mode at the n-th detector location. We now assume that all the input photons are in the lowest order plane-polarized transverse mode (j=1). It is clear that as long as the other transverse modes and polarizations are in the vacuum state, then no counts will result from their appearance in (2.1). Accordingly, we drop these terms completely from now on, writing $\hat{\Phi}_n(t) = \hat{\Phi}_{n1}(t,x_n^o)$. Here x_n^o is the average x-coordinate of the detector active volume. We also assume that the detector is relatively thin, and covers an area A_n of uniform efficiency, so that:

$$\int_{A_n} |\mathbf{u}_{n1}(x_n^o,\mathbf{r})|^2 \, d^2\mathbf{r} = 1 . \tag{2.6}$$

In order to express the count-rate in terms of $\hat{\Phi}_n$, we now suppose that the detector can count nearly all the photons in the fundamental mode that arrive in a given time-interval. More generally, there will be an overall efficiency factor $\eta_n \leq 1$. From the definition of photon density, the number of photons per unit length is equal to $\langle\hat{\Phi}_n^\dagger\hat{\Phi}_n\rangle/c$. Hence the counting probability must be:

$$P^{(N)}_{\{v\}}(t_1, \dots t_N) = \prod_{n=1}^{N}(\eta_n \Delta t) \langle \hat{\Phi}_{v(1)}^\dagger(t_1) \dots \hat{\Phi}_{v(N)}^\dagger(t_N)\, \hat{\Phi}_{v(N)}(t_N) \dots \hat{\Phi}_{v(1)}(t_1)\rangle$$

$$\text{where: } t_1 \leq t_2 \dots \leq t_N . \tag{2.7}$$

The effective detected photon intensity is:

$$\hat{I}_n(t) = \eta_n \hat{\Phi}_n^\dagger(t)\, \hat{\Phi}_n(t) . \tag{2.8}$$

The time-ordered, normally ordered symbols (: :) can now be employed to re-express (2.7) in terms of coincidence rates $G^{(N)}_{\{v\}}(t_1, \dots t_N)$ as:

$$G^{(N)}_{\{v\}}(t_1, \dots t_N) = P^{(N)}_{\{v\}}(t_1, \dots t_N)/\Delta t^N = \langle : \hat{I}_{v(1)}(t_1) \dots \hat{I}_{v(N)}(t_N) : \rangle . \tag{2.9}$$

Here it is assumed that the fields are relatively narrow-band compared to the photo-detector spectral response, in order that all the input photons are detected with uniform efficiency at all relevant frequencies.

3. LOCAL OSCILLATORS

The fields that are incident on the detectors will be composed of local oscillator inputs and signal inputs. In general, there can usefully be up to L input local oscillators or signals $\hat{\Phi}_m^{IN}$ for L photo-detectors. These are combined using beam-splitters, interferometers or other linear devices to generate the L detectable photon intensities. Thus, the detected fields can be written as a linear combination of all the input fields:

$$\hat{\Phi}_m(t) = \int \sum_n g_{mn}(t-t') \, \hat{\Phi}_n^{IN}(t') \, dt' . \tag{3.1}$$

Here the meaning of g_{mn} is of an amplitude transmission. We note that this formalism is closely related to the definition of the time-dependent spectrum [10], except that it is generalized to treat L traveling wave modes. It can also be extended to include linear devices that cause squeezing through phase-conjugation, as in the input-output formulation [10]. However, the intention here is to focus on the usual, essentially noiseless, filters employed in the detection process. For simplicity, we include any propagation delays in g_{mn}. An overall phase factor due to the dielectrics is also included. An input term $\hat{\Phi}_o^{IN}(t')$ can be used to represent the loss-reservoirs of the combining media where necessary. It needs to be included for the commutation relations, and is assumed to represent low temperature reservoirs with negligible thermal photon numbers. Because of normal-ordering, it does not come into our final results, although it would be important at low frequencies. There would be a corresponding output photo-detector signal $\hat{\Phi}_o(t)$ which is not observed.

We can now define the average detected photon intensity at the n-th photo-detector as :

$$I_n(t) = G_n^{(1)}(t) = \eta_n \langle \hat{\Phi}_n^\dagger(t) \, \hat{\Phi}_n(t) \rangle . \tag{3.2}$$

Clearly, the coincidence rate for two photon counts is just:

$$G_{mn}^{(2)}(t,t') = \langle : \hat{I}_m(t) \, \hat{I}_n(t') : \rangle . \tag{3.3}$$

However, the quantity most often measured in experiments is the output current correlation function. The output current is composed of current pulses that occur at each photon arrival. We suppose that these can overlap, but do not interfere with each other:— the detectors are not saturated, and possible dead-time effects are neglected. The current pulse for each photon is $J_n^o(t-t_p)$ for a photon at the n-th detector, arriving at time t_p. The total output current at the n-th detector is thus:—

$$J_n(t) = \sum_{\{t_{jn}\}} J_n^o(t-t_{jn}) . \tag{3.4}$$

Here $\{t_{jn}\}$ is the set of photon arrival times for the n-th detector. An average output current correlation can also be calculated, with:

$$G^{J}_{mn}(t,t') = \langle J_m(t)\, J_n(t')\rangle \, . \tag{3.5}$$

The expected value of the observed current correlation function is easily calculated using the coincidence rate $G^{(2)}_{mn}(t,t')$. We first note that the expression (3.5) can be rewritten as:

$$\begin{aligned} \langle J_m(t)\, J_n(t')\rangle \;=\; & \langle \delta_{mn} \sum_{\{t_{jn}\}} J^{o}_{n}(t-t_{jn})\, J^{o}_{n}(t'-t_{jn}) \rangle + \\ & + \langle \sum_{\{t_{jm}\}\neq\{t_{kn}\}} \sum J^{o}_{m}(t-t_{jm})\, J^{o}_{n}(t-t_{kn}) \rangle \, . \end{aligned} \tag{3.6}$$

This reduces the expression to a sum over terms in which the photons are the same in $J_m(t)$ and $J_n(t')$, together with a sum over terms corresponding to coincidence measurements of different photons. The probability of a single photon count is $G^{(1)}_{n}\cdot\Delta t$, while the probability of a coincidence is $G^{(2)}_{mn}(t,t')\cdot\Delta t^2$. Accordingly, the sums can be replaced by time integrals, with:

$$\begin{aligned} G^{J}_{mn}(t,t') \;=\; & \delta_{mn} \int G^{(1)}_{n}(t_1)\, J^{o}_{n}(t-t_1)\, J^{o}_{n}(t'-t_1)\, dt_1 \;+ \\ & + \iint G^{(2)}_{mn}(t_1,t_2)\, J^{o}_{m}(t-t_1)\, J^{o}_{n}(t'-t_2)\, dt_1\, dt_2 \, . \end{aligned} \tag{3.7}$$

This expression is correct for arbitrary current pulse-shapes or local-oscillator envelopes, provided the earlier assumptions about the photo-detector response are valid. It is useful to simplify this somewhat. Thus, it is often the case that the output current pulses are short compared to the correlation time-scales of interest. This is even possible when the local oscillator is modulated more rapidly than the photo-detector response time, as could happen with picosecond input time-scales.

We therefore turn to the photo-current spectrum. Defining:

$$G^{J}_{mn}(\omega,\omega') \equiv \frac{1}{2\pi}\iint e^{i\omega t+i\omega t'}\, G^{J}_{mn}(t,t')\, dt\, dt'$$

$$G^{(2)}_{mn}(\omega,\omega') \equiv \frac{1}{2\pi}\iint e^{i\omega t+i\omega' t'}\, G^{(2)}_{mn}(t,t')\, dt\, dt'$$

and:

$$J^{o}_{n}(\omega) \equiv \frac{1}{\sqrt{2\pi}}\int e^{i\omega t}\, J^{o}_{n}(t)\, dt$$

$$G^{(1)}_{n}(\omega) \equiv \frac{1}{\sqrt{2\pi}}\int e^{i\omega t}\, G^{(1)}_{n}(t)\, dt \tag{3.8}$$

we obtain the photo-current spectrum normalized by the spectral response:

$$\frac{G^J_{mn}(\omega,\omega')}{J^o_m(\omega)\,J^o_n(\omega')} = \sqrt{2\pi}\delta_{mn}\,G^{(1)}_n(\omega+\omega') + 2\pi G^{(2)}_{mn}(\omega,\omega') \,. \tag{3.9}$$

This general result includes all the spectral properties of the local-oscillators and the photo-detector response. The first term is known as the shot-noise term, while the next includes the nonclassical correlation effects.

4. QUADRATURE PHASE DETECTION

In practical situations of local oscillator measurement, the local oscillator field is intense, while the signal field is relatively weak. This allows the expression of (3.9) to be greatly simplified. We first define a fluctuation operator:

$$\delta\hat{\Phi}_n(t) = \hat{\Phi}_n(t) - \langle\hat{\Phi}_n(t)\rangle \tag{4.1}$$

Next, in evaluating the correlation spectrum it is useful to define quadrature phase operators. We therefore introduce a normalized envelope function $u_n(t)$ for the local oscillator incident at the n-th detector. This is evaluated in terms of the coherent part of the total input field. That is, the local oscillator is taken to include the coherent part of any input signal as well as coherent local oscillator signal. In reality, the average amplitude of any input signal is changed by the existence of phase diffusion [11], which can strongly modify the squeezing spectrum. These effects are neglected in the present treatment. The coherent part of the total input field is:

$$\phi_n(t) = \langle\hat{\Phi}_n(t)\rangle = |\phi_n(t)|\, \exp\,(-i\theta(t)). \tag{4.2}$$

This generates a coherent detected photon intensity $I^c_n\,(t)$ of:

$$I^c_n\,(t) = \eta_n|\phi_n(t)|^2. \tag{4.3}$$

In a time interval T, the average intensity in the coherent signal is then $\bar{I}^c_n$, where:

$$\bar{I}^c_n = \int_{-T/2}^{T/2} I^c_n\,(t)\; dt/T. \tag{4.4}$$

This intensity will be used to normalize the coherent part u_n of the n-th local oscillator envelope. The phase angle of the local oscillator field is $\theta(t)$, which is assumed to be experimentally adjustable without altering the fluctuation field. The input coupling term g_{mn} implicitly includes this phase angle. Let:

$$u_n(t) = \eta_n\,|\phi_n(t)|\,(\bar{I}^c_n)^{-1/2}. \tag{4.5}$$

The quadrature field $\hat{X}^\theta_n(t)$ at the n-th detector can now be defined in a straightforward way, relative to the phase angle θ_n and the phase independent temporal field envelope $u_n(t)$. An alternative way of understanding this is that the time-dependent local-oscillator detector is sensitive to quadrature components whose longitudinal mode envelope in the x direction is just $u_n(t-x/c)$. This is an obvious extension of the usual c.w. local oscillator case, where the longitudinal mode envelope is $\exp[-i\omega(t-x/c)]$. In terms of the input fields, we let:

$$\hat{X}^{\theta_m}_m(t) = u^*_m(t)\, e^{i\theta_m}\, \delta\hat{\Phi}_m(t) + \text{h.c.} = \int u^*_m(t)\, e^{i\theta_m} \sum_n g_{mn}\,(t-t')\; \delta\hat{\Phi}^{IN}_n(t')\, dt' + \text{h.c.}$$

where:

$$\delta\hat{\Phi}^{IN}_n(t) \;=\; \hat{\Phi}^{IN}_n(t) \;-<\hat{\Phi}^{IN}_n(t)> . \tag{4.6}$$

Expanding out the correlation function $G^{(2)}_{mn}(t,t')$ in term of these quadrature operators gives to lowest non-vanishing order in the fluctuation operators:

$$G^{(2)}_{mn}(t,t') = G^{(1)}_m(t)\, G^{(1)}_n(t') + \left(\bar{I}^c_m\, \bar{I}^c_n\right)^{1/2} <:\hat{X}^{\theta_m}_m(t)\, \hat{X}^{\theta_n}_n(t'):> . \tag{4.7}$$

There are also terms of order $<|\delta\hat{\Phi}^{IN}|^4>$ which can be neglected here provided there is a strong coherent local oscillator signal. The result of (4.7) can now be utilized in the Fourier transform relation for the current spectrum. First it is necessary to define a Fourier transform quadrature operator. This is defined so that it will satisfy the usual commutation relations in frequency space in the limit of a time-independent local oscillator:

$$\hat{X}^{\theta_n}_n(\omega) \;=\; \frac{1}{\sqrt{2\pi}} \int e^{i\omega t}\, \hat{X}^{\theta_n}_n(t)\, dt . \tag{4.8}$$

Substituting (4.7) and (4.8) into the earlier result for the current correlation function gives:

$$\frac{G^J_{mn}(\omega,\omega')}{J^o_m(\omega) J^o_n(\omega')} \;=\; \sqrt{2\pi}\,\delta_{mn}\, G^{(1)}_n(\omega+\omega') + 2\pi\left[\, G^{(1)}_m(\omega)\, G^{(1)}_n(\omega') + \left(\bar{I}^c_m\, \bar{I}^c_n\right)^{1/2} <:\hat{X}^{\theta_m}_m(\omega)\, \hat{X}^{\theta_n}_n(\omega'):>\right] \tag{4.9}$$

A simplified version of (4.9) is obtained if the power spectrum, $G^J_{mn}(\omega) \equiv (2\pi/T) G^J_{mn}(-\omega,+\omega)$ is calculated. In addition, as stated earlier, it is often the case that the current pulse spectrum $J^o(\omega)$ is frequency-independent out to some cutoff of the photo-detector response at ω_r. For frequencies below this limit, $J^o_n(\omega)$ can be replaced by $(Z_n e)/\sqrt{2\pi}$, where Z_n is the total number of output electrons per current pulse. The resulting power spectrum is then:—

$$\frac{G^J_{mn}(\omega)}{Z_m Z_n e^2} \;=\; \delta_{mn}\, \bar{I}^c_m + \frac{2\pi}{T}\left\{ G^{(1)}_m(-\omega) G^{(1)}_n(\omega) + \left(\bar{I}^c_m\, \bar{I}^c_n\right)^{1/2} <:\hat{X}^{\theta_m}_m(-\omega)\, \hat{X}^{\theta_n}_n(\omega):>\right\}. \tag{4.10}$$

In this equation, the shot noise term of order $<|\delta\hat{\Phi}^{IN}|^2>$ is neglected relative to the large coherent term $\bar{I}^c_m$.

The new feature in this equation compared to that for c.w. experiments is clearly the terms like $|G_n^{(1)}(\omega)|^2$, which correspond to the local-oscillator spectrum. These terms would normally have a delta-function like behaviour at zero frequency, in c.w. experiments. In a mode-locked laser, they become a comb of delta-functions at discrete intervals that correspond to the laser cavity mode spacing. The local-oscillator spectral terms can be reduced by employing balanced homodyne detectors [7,12]. This technique simply involves taking the difference of the output current at two detectors to cancel the local-oscillator term. However, it is difficult in practical terms to obtain a complete cancellation when the local oscillator is intense. It is then necessary to investigate the interesting quantum fluctuation behaviour of $\hat{X}_n(\omega)$, by examining the regions of frequency space where the local-oscillator spectrum vanishes[8], which simplifies the result of (4.9) even further. In this region it is useful to define a normalized squeezing variance or correlation $V_{mn}(\omega)$ by normalizing with the geometric mean of the shot-noise in each channel, so that:

$$V_{mn}(\omega) = \left(\frac{G^{J}_{mn}(-\omega,\omega)}{Z_m Z_n e^2 \left(\bar{I}^c_m \, \bar{I}^c_n\right)^{1/2}} \right) . \tag{4.11}$$

In the relevant region of zero local-oscillator spectral intensity, this reduces to:

$$V_{mn}(\omega) = \delta_{mn} + S_{mn}(\omega)$$

where:
$$S_{mn}(\omega) = \frac{2\pi}{T} \langle : \hat{X}^{\theta_m}_m(-\omega) \, \hat{X}^{\theta_n}_n(\omega) : \rangle . \tag{4.12}$$

Here the squeezing spectrum $S_{mn}(\omega)$ is the departure from coherent shot-noise behaviour. In fact, according to GLAUBER[9], the definition of a time-dependant coherent field is that there is a factorization of $G^{(2)}_{mn}(t,t')$ into a product of the form $G^{(1)}_m(t)\, G^{(1)}_n(t')$, which clearly can only occur when $S_{mn}(\omega) = 0$. In the next section, we show that squeezing results occur when $S_{mn}(\omega) < 0$, while Cauchy-Schwarz violations of classical correlation inequalities occur when $S_{nn}(\omega)S_{mm}(\omega) < |S_{mn}(\omega)|^2$. All of these results are dependent on the phase-angle of the quadratures $\hat{X}^{\theta_n}_n(\omega)$, which are included in the transmission amplitudes $g_{mn}(t-t')$.

5. CORRELATIONS AND SQUEEZING

The quadrature operator $\hat{X}^{\theta_n}_n$ introduced here has a novel appearance when compared to the usual form of this operator. Accordingly, we wish to demonstrate how this is related to quantum correlations and squeezing. From the definitions of $\hat{\Phi}^{IN}_n(t)$, $\hat{\Phi}_n(t)$, it is clear that:

$$[\hat{\Phi}_m(t), \hat{\Phi}^{\dagger}_n(t')] = \delta_{mn}\, \delta(t-t')$$

$$[\, \hat{\Phi}^{IN}_m(t), [\hat{\Phi}^{IN}_n]^{\dagger}(t')] = \delta_{mn}\, \delta(t-t'). \tag{5.1}$$

This implies that $g_{nm}(\tau)$ is both unitary and causal:

$$g_{nm}(\tau) = 0 \qquad (\tau < 0)$$

$$\sum_n \int g_{mn}(t_1-\tau)\, [g_{m'n}(t_2-\tau)]^* d\tau = \delta_{mm'}\, \delta(t_1-t_2)\,. \tag{5.2}$$

The commutation properties of the $\hat{X}$ operators amongst themselves can also be easily calculated, as:

$$[\hat{X}^{\theta}_m(t), \hat{X}^{\theta'}_n(t')] = 2i \sin(\theta(t) - \theta'(t))|u_n(t)|^2\, \delta_{mn}\, \delta(t-t')\,. \tag{5.3}$$

It is clear that the arbitrary time-dependent filters $g_{mn}(t)$ have little effect on these commutators. However, the local oscillator time-dependence does change the commutation relations of the quadrature Fourier transforms. These are now found to be (for constant $\theta(t)$):—

$$[\hat{X}^{\theta}_m(\omega), \hat{X}^{\theta'}_n(\omega')] = i\sin(\theta - \theta')\, \delta_{mn} \int |u_n(t)|^2\, e^{i(\omega+\omega')t}\, dt/\pi \tag{5.4}$$

In particular, if we choose $\omega = -\omega'$, $\theta-\theta' = \pi/2$, the following result is obtained:

$$[\hat{X}^{\pi/2}_m(-\omega), \hat{X}^{0}_n(\omega)] = i\, \eta_n\, \delta_{mn}\, (T/\pi). \tag{5.5}$$

This identifies the operators $\hat{X}^{\pi/2}_m(-\omega)$, $\hat{X}^{0}_n(\omega)$ as having an identical commutator algebra to the $(\hat{x},\hat{p})$ commutator of the Heisenberg uncertainty principle. However, note that $\hat{X}^{\theta}_n(-\omega)$, $\hat{X}^{0}_n(\omega')$ do *not* commute even at different frequencies. Another difficulty with the use of a Fourier expansion basis is that the operators $\hat{X}^{\theta}_n(\omega)$ are not Hermitian. For C.W. local oscillators, it is known that these operators can be divided up into real and imaginary parts which commute, and hence are independently measurable quantum mechanical observables. This is not the case for time-dependent local oscillators, as is easily verified by considering the effect of sin and cosine transformations in the integrand of (5.4). It is possible, instead, to expand the Hermitian quadrature operators in an orthonormal series $\hat{x}^{\theta}_{kn}$:

$$\hat{x}^{\theta}_{kn} = \int_{-T/2}^{T/2} u_{kn}(t)\, \hat{X}^{\theta}_n(t)\, dt$$

where: $$\int u_{kn}(t)\, u_{jn}(t)\, |u_n(t)|^2\, dt = \delta_{kj}\,. \tag{5.6}$$

Given this choice, it is clear that:

$$[\hat{x}^{\theta}_{kn}, \hat{x}^{\theta'}_{k'n'}] = 2i \sin(\theta-\theta')\, \delta_{kk'}\delta_{nn'}\,. \tag{5.7}$$

Having identified the commutators of the quadrature operators, we can now investigate the types of nonclassical behaviour in these experiments. This is easily obtained, due to the fact that a field with classical statistics has a GLAUBER-SUDARSHAN P-representation that is positive [9]. The following *classical* inequalities are obtained, independent of detector efficiency:

$$\text{(i)} \quad S_{nn}(\omega) \geq 0$$
$$\text{(ii)} \quad S_{nn}(\omega)S_{mm}(\omega) \geq |S_{mn}(\omega)|^2 . \qquad \text{[CLASSICAL]} \qquad (5.8)$$

Violation of these inequalities will be called squeezing and quantum correlation behaviour respectively.

These inequalities can be violated by the full quantum operators, although there are corresponding *quantum* inequalities:

$$\text{(i)} \quad \langle \hat{X}_n^{\theta}(-\omega)\hat{X}_n^{\theta}(\omega)\rangle \geq 0$$
$$\text{(ii)} \quad \langle \hat{X}_m^{\theta}(-\omega)\hat{X}_m^{\theta}(\omega)\rangle\langle \hat{X}_n^{\theta'}(-\omega)\hat{X}_n^{\theta'}(\omega)\rangle \geq |\langle \hat{X}_m^{\theta}(-\omega)\hat{X}_n^{\theta'}(\omega)\rangle|^2 . \ \text{[QUANTUM]} \quad (5.9)$$

These full operator relations can be transformed into relations for the observed normally-ordered variances by using the commutation relations to order the operators in (5.9), hence obtaining:

$$\text{(i)} \quad S_{nn}(\omega) + \eta_n \geq 0$$
$$\text{(ii)} \quad [S_{mm}(\omega) + \eta_m][S_{nn}(\omega) + \eta_n] \geq |S_{mn}(\omega) + \delta_{mn}\eta_n|^2. \quad \text{[QUANTUM]} \quad (5.10)$$

Similar inequalities hold for the squeezing variance, which is related to the non-ordered quadrature operators by:

$$V_{mn}(\omega) = \delta_{mn}(1-\eta_n) + \frac{2\pi}{T}\langle \hat{X}_m^{\theta_m}(-\omega)\, \hat{X}_n^{\theta_n}(\omega)\rangle. \qquad (5.11)$$

A complete suppression of the shot noise, or a perfect correlation between the operators, is only possible when the detectors are perfectly efficient ($\eta_n = 1$). This is independent of the input fields or any effects of the beam-splitters and filters.

6. SUMMARY

A normally ordered theory of time-dependent intensity spectral measurements was presented. Here, the quadrature operators are defined relative to the envelope of the average amplitude of the total field at the detector. This formalism includes the situation where the input field is itself used as the local oscillator, possibly after spectral filtering. The time-dependence of the average amplitude at the detector is regarded here as the local oscillator, regardless of its origin. The time-dependant properties of the local oscillator relative to those of the fluctuations crucially determines the extent of squeezing or quantum correlations that are observed.

Since the time-dependence of the local-oscillator itself determines the quadrature operator that is detected, it is often necessary to mode-match in the time-domain between the input signal with the local-oscillator. The duration of the local oscillator should correspond to the duration of the squeezed signal for optimal results. Unlike the c.w. case, the different Fourier components of the output current are not independent observables. At different phase angles, the frequency domain quadrature operators do **not** commute when their frequencies are distinct. To obtain a set of orthogonal quadratures, it is necessary to use generalized transforms that are defined relative to a kernel given by the local-oscillator intensity. However, it is still possible to obtain inequalities for the Fourier components, whose violation signals the existence of squeezing and quantum correlations.

REFERENCES

1. H. Takahashi: Adv. in Comm. Syst. **1**, 227 (1965).
   D.R. Robinson: Comm. Math. Phys. **1**, 159 (1965).
   D. Stoler: Phys. Rev. **D1**, 3217 (1970); **D4**, 1925 (1971).
   H.P. Yuen: Phys. Rev. **A13**, 2226 (1979).
   For a review, see: R. Loudon and P.L. Knight: J. Mod. Opt. **34**, 709 (1987).
2. R.E. Slusher, L.W. Hollberg, B. Yurke, J.S. Mertz and J.F. Valley: Phys. Rev. Lett. **55**, 2409 (1985).
   R.M. Shelby, M.D. Levenson, S.H. Perlmutter, R.S. Devoe and D.F. Walls: Phys. Rev. Lett. **57**, 691 (1986).
   L. Wu, H.J. Kimble, J.L. Hall and H. Wu: Phys. Rev. Lett. **57**, 2520 (1986).
   M.W. Maeda, P. Kumar and J.M. Shapiro: Opt. Lett. **12**, 161 (1987).
   B.L. Schumaker, S.H. Perlmutter, R.M. Shelby and M.D. Levenson: Phys.Rev. Lett. **58**, 357 (1987).
3. M.D. Reid and P.D. Drummond: Phys. Rev. Lett. **60**, 2731 (1988).
4. A. Einstein, B. Podolsky and N. Rosen: Phys. Rev. **47**, 777 (1935).
   J.S. Bell: Physics **1**, 195 (1964).
5. S.J. Carter, P.D. Drummond, M.D. Reid and R.M. Shelby: Phys. Rev. Lett. **58**, 1841 (1987).
6. C.M. Caves: Phys.Rev. **D26**, 1817 (1980).
7. H.P. Yuen and J.H. Shapiro: I.E.E.E. Trans. Inf. Theory **IT-26**, 79 (1980).
   J.H. Shapiro: I.E.E.E. J. Quant. Elect. **QE-21**, 237 (1985).
   Z.Y. Ou, C.K. Hong and L. Mandel: J.O.S.A. **B4**, 1574 (1987).
   H.J. Carmichael: J.O.S.A. **B4**, 1588 (1987).
   M.J. Collett and R. Loudon: J.O.S.A. **B4**, 1525 (1987).
8. B. Yurke, P. Grangier, R.E. Slusher and M.J. Potasek: Phys. Rev. **A35**, 3586 (1987).
   R.E. Slusher, P. Grangier, A. LaPorta, B. Yurke and M.J. Potasek: Phys. Rev. Lett. **59**, 2566 (1987).
9. R.J. Glauber: Phys. Rev. **130**, 2529; **131**, 2766 (1963).
   E.C.G. Sudarshan: Phys. Rev. Lett. **10**, 277 (1963).
10. J.H. Eberly and K. Wodkiewicz: J.O.S.A. **67**, 1252 (1977).
    C.W. Gardiner and M.J. Collett: Phys. Rev. **A31**, 3761 (1985).
11. P.D. Drummond and M.D. Reid: Phys. Rev. **A37**, 1806 (1988).
12. H.P. Yuen and V.W.S. Chen: Opt. Lett. **8**, 177 (1983).
    B.L. Schumaker: Opt. Lett. **9**, 189 (1984).

Part II

Quantum Measurement Problems in Optics

Backaction Evasion Experiments

R.E. Slusher, A. La Porta, and B. Yurke

AT & T Bell Laboratories, Murray Hill, NJ 07974, USA

1. Introduction

Measurements of an optical field typically involve using a beamsplitter to split off and detect a small fraction of a beam. This process attenuates the field being measured. The small value of the measured portion of the field together with the quantum uncertainty of the measurement process results in a signal-to-noise ratio much smaller than that intrinsic to the incident field and its quantum uncertainty. One can think of this uncertainty as resulting from vacuum fluctuations entering the unused port of the beamsplitter. These fluctuations combine with the split off beam and increase the uncertainty in the measurement. This noise increase is called "backaction" noise.[1] For measurements of the quadrature components of the optical field, it is possible to "evade" this backaction noise by using a combination of linear and nonlinear optical interactions. We will describe in detail a particularly simple scheme using only two polarization rotators and an optical parametric amplifier based on parametric downconversion.[2,3] Using this simple arrangement, we demonstrate experimentally that one quadrature component of a coherent state input field is measured without changing either its initial magnitude or its variance.

A simple beamsplitter with squeezed vacuum entering the normally dark input port can also serve as a backaction evading device, as pointed out by Shapiro.[4] However, in this case, in order to obtain a transmitted signal equal in magnitude to the input signal, the fraction of the beam split off must approach zero. At the same time the degree of squeezing at the input must be very large. This squeezing would typically be achieved by using a parametric amplifier. This paramp is used in a more efficient manner in the scheme described here. The coupling to the backaction evader we describe can be large, of order unity, allowing the most efficient use of the paramp gain.

Our backaction evader can be described by a simple dual parametric amplifier model outlined in the next section. Details of the present experimental arrangement are given in the third section. We conclude with some suggestions for future experiments including a rigorous quantum nondemolition experiment and a new idea for a machine to generate small optical Schrödinger cats.

2. Dual Parametric Amplifier Mode

A backaction evasion apparatus can be thought of as a black box with signal input/output and readout input/output channels. The signal and readout channels can be separate spatial modes, separate frequencies or orthogonal polarization channels. Yurke[2] has described designs for several variations of backaction evaders utilizing strong coupling of both channels to a paramp. Polarization channels for these strong coupling evaders have been suggested by two other groups.[5,6] We have demonstrated a scheme using polarization channels and a single pass through a polarization rotator, a paramp and a second polarization rotator.[3] The paramp is a pair of KTP crystals pumped by a doubled Nd:YAG laser in the type II phase matching geometry.

Springer Proceedings in Physics, Vol. 41 **Quantum Optics V**
Editors: J.D. Harvey and D.F. Walls

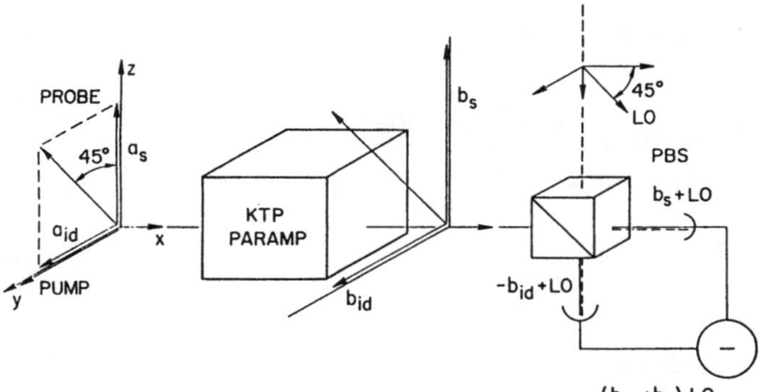

Fig. 1 - A schematic diagram of the experimental geometry and polarization directions used to measure the parametric gain of a KTP crystal. The probe polarization is at 45° with respect to the z axis of the KTP crystal. It has polarization components along the signal and idler axes where field operators are designated as a_s and a_{id}. The idler and pump polarizations are along the y axis corresponding to type II phase matching. A measurement of the amplified fields (operators b_s and b_{id}) is accomplished by a polarization beam splitter and balanced homodyne detectors.

It is particularly simple to model our evader using the dual paramp model analyzed in section V of Yurke's article.[2] The two paramps correspond to the two polarization directions at 45° with respect to the z axis of the KTP crystals. These axes are also used for the classical gain measurements as shown in Fig. 1.

A probe beam is incident on the paramp with its axis at 45° with respect to the z axis of the KTP crystals. It has equal components along the signal and idler polarization axes. Field annihilation operators along these two axes are denoted by a_s and a_{id}. After passing through the paramp both signal and idler fields are transformed to the operators b_s and b_{id}, and the amplified probe polarization remains at 45°. The amplitude of this amplified probe is measured by a balanced homodyne detector as shown in Fig. 1. The measured gain, G_f, is the ratio of the output/input probe amplitude. No polarization rotation occurs during the amplification process, indicating that the two channels at 45° with respect to the z axis are independent. The independence of the two channels is easily seen by considering the standard paramp transformation

$$b_s = Ga_s + \sqrt{G^2-1}\ a_{id}^\dagger \tag{1}$$

and

$$b_{id} = \sqrt{G^2-1}\ a_s^\dagger + Ga_{id}\ , \tag{2}$$

where G is the gain used in Yurke's analysis.[2] The field operators along the 45° axes, denoted a_I and a_{II} are related to the signal and idler fields by

$$a_I = \frac{a_s + a_{id}}{\sqrt{2}} \tag{3}$$

and

$$a_{II} = \frac{a_s - a_{id}}{\sqrt{2}}\ . \tag{4}$$

Combining Eqs. (1-4), the independence of the polarization channels I and II is immediately obtained

$$b_I = Ga_I + \sqrt{G^2-1}\ a_I^\dagger \tag{5}$$

and

$$b_{II} = Ga_{II} - \sqrt{G^2-1}\ a_{II}^\dagger\ . \tag{6}$$

One can show that the measured gain G_f is related to G by

$$G_f = (G + \sqrt{G^2-1}) \tag{7}$$

or

$$G = (G_f + G_f^{-1})/2\ . \tag{8}$$

The complete backaction transformation can now be written for the polarization channels I and II following the diagram in Fig. 2.

Input signal field operators are denoted by s_{1i} and s_{2i}. This input field is vertically polarized along the KTP z axis so that

$$a_{Ii} = (s_{1i} + s_{2i})/\sqrt{2} \tag{9}$$

and

$$a_{IIi} = (s_{2i} - s_{1i})/\sqrt{2}\ . \tag{10}$$

The first polarization rotation by an angle θ corresponds to the transformation

$$a_I = Ka_{Ii} - \sqrt{1-K^2}\ a_{IIi} \tag{11}$$

and

$$a_{II} = \sqrt{1-K^2}\ a_{Ii} + Ka_{IIi}\ , \tag{12}$$

where $K = \cos\theta$. Equations (5) and (6) describe the action of the paramp and the second polarization rotation is represented by

$$b_{Io} = Kb_I - \sqrt{1-K^2}\, b_{II} \tag{13}$$

and

$$b_{IIo} = \sqrt{1-K^2}\, b_I + Kb_{II}\ . \tag{14}$$

Finally, we transform back 45° to the initial polarization reference frame

$$s_{1o} = (b_{Io} - b_{IIo})/\sqrt{2} \tag{15}$$

and

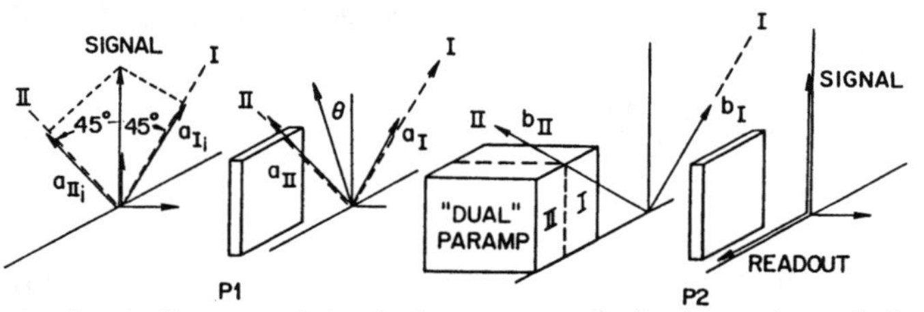

Fig. 2 - A diagram of the dual paramp analysis shows the polarization orientations of independent channels I and II and the corresponding field operators. Polarization rotators P1 and P2 each rotate the polarization by an angle θ.

$$s_{2o} = (b_{Io} + b_{IIo})/\sqrt{2} \; . \qquad (16)$$

to obtain the output signal operators s_{1o} and s_{2o}. The output field is related to the input by

$$\begin{aligned} 2s_{jo} = {} & \left[(2K^2-1)G \mp 2KG\sqrt{1-K^2}\,\right](s_{1i} + s_{2i}) \qquad (17) \\ & + \left[\mp(2K^2-1)G - 2KG\sqrt{1-K^2}\,\right](s_{2i} - s_{1i}) \\ & + \sqrt{G^2-1}\,(s^\dagger_{1i} + s^\dagger_{2i}) \pm \sqrt{G^2-1}\,(s^\dagger_{2i} - s^\dagger_{1i}) \, , \end{aligned}$$

where j = 1 for the upper signs and j = 2 for the lower signs. As shown in Yurke's analysis, in order to replicate one quadrature component of the field in transmission through the evader, i.e., to obtain backaction evasion, the output field operators must be of the following form

$$s_{1o} = s_{1i} + \sqrt{G^2-1}\,(s^\dagger_{2i} - s_{2i}) \qquad (18)$$

and

$$s_{2o} = s_{2i} + \sqrt{G^2-1}\,(s_{1i} + s^\dagger_{1i}) \, . \qquad (19)$$

Comparing Eqs. (18) and (19) with Eq.(17), a simple evasion angle condition

$$\theta_e = \cos^{-1}\left[\sqrt{\frac{(G+1)}{2G}}\right] = \cos^{-1}\left[\frac{G_f+1}{\sqrt{2(G_f^2+1)}}\right] \qquad (20)$$

is obtained.

This is a remarkable result. It is certainly not intuitively clear that the two rotations by the same angle θ_e will lead to backaction evasion. The dual paramp model can help to clarify thinking about the process since one has the possibly of combining arbitrary amounts of the fields from these two independent paramps so that the backaction noise coupled into the signal channel can cancel for one quadrature. It is also clear from this model that the process works over a broad range of frequencies. Each of the paramp transformations is well known to apply over a broad frequency range.[7] An operator at frequency ω is always coupled with the conjugate of that operator at frequency -ω, where ω is the frequency shift from the degenerate condition. Of course the polarization rotators are broadband. Thus all of the equations above carry through for the broadband case. The result is quantum correlation of the readout and replicated signal beams over a broad frequency range around the degenerate frequency. It also means the process works well for the short pulse conditions obtained in the present experiments.

In summary, we write the backaction evasion equations for the quadrature components of the signal and readout fields[3] corresponding to the operator equations (18) and (19)

$$S_{1o} = S_{1i} \qquad (21)$$

$$S_{2o} = S_{2i} + \left(G_f - G_f^{-1}\right)R_{2i} \qquad (22)$$

$$R_{1o} = R_{1i} - \left(G_f - G_f^{-1}\right)S_{1i} \qquad (23)$$

$$R_{2o} = R_{2i} \, . \qquad (24)$$

The "1" quadrature component of the incident signal field S_{1i} is replicated at the output as S_{1o} and a quantum correlated copy is generated in the "1" quadrature of the readout field R_{1o}. Note that the notation has been changed. The 1 and 2 now correspond to quadrature components instead of polarization channels. The quantum correlation of R_{1o} and S_{1o} can be complete by making G_f large; i.e., $G_f S_{1i}/R_{1i}$ can be very large since R_{1i} is at the vacuum fluctuation level. The backaction noise is evaded and appears in the S_{2o} component. Each of the levels and variances in Eqs. (21) through (24) will be verified in the experiments.

Experimently the correlation between the signal and the idler is demonstrated by summing the photocurrents from balanced homodyne detectors measuring the signal quadrature components and the readout quadratures. The ratio of the gains of these two detectors is g. Consider the case when measuring quadrature 1. The sum of the photocurrents is then

$$g i_{S_{1o}} + i_{R_{1o}} = C\left[gS_{1i} + R_{1i} - (G_f - G_f^{-1})S_{1i}\right], \tag{25}$$

where C is a constant relating the currents and the quadrature fields. For a coherent state signal input and vacuum readout input the variances or noise power for the fields S_{1i} and R_{1i} are equal and uncorrelated. When the photodetector gain ratio is chosen at the optimum value

$$g = (G_f - G_f^{-1}), \tag{26}$$

the ratio of the summed current noise power drops below the combined shot-noise level by a factor

$$F = \frac{1}{(G_f - G_f^{-1})^2 + 1}\ . \tag{27}$$

This correlation effect in the summed currents is shown in Fig. 3 along with the correlations expected in a squeezing experiment.

Note that at low classical gain the backaction evasion correlation is considerably less than the squeezing correlation. It is also clear that losses degrade both squeezing and backaction evasion is a similar fashion.

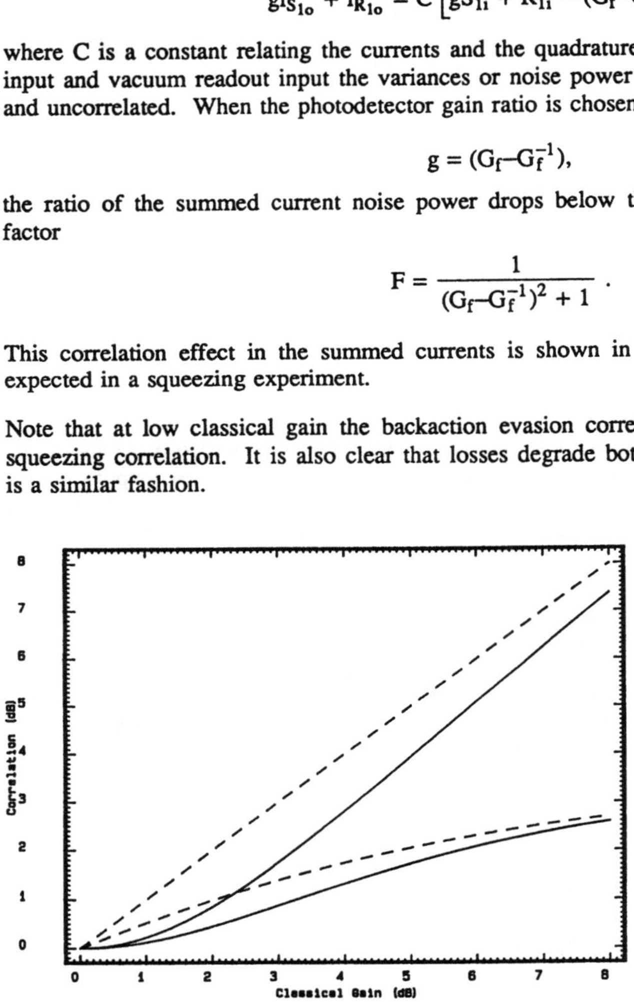

Fig. 3 - The correlation of the signal and read channels is shown as the drop in dB below the summed photocurrent shot-noise as a function of classical gain G_f. Solid curves correspond to the backaction evasion experiment and the dashed curves to an optimized squeezing experiment. In each case the upper curves correspond to $\eta = 1$ and the lower lines to $\eta = 0.7$.

3. Experiments

Experimentally, the evader is similar to the pulsed squeezing apparatus used in earlier studies.[8] A schematic diagram of the apparatus is shown in Fig. 4.

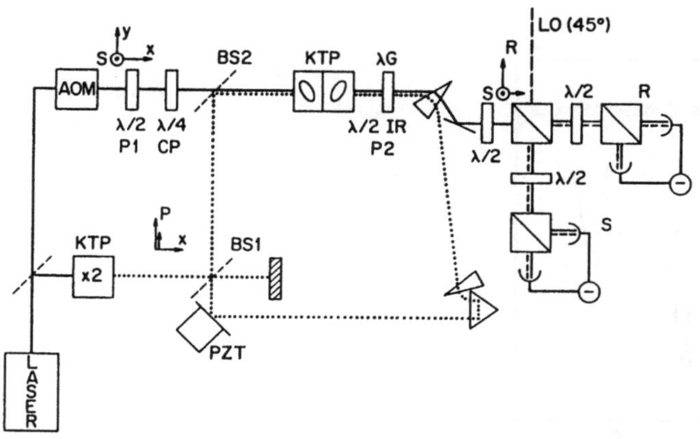

Fig. 4 - A schematic diagram is shown for the backaction evasion apparatus. A Nd:YAG mode-locked laser is doubled in a KTP crystal to form the green pump beam (dotted line) that enters a ring cavity at beamsplitter BS1. A 15 cm focal length lens just before BS2 focuses the pump into the KTP and a similar one recollimates the beam immediately after the KTP. The polarization rotator P2 is a fullwave plate in the green and does not affect the pump polarization. Astigmatism introduced by the first prism is corrected by a second prism. The ring is locked with a portion of the beam which leaks out the back of a BS2 where a careful measurement of the beamwaist is also made. The signal beam is shifted 65 MHz by the acoustooptic modulator AOM. Polarization orientations of the pump P, the signal S and the readout R are indicated. The LO (dashed line) is also split off from the initial unshifted laser beam.

The pump, probe and local oscillator beams are all derived from a standard Nd:YAG mode-locked laser operating at an average power of 10W, a pulse repetition rate of 200 MHz, and pulse widths near 140 ps. A major portion of the infrared beam is doubled to a wavelength of 0.532m using a KTP crystal. This green pump light is injected in a ring resonator to increase the peak power in the green pulses to a level of 200W corresponding to a peak intensity of 10MW/cm^2 in the focal volume at the KTP paramp with a beam waist of 25 μm. A small portion of the leaked light from this ring is used in a PZT mounted mirror feedback loop to stabilize the power in the loop within a $\pm 1\%$ range of the peak value.

Two KTP crystals are used for the parametric amplifier. They are cut with phase matching angles near 24°, one at +24° and the other at -24°, with respect to the y axis so that a compensation for walkoff of light with polarization along the y axis is obtained.[9] The pump polarization is along the y axis. Using a probe beam as described in Fig. 1, the gain G_f is measured to be 1.33 (2.5 dB in power). This corresponds to an evasion angle $\theta_e = 8°$. Birefringence in the KTP crystals is compensated for by the compensating plate, CP, in Fig. 4. Two halfwave plates rotate the polarization at the input and output of the paramp by the 8° evasion angle. The polarization rotator P2 rotates the polarization of the infrared to a horizontal orientation for lossless transmission through a prism that separates the infrared and green light. After this prism the second evasion angle rotator completes the backaction evasion process.

A vertically-polarized signal beam, shifted 65 MHz from the laser frequency by an acoustooptic modulator, is injected into the evader. After the second evasion angle rotation, the vertically-polarized signal beam and the horizontally-polarized read beam are separated by a polarization beam splitter and directed toward two phase-sensitive homodyne detectors. This separating beamsplitter is also used to introduce the local oscillator beam. The LO is split equally between the two detectors by orienting its polarization at 45° with respect to the

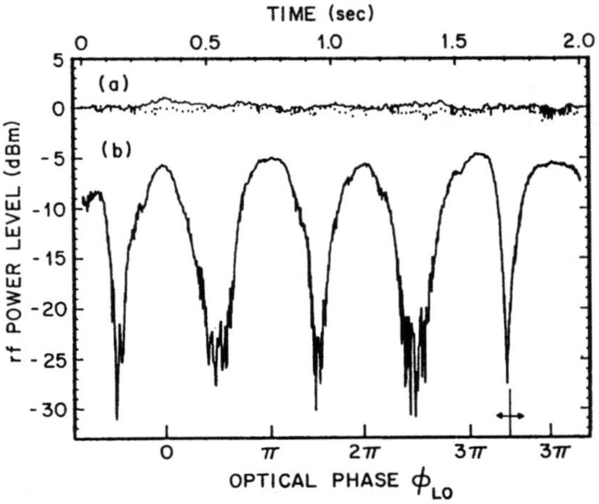

Fig. 5 - Levels of the (a) signal and (b) readout are shown as a function of local oscillator phase ϕ_{LO} and time. The rf level of the 65-MHz beat between LO and signal or readout is measured on a spectrum analyzer as ϕ_{LO} is swept. In (a) the dotted line is the signal input level measured with the paramp off, and the solid line is the signal output level measured with the paramp on. The readout level in (b) is only present with the paramp on and is large for only one quadrature component. The rf bandwidth is 10 kHz and the video bandwidth is 100 Hz for these data.

beamsplitter axis. A halfwave plate at the input to each detector rotates the polarization of the combined LO and signal or read beams by 45° so that equal portions of the combined beams are separated by final polarization beamsplitter and directed to a pair of balanced detectors for both the signal and read channels. Both signal and read photocurrents from the balanced detectors are subtracted and amplified to yield currents proportional to the signal and read fields (see Fig. 1). The levels of these fields are shown in Fig. 5 as a function of the LO phase.

The level of the signal beam with the evader on or off is identical within the ±0.5 dB fluctuations due to low-frequency technical noise, while the read level rises to a value near the signal level for only one phase quadrature. At the other phase quadrature, the read level drops nearly 25 dB to a value limited by rf pickup from the 65 MHz AOM. These measured levels are in excellent agreement with Eqs. (21) through (24) using the measured gain G = 1.33.

Next we measure the variances of the read and signal levels at frequencies near the shifted signal frequency. We choose a narrow band of frequencies 740 kHz down from 65 MHz where the noise level is at the shot-noise level with the paramp off. With the paramp on and the evasion angle adjusted to 8°, the variances as a function of LO phase for both signal and read channels are shown in Fig. 6.

Good qualitative agreement is again obtained with the backaction evasion predictions. When the signal variance falls to the shot-noise value, indicating that the signal variance is replicated at the output for one quadrature, the read variance is a maximum, as predicted by Eqs. (21) and (23). Similarly, the measured variances at the other quadrature are in agreement with Eqs. (22) and (24). In order to obtain quantitative agreement, we must take the net quantum efficiency η into account. The measured efficiency is $\eta = 0.7$ due to a detector efficiency of 0.9, a mode matching efficiency of 0.95 and losses in the optics of 0.9. Using $\eta = 0.7$, the noise rise above the shot-noise level for both signal and read channels is predicted to be 0.9 dB compared with the observed value of 0.7 dB. This discrepancy may be accounted for by the difference in the pulse lengths of the LO and pump pulses.

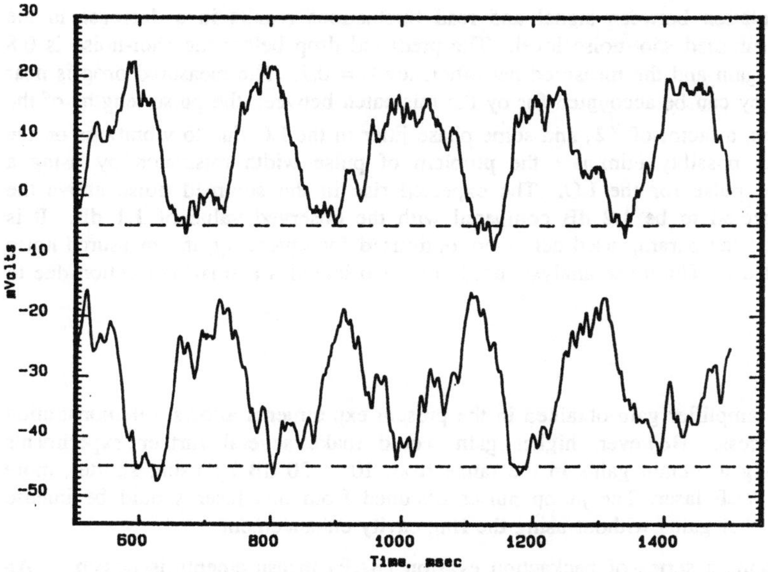

Fig. 6 - A simultaneous scan of the signal and read variances is shown as a function of time as the local oscillator phase is swept. The phase shift between minima in the noise is π. Shot-noise levels for signal (upper trace) and read (lower trace) is shown as a dashed line.

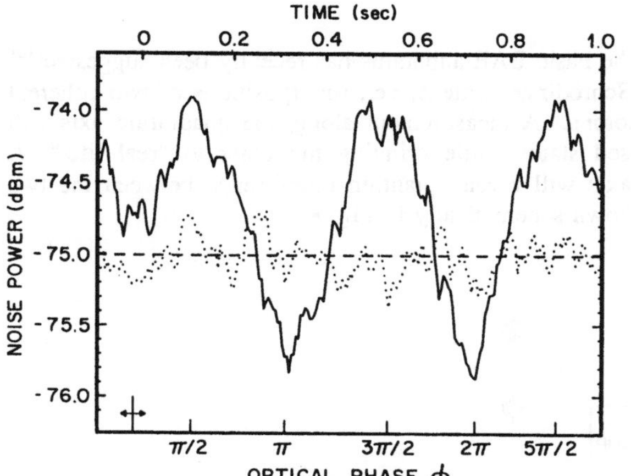

Fig. 7 - The sum of the amplified signal and readout noise levels is shown as ϕ_{LO} is swept. Dotted data correspond to the paramp off and the level is identical with the randomly added shot-noise levels of signal and readout detectors (dashed average line). With the paramp on (solid line) correlation of the noise in the amplified readout and signal results in a reduction of the noise in one quadrature 0.6 dB below the combined shot-noise level. The rf bandwidth is 30 kHz and the video bandwidth is 10 Hz.

Finally, we confirm the backaction evasion process by measuring the correlation of the quantum fluctuations in the signal and read photocurrents. As described in the previous section, this is accomplished by summing the signal and read photocurrents with a relative amplification gain of $g = (G_f - G_f^{-1}) = 0.58$. The results of this summing process are shown in Fig. 7.

As predicted, the correlation between signal and read fluctuations results in a decrease in the noise level below the summed shot-noise level. The predicted drop below the shot-noise is 0.8 dB using the measured gain and the measured net efficiency $\eta = 0.7$. The measured drop is near 0.6 dB. This discrepancy can be accounted for by the mismatch between the pulse lengths of the LO and the pump pulses, a factor of $\sqrt{2}$, and some phase jitter in the LO due to vibrations on the optical table. One can possibly eliminate the problem of pulse width mismatch by using a parametrically-amplified pulse for the LO. The expected rise in the summed noise above the shot-noise level is expected to be 1.2 dB compared with the observed value of 1.1 dB. It is interesting to note that if the paramp used here were optimized for squeezing, the measured noise reduction would be 1.4 dB. The noise analysis used here also includes a small correction due to amplifier noise.

4. Future

The modest parametric amplifier gain obtained in the present experiments allows a demonstration of the backaction process. However, higher gain would make several further experiments possible. We are hoping to obtain gains in the range from 10 to 20 dB by using shorter, more intense pulses from a YLF laser. The pump pulses obtained from this laser should be intense enough to obtain the higher gains without using the ring cavity enhancement.

At these higher gains a series of backaction evasion (BAE) measurements is possible. An initial BAE measurement leaves the signal in a particular eigenstate for one quadrature field due to the nature of the measurement process. A second BAE would measure this eigenvalue without perturbing it. As a result of this pair of BAE measurements, a rigorous quantum nondemolition measurement is made. A known quantum state of a system is measured without perturbing one quadrature component of that state.

A second interesting modification of the basic BAE apparatus has recently been suggested[10] to construct a machine that will generate Schrödinger kittens, i.e., superpositions of two coherent states separated by a small number of photons. A measurement along one quadrature axis will result in "killing" one of the superimposed states while bringing the other to "real life." A measurement along the other quadrature axis will reveal quantum interference between the two states. The kitten generating machine is shown schematically in Fig. 8.

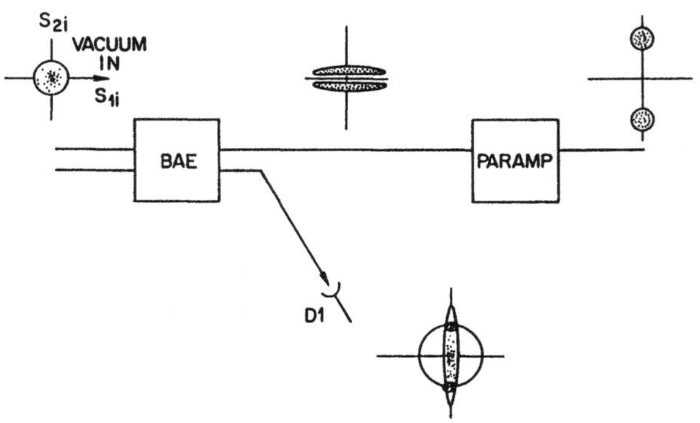

Fig. 8 - A schematic diagram of a Schrödinger kitten generator is shown utilizing a backaction evader BAE and a parametric amplifier. A direct detector D1 at the read channel of the BAE measures photon number corresponding to the circle. The resulting state is in the cross-hatched regions above or below the horizontal axis. As a result, the signal beam probability distribution is fixed as shown in the diagram above D1. The paramp amplifies the vertical quadrature and deamplifies the horizontal quadrature resulting in the "kitten" state shown at the output.

The input state is a vacuum field. In this case, the first polarization rotation is not required. The read beam from the BAE is measured by a direct detector that can be thought of as measuring photon number represented by the circle in the figure. At the instant of measurement the quantum state of the read channel is determined to be either of the cross-hatched regions formed by the intersection of the circle and the squeezed vacuum. Since the transmitted signal is strongly correlated with this read state, its probability distribution is forced into the shape shown in the figure by the direct measurement process. This signal distribution can be reformed into the desired output state using a paramp that amplifies the vertical quadrature and deamplifies the horizontal quadrature. A Schrödinger cat state is thus produced at the output. Losses in the optical system will limit the magnitude of the separation between the two quantum fuzz balls to $\sim 1/(1-\eta)$ photons.

References

1. C. M. Caves, K. S. Thorne, P. W. P. Drever, V. D Sandberg, M. Zimmermann: Rev. Mod. Phys. *52*, 341 (1980).
2. B. Yurke: J. Opt. Soc. Am. B2, 732 (1985).
3. A. La Porta, R. E. Slusher, B. Yurke: Phys. Rev. Lett. *62*, 28 (1989).
4. J. H. Shapiro: IEEE J. Quantum Electron. *21*, 237 (1985).
5. R. M. Shelby, M. D. Levenson: Opt. Commun. *64*, 553 (1987).
6. S. F. Pereira, H. J. Kimble, P. Alsing, D. F. Walls: J. Opt. Soc. Am. A*4*, 24 (1987).
7. B. Yurke: Phys. Rev. A*32*, 300 (1985).
8. R. E. Slusher, P. Grangier, A. La Porta, B. Yurke, M. Potasek: Phys. Rev. Lett. *59*, 2566 (1987) and R. E. Slusher, A. La Porta, P. Grangier, B. Yurke: in *Squeezed and Nonclassical Light*, ed by P. Tombesi and E. R. Pike (Plenum Press, 1988).
9. A. La Porta: submitted to J. Opt. Soc. Am.
10. S. Song, C. M. Caves, B. Yurke: Proc. of Sixth Rochester Conference on Coherence and Quantum Optics, June 26-28, 1989.

Squeezing and Quantum Nondemolition Measurement Using Self-Induced Transparency Solitons

K. Watanabe[1], *H. Nakano*[1], *A. Honold*[2], *Y. Yamamoto*[1], *H.A. Haus*[2], *and Y. Lai*[2]

[1]NTT Basic Research Laboratories, Musashino-Shi, Tokyo 180, Japan
[2]Department of Electrical Engineering and Computer Science, Massachusetts Institute of Technology, Cambridge, MA 02139, USA

1.Introduction

A number-phase squeezed state is produced by self-phase modulation (or quantum phase diffusion) in a Kerr nonlinear medium [KITAGAWA]. A quantum nondemolition (QND) measurement of photon number is also realized in a Kerr medium [MILBURN][IMOTO] by mutual-phase modulation of a signal and a probe wave.

Conventional optical Kerr nonlinearities are of two kinds: nonresonant-coherent and resonant-incoherent. When the frequency of the exciting field is detuned far from the atomic transition frequency, the atoms are virtually (coherently) excited so that the response time is fast and the absorption loss is negligible but the Kerr nonlinearity coefficient is usually small. When the field frequency is close to or just on the atomic transition frequency, the atoms are actually (incoherently) excited so that the Kerr nonlinearity coefficient is resonantly enhanced but the response time becomes slow and the absorption loss is high. These trade-offs between the Kerr nonlinearity and the absorption loss coefficient place a serious limit on the squeezed state generation and the QND measurement, because loss processes introduce vacuum field fluctuations.

Self-induced transparency (SIT) is the resonant and still coherent coupling process between a field and a two-level atomic system [MCCALL]. The absorbing atoms are excited from the ground state in the leading edge of the pulse and are down-converted by the stimulated process to the ground state in the trailing edge of the pulse. The field does not lose its energy at all but still excites the atoms efficiently. Therefore, a large Kerr nonlinearity coefficient and a small absorption loss coefficient are expected for this optical nonlinearity that is both resonant and coherent. This paper studies the application of the SIT to squeezed state generation and QND measurements.

In Sec. 2 we review the classical nonlinear Schrödinger (NLS) equation and the inverse scattering solution for an optical soliton in a fiber. Sec. 3 describes the quantization of an optical solitons. In Sec. 4 we demonstrate that the Maxwell-Bloch equations for the SIT has similar properties as the fiber soliton. The squeezing due to the quantum phase diffusion is treated in Sec. 5. Sec. 6 demonstrates that the collision of the two solitons realizes the QND measurements of photon number. In Sec. 7 we evaluate the Kerr nonlinearity coefficients both for self-phase modulation and mutual-phase modulation, and the absorption coefficient due to finite energy and phase decay time constants for a bound exciton system in CdS semiconductors. In Sec. 8 we finally demonstrate that very efficient squeezed state generation and QND measurements can be realized with SIT solitons.

Springer Proceedings in Physics, Vol. 41 **Quantum Optics V**
Editors: J.D. Harvey and D.F. Walls

2. Classical Inverse Scattering Method and Classical Solitons

The analysis of a pulse in a nonlinear dispersive waveguide is complicated, when expressed in terms of the Fourier component modes $A(z,\beta)$ at time z and wavenumber β as shown in Fig. 1, because the Kerr nonlinearity causes mode coupling (four wave mixing). If the local modes $\alpha(z,x)$ at "time" z and position x, which are related to $A(z,\beta)$ by a linear Fourier transform, are adopted as shown in Fig. 1, the analysis leads to the NLS equation.

$$i\frac{\partial\alpha(z,x)}{\partial z} = \frac{1}{2}D\frac{\partial^2\alpha(z,x)}{\partial x^2} - \frac{1}{2}\kappa|\alpha(z,x)|^2\alpha(z,x). \tag{1}$$

Here $D = (\frac{d^2k}{d\omega^2})/(\frac{dk}{d\omega})^2$ is a diffusion constant, k is a propagation constant, ω is an angular frequency and κ is the Kerr nonlinearity coefficient. The nonlinear Fourier transform or inverse scattering method (ISM) converts the NLS equation into the linear evolution equations [ZAKHAROV]. The solution of (1) can be constructed as a scattering potential for two waves v_1 and v_2, which satisfy the following scattering equations [ZAKHAROV],

$$\frac{\partial v_1}{\partial x} = -i\zeta v_1 + \sqrt{\frac{\kappa}{2|D|}}\alpha(0,x)v_2, \tag{2}$$

$$\frac{\partial v_2}{\partial x} = i\zeta v_2 + \sqrt{\frac{\kappa}{2|D|}}\alpha(0,x)v_1, \tag{3}$$

and

$$i\frac{\partial v_1}{\partial z} = (|D|\zeta^2 - \frac{\kappa}{2}|\alpha|^2)v_1 + \kappa|D|(i\zeta\alpha - \frac{1}{2}\frac{\partial\alpha}{\partial x})v_2, \tag{4}$$

$$i\frac{\partial v_2}{\partial z} = -\kappa|D|(i\zeta\alpha^* - \frac{1}{2}\frac{\partial\alpha^*}{\partial x})v_1 - (|D|\zeta^2 - \frac{\kappa}{2}|\alpha|^2)v_2. \tag{5}$$

When D is negative, the scattering equations (2)-(5) possess a bound solution for a particular α, which corresponds to a soliton solution. The scattering data are expressed, in part, in terms of the reflection coefficient spectrum

$$\rho(\xi) \equiv b(\xi)/a(\xi) \tag{6}$$

which is the ratio of the reflected wave

$$v_2 = b(\xi)e^{i\xi x} \tag{7}$$

to the incident wave

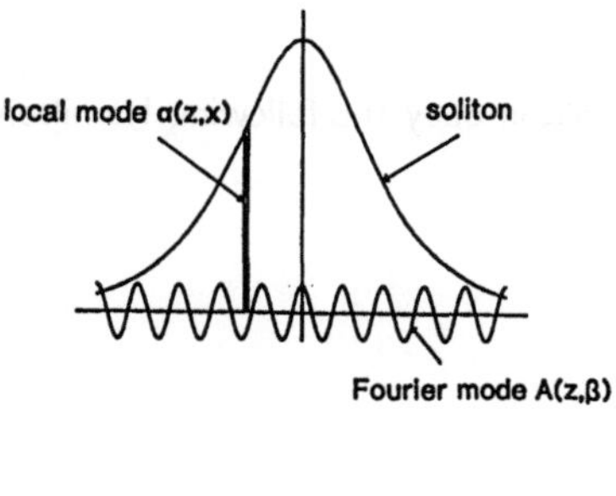

Fig. 1

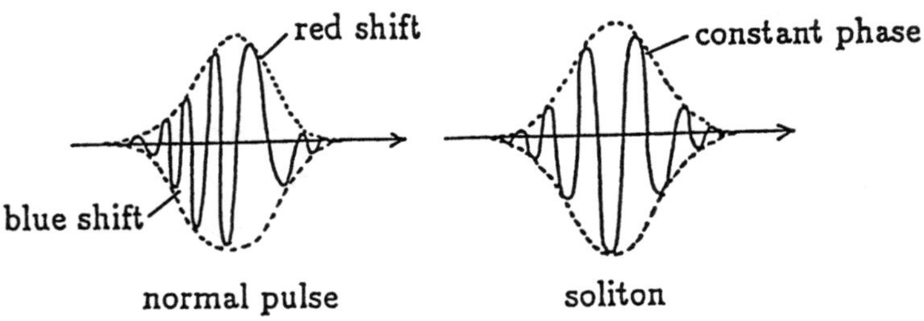

Fig. 2

$$v_1 = a(\xi)e^{-i\xi x} \tag{8}$$

at $x \to \infty$, with $\zeta = \xi = \text{real}$.

The scattering data for a soliton are contained in the poles of $\rho(\zeta)$ in the upper half of the ζ -plane, $\zeta = \xi + i\eta$, and the value of the reflected wave, b(ζ) at a zero of a(ζ). The evolution equation for b(ζ) is

$$\frac{\partial \ln b}{\partial z} = 2i\frac{d\omega}{dk}|D|\zeta^2. \tag{9}$$

The soliton solution of (1) is constructed from (9) as

$$\begin{aligned} \alpha(z,x) &= 2\sqrt{\frac{2|D|}{\kappa}}\eta \text{sech}[2\eta(x - q_0 + 2|D|\xi z)] \\ &\quad \times \exp[-2i\xi x - 2i|D|(\xi^2 - \eta^2)z + i\phi_0]. \end{aligned} \tag{10}$$

The soliton solution (10) can be uniquely determined by the four independent parameters: the momentum $p = -2\hbar\xi$, the initial position $q_0 = \frac{1}{2\eta}|D|\ln\frac{|b(0)|}{2\eta}$, the initial phase $\phi_0 = -arg\sqrt{b(0)}$, the photon number $N = 8|D|\eta/\kappa$. The two parameters p and q_0 determine the particle property (or the mode function) of the soliton and the other two parameters ϕ_0 and N determine the wave property (or the excitation) of the soliton. The soliton solution (10) is a "normal mode" in a nonlinear dispersive waveguide as shown in Fig. 1, even though the Fourier modes and the local modes are "coupled modes".

The physical picture of the classical soliton (10) is simple. As shown in Fig. 2, the leading edge of the pulse is red-shifted due to the Kerr nonlinearity , and the trailing edge of the pulse is blue-shifted. If the waveguide has a negative dispersion, the red-shifted leading edge is delayed with respect to the blue-shifted trailing edge. The Kerr nonlinearity and the negative disperss counteract each other and a stable soliton with fixed pulse envelope and phase is formed.

3. Quantization of Optical Solitons

The four parameters that uniquely determine the classical soliton obey the following Hamilton equations [KAUP][HAUS].

$$\begin{aligned} \frac{\partial q}{\partial t} &= \frac{\partial H}{\partial p}, \\ \frac{\partial p}{\partial t} &= -\frac{\partial H}{\partial q} = 0, \end{aligned}$$

$$\begin{aligned} \frac{\partial n}{\partial t} &= \frac{\partial H}{\partial \phi} = 0, \\ \frac{\partial \phi}{\partial t} &= -\frac{\partial H}{\partial n}. \end{aligned} \tag{11}$$

Here the Hamiltonian consists of the kinetic energy and the potential energy:

$$H = \frac{8\hbar v D^2}{\kappa}(2\xi^2\eta - \frac{2}{3}\eta^3), \tag{12}$$

where $v \equiv d\omega/dk$ is the group velocity. We use lower case n for the photon number for consistency of notation, contrary to the notation of [HAUS].

A photon acquires a mass $m = 1/\hbar v|D|$ due to the dispersion and has an associated kinetic energy $p^2/2m$. The conventional cannonical quantization is applied to the optical soliton by replacing the two Poisson brackets corresponding to relations (8) by the two commutator brackets,

$$[\hat{q}, \hat{p}] = i\hbar, \tag{13}$$

$$[\hat{n}, \hat{\phi}] = i. \tag{14}$$

A quantized soliton is characterized by two pairs of observables, photon number and phase, position and momentum. It resembles a particle (boson), with two internal degrees of freedom (oscillator). Thus it has an uncertainty in position and momentum, and an uncertainty in photon number and phase.

The center position and the momentum are now conjugate observables and cannot be determined simultanuously beyond the accuracy dictated by the uncertainty principle, $< \Delta\hat{q}^2 >< \Delta\hat{p}^2 > \geq \frac{\hbar^2}{4}$. The photon number and the phase are also under the constraint, $< \Delta\hat{n}^2 >< \Delta\hat{\phi}^2 > \geq \frac{1}{4}$.

The building block for the description of quantum optical solitons is an eigenstate $|n, p>$ of a photon number operator and a momentum operator. It is constructed from a vacuum state by the position dependent photon creation operator [BETHE][WADATI],

$$|n, p> = \frac{1}{\sqrt{n!}} \int f_{n,p}(x_1, \cdot, \cdot, x_n) \hat{\phi}^\dagger(x_1) \cdot \cdot \hat{\phi}^\dagger(x_n) dx_1 \cdot \cdot dx_n |0> \tag{15}$$

where

$$f_{n,p}(x_1, \cdot, \cdot, x_n) = N_n \exp[i\frac{k}{\hbar}(\sum_{j=1}^{n} x_j) + \frac{\kappa}{2D} \sum_{1 \leq i < j \leq n} |x_j - x_i|]. \tag{16}$$

The quantum soliton in a coherent state is constructed by the linear superposition of $|n, p>$ eigenstate [LAI],

$$|\alpha> = \sum_n \frac{\alpha_0^n}{\sqrt{n!}} e^{-\frac{1}{2}|\alpha_0|^2} \int \frac{1}{\sqrt{\Delta p}\pi^{\frac{1}{4}}} e^{-\frac{1}{2}\frac{|p-p_0|^2}{(\Delta p)^2}} e^{-inpx_0/\hbar} e^{-iP(n,p)^2 Dz/2\hbar^2} |n, p> dp, \tag{17}$$

and

$$P(n, p)^2 = np^2 - \frac{\hbar^2\kappa^2}{12|D|^2} n(n^2 - 1). \tag{18}$$

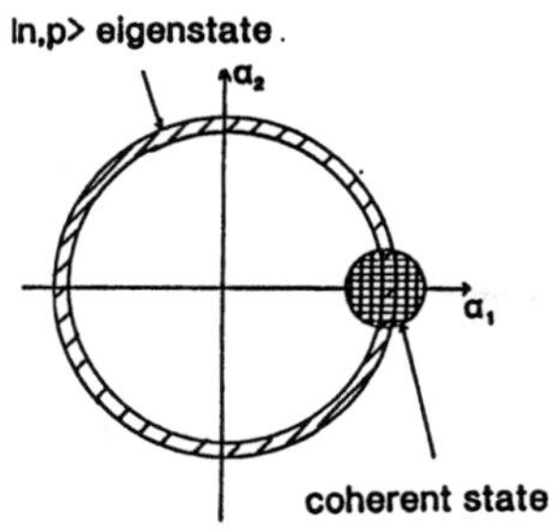

Fig. 3(a)

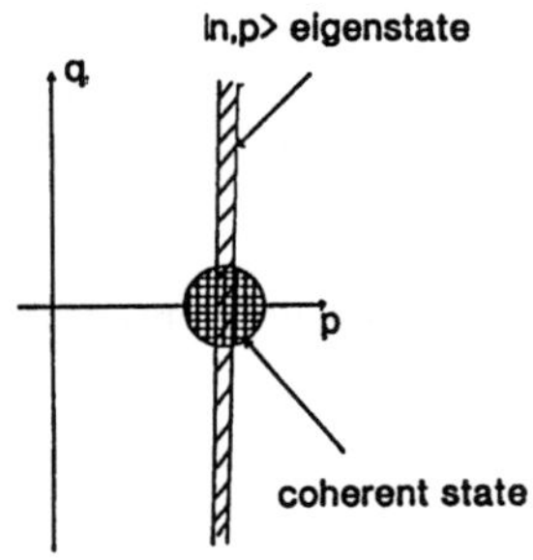

Fig. 3(b)

If the coherent state soliton is projected onto the two quadrature phase amplitudes phase space (α_1, α_2), the quasi-probability density is localized, possessing an average phase at a sacrifice of an uncertainty of photon number as shown in Fig. 3(a). If the coherent state soliton is projected onto the momentum-position phase space (p, q), the quasi-probability density is localized, possessing an average position at the expense of a finite uncertainty of momentum as shown in Fig. 3(b).

4. Self Induced Transparency Solitons

The optical field amplitude $\tilde{E}$ and the complex excitation amplitudes w_1 and w_2 for the upper and lower levels of the atomic system obey the Maxwell-Bloch equations [LAMB],

$$(\frac{v}{\Omega})^2\frac{\partial\tilde{E}}{\partial z}+\frac{v}{\Omega^2}\frac{\partial\tilde{E}}{\partial t} = < w_1 w_2^* >, \tag{19}$$

$$\frac{dw_1}{dt} = i\frac{\delta}{2}w_1+v\tilde{E}w_2, \tag{20}$$

$$\frac{dw_2}{dt} = -i\frac{\delta}{2}w_2-v\tilde{E}^*w_1,. \tag{21}$$

Here $\delta = \omega_0 - \omega_{21}$ is the detuning between the field frequency ω_0 and the atomic frequency ω_{21}, $\tilde{E} = \frac{Ep_{21}}{i\hbar v}, \Omega^2 = \frac{\omega N p_{21}^2}{2\hbar\varepsilon}$, E is the field envelope function, p_{21} is the dipole moment, v is a nonresonant light velocity, ε is the dielectric constant, and N is the atomic density. In the moving frame coordinate, the above equations are reduced to the Zakharov-Shabat scattering equations (2)-(5) of the ISM.

When the optical pulse satisfies the relation, $\frac{p_{21}}{\hbar}\int Edx = 2\pi$, it propagates stably like a soliton. The field envelope function for such a 2π-soliton is expressed as

$$E=\frac{2\hbar}{p_{21}}2\eta\mathrm{sech}[2\eta(x-q_0-\frac{\Omega^2}{4v^2\kappa_1}z)]\exp[-2i\xi x+\frac{i\Omega^2\xi z}{2v^2\kappa_0}+i\phi_0]. \tag{22}$$

where the factor $\exp[i(\omega_0 t - k_0 z)]$ has been dropped.

Here $\xi = (\omega - \omega_0)/2v$ represents the nonlinear frequency shift and, the pulse height η is propotional to the photon number n_p, i.e. $\eta = \omega p_{21}^2 n_p / 16\hbar v^2 \varepsilon A_p$, and determines the pulse duration $\tau = 1/2\eta v$ and, A_P is the area of the pulse. If the lineshape function of the atomic system is assumed to be a delta function $\delta(\omega - \omega_{21})$, the coefficients κ_0 and κ_1, which determine the group velocity V, and the nonlinear phase shift ϕ, are written as

$$V \equiv \frac{v}{1 + \frac{\Omega^2}{4v^2\kappa_1}} = \frac{v}{1 + \frac{\Omega^2}{(\omega-\omega_{21})^2 + 1/\tau^2}}, \tag{23}$$

and

$$\phi \equiv \frac{\Omega^2 \xi z}{2v^2\kappa_0} = \frac{\Omega^2(\omega - \omega_{21})\frac{z}{v}}{(\omega - \omega_{21})^2 + 1/\tau^2}. \tag{24}$$

The fiber soliton and the SIT soliton are compared in Table I. Because of the mathematical similarity of the two solitons, the quantization of the SIT soliton can be done according to the same procedure of Sec. 2.

Table I. Comparison of fiber solitons and SIT solitons

	Fiber Soliton	SIT Soliton
Physical mechanism	counter balance of nonlinearity and dispersion (nonresonant)	Rabi flopping between field and atoms (resonant)
Basic equation	nonlinear Schrödinger eq.	Maxwell-Bloch eq.
Squeezing	$<\Delta\hat{a}_1^2> = \frac{1}{4}\exp(-\frac{2\chi^{(3)}\hbar\omega^2 <\hat{n}> z}{\epsilon^2 v^2 A\tau})$	
	$\chi^{(3)}_{\text{self}} = \text{const.}(= 10^{-14}\text{esu})$	$\chi^{(3)}_{\text{self}} = \frac{p_{21}^4 N\tau^2}{32\hbar^3}\frac{\omega-\omega_{21}}{(\omega-\omega_{21})^2+1/\tau^2}$
QND measurement	$<\Delta\hat{n}_1^2> = [(\frac{4\hbar\omega^2\chi^{(3)}z_{\text{eff}}}{v^2\epsilon^2 A\tau})^2 <\hat{n}_2>]^{-1}$	
	$\chi^{(3)}_{\text{mutual}} = \text{const.}(= 10^{-14}\text{esu})$	$\chi^{(3)}_{\text{mutual}} = \chi^{(3)}_{\text{self}}$

5. Squeezing of Optical Solitons

As shown in (10) and (22), both the fiber soliton and the SIT soliton have phase factors dependent on the photon number. Therefore, the coherent state soliton, consisting of the linear superposition of different photon number states, experiences quantum phase diffusion. The quasi-

Fig. 4

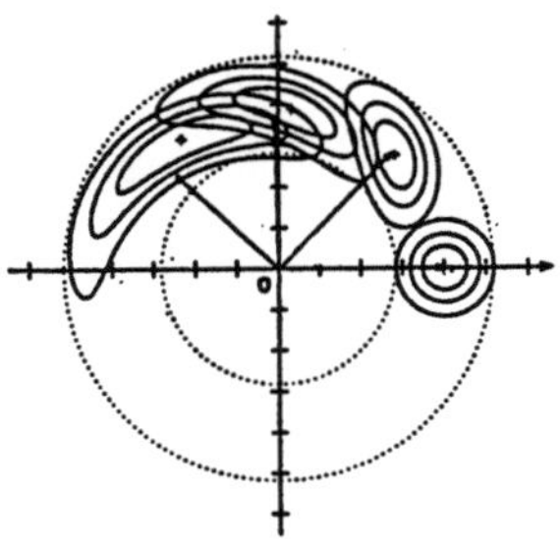

probability density becomes bean-shaped as shown in Fig. 4, which is nothing but the generation of a number-phase squeezed state from a coherent state. The quantum phase diffusion process does not alter the photon number noise but it establishes the quantum correlation between the photon number and the phase, so that the rotation of the coherent excitation results in the reduction of the photon number noise [KITAGAWA].

When the average photon number is big and the quadrature phase diffusion is moderate, the squeezed photon number noise (or the squeezed quantum phase component noise) is approximately given by

$$< \Delta\hat{a}_1^2 >= \exp(-\frac{2\chi^{(3)}\hbar\omega^2 < \hat{n} > z}{\epsilon^2 v^2 A\tau}), \tag{25}$$

The quantum phase diffusion under the above approximation of the fiber soliton is exactly the same as that of a pulse in a nondispersive Kerr medium. Therefore, the $\chi^{(3)}$ coefficient of the fiber material determines the degree of squeezing for the soliton. In a similar way, the quantum phase diffusion of the SIT soliton is determined by an effective $\chi^{(3)}$ coefficient,

$$\chi_{self}^{(3)} = \frac{p_{21}^4 N\tau^2}{16\hbar^3}\frac{\omega - \omega_{21}}{(\omega - \omega_{21})^2 + 1/\tau^2}. \tag{26}$$

$\chi_{self}^{(3)}$ is zero both just on resonance $\omega = \omega_{21}$ and approaches zero far from resonance $|\omega - \omega_{21}| \gg \frac{1}{\tau}$. The maximum $\chi_{self}^{(3)}$ occurs at the optimum frequency detuning, $\Delta\omega_{opt} = |\omega - \omega_{21}| = \frac{1}{\tau}$, and is proportional to p_{21}^4, N, and τ^3.

6. QND Measurement of Soliton Photon Number

The soliton photon number is preserved during the propagation and also in the collision with another soliton. However, the phase is changed due to the collision process. According to the J=2 soliton solution [ZAKHAROV], the phase shift of the soliton 2 due to the collision with the soliton 1 is

$$\hat{\phi}_2 = 2\{\tan^{-1}[(\hat{\eta}_2 + \hat{\eta}_1)(\hat{\xi}_2 - \hat{\xi}_1)^{-1}] - \tan^{-1}[(\hat{\eta}_2 - \hat{\eta}_1)(\hat{\xi}_2 - \hat{\xi}_1)^{-1}]\}. \tag{27}$$

If $| < \hat{\xi}_1 - \hat{\xi}_2 > | \gg < \hat{\eta}_1 >$ and $< \hat{\eta}_2 >$, then the linearization, such as $\hat{\phi}_2 =< \hat{\phi}_2 > +\Delta\hat{\phi}_2$ and $\hat{\eta}_1 =< \hat{\eta}_1 > +\Delta\hat{\eta}_1$, leads to the relation

$$\Delta\hat{\phi}_2 = \frac{4}{|<\hat{\xi}_1 - \hat{\xi}_2>|}\Delta\hat{\eta}_1 = \sqrt{F'}\Delta\hat{n}_1 \tag{28}$$

Here $\sqrt{F'} = 4\hbar\omega^2\chi^{(3)}z_{\rm eff}/v^2\varepsilon^2 A\tau$ and the effective interaction length $z_{\rm eff}$ is the collision length of the two solitons, which is calculated by

$$z_{\rm eff} = \frac{1}{2|D||<\hat{\xi}_1 - \hat{\xi}_2>|<\eta>} = \frac{4}{\kappa<\hat{n}>|<\hat{\xi}_1 - \hat{\xi}_2>|}, \tag{29}$$

for the fiber soliton. Here we assume the signal and the probe soliton have the same average photon numbers and probe durations. From the above argument, the signal soliton photon number $\hat{n}_1$ is measured by the probe soliton phase $\hat{\phi}_2$ in a QND manner. If the probe soliton is in a coherent state, the measurement error is

$$<\Delta\hat{n}_1^2>_{\rm meas} = \frac{1}{4F'<\hat{n}_2>}. \tag{30}$$

The mutual phase modulation for the SIT soliton is independent of the frequency detuning from ω_{12}, but is dependent on the frequency difference $\omega_1 - \omega_2 = 2v(\xi_1 - \xi_2)$ as indicated by (28). The mutual phase modulation increases with decreasing $|\omega_1 - \omega_2|$, but the collision length also increases by the relation,

$$z_{\rm eff} = \frac{2v}{\Omega^2}(\tau_s + \tau_p)\frac{[(\omega_s - \omega_{21})^2 + 1/\tau_s^2][(\omega_p - \omega_{21})^2 + 1/\tau_p^2]}{|(\omega_s - \omega_p)(\omega_s + \omega_p - \omega_{21}) + (\frac{1}{\tau_s} - \frac{1}{\tau_p})(\frac{1}{\tau_s} + \frac{1}{\tau_p})}. \tag{31}$$

The effective $\chi^{(3)}$ coefficient for mutual phase modulation is evaluated by comparing (28) to the mutual phase modulation between the two normal pulses in the same interaction length $z_{\rm eff}$,

$$\chi^{(3)}_{\rm mutual} = \chi^{(3)}_{\rm self}. \tag{32}$$

$\chi^{(3)}_{\rm mutual}$ is not zero even if the probe soliton is just on resonance $\omega_2 = \omega_{21}$, where $\chi^{(3)}_{\rm self}$ is zero. This is the important feature of SIT solitons for optical switching and QND measurement applications, because the phase shift of the probe soliton is only dependent on the photon number of the signal soliton.

7. Numerical Examples for Self Induced Transparency Solitons

Next let us consider the SIT soliton in the excitonic spectral range in semiconductors. An exciton bound to a neutral donor(the I_2 line) in CdS at 2 K has an absorption peak wavelength of $\lambda = 487$ nm, a T_1 time constant of 135ps, a T_2 time constant between 40 ps and 270ps, an exciton density of $N = 1 \times 10^{21}$ m^{-3}, and an oscillator strength of $f_0 = 26$, that corresponds to the dipole moment $p_{21} \equiv e\sqrt{\frac{f_0\hbar}{2m_e\omega}} = 1 \times 10^{-28}(C \cdot m)$ [DAGENAIS]. When the beam radius r is assumed to be 30 μm and 0.1μm respectively, the photon number and the peak power to produce the 2π-soliton of 1 ps pulse duration are 5.0×10^8 and 200 W, and 5×10^3 and 2mW, respectively. The $\chi^{(3)}_{\rm self}$ coefficient vs.

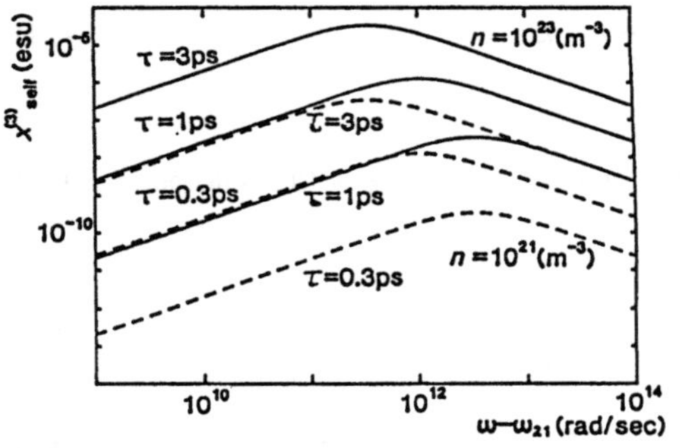

Fig. 5(a)

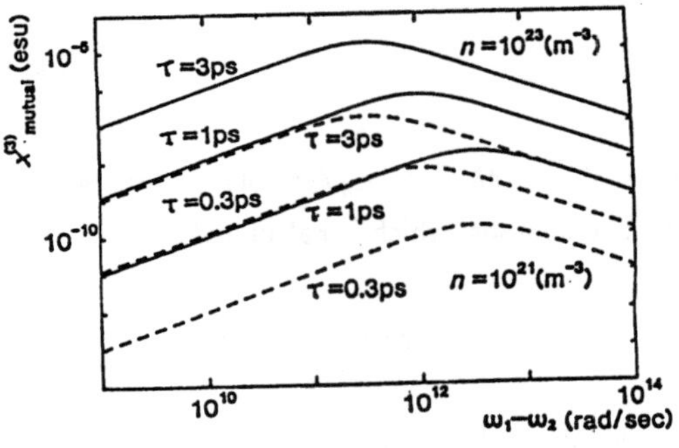

Fig. 5(b)

the frequency detuning $\omega - \omega_{21}$ is plotted in Fig. 5(a) as a function of pulse duration. For the pulse durations of $\tau = 1$ ps and $\tau = 3$ ps, the $\chi^{(3)}_{\text{self}}$ values of 10^{-8} esu and $10^{-6.5}$ esu are obtained at the optimum frequency detuning $\Delta\omega_{\text{opt}} = 10^{12}$ Hz and 3×10^{11} Hz. If the exciton density increases to N=1 $\times 10^{23} m^{-3}$, which is still well below the maximum density determined by the exciton Bohr radius, the $\chi^{(3)}_{\text{self}}$ values are 10^{-6} esu and $10^{-4.5}$ esu, respectively. The $\chi^{(3)}_{\text{mutual}}$ coefficient vs. the frequency difference $\omega_1 - \omega_2$ of the two 2π-solitons is plotted in Fig. 5(b). Here we have assumed $\omega_2 = \omega_{21}$. The $\chi^{(3)}_{\text{mutual}}$ values are of the same order as the $\chi^{(3)}_{\text{self}}$ values. Linear absorption occurs even for the SIT soliton due to finite energy and phase decay time constants T_1, T_2, and the pulse energy decreases according to the following equation[MACALL],

$$\frac{d(N_p/A_p)^2}{dz} = -64\frac{n\hbar\nu\varepsilon}{\omega p_{21}^2}\{\frac{1}{T_2}\frac{1}{1+(\omega-\omega_{21})^2\tau^2} + \frac{2}{3}(\frac{1}{T_1}-\frac{1}{T_2})[\frac{1}{1+(\omega-\omega_{21})^2\tau^2}]^2\}. \quad (33)$$

From (33), an effective linear absorption coefficient α_{eff} can be defined when the energy loss is small,

$$\alpha_{\text{eff}} = \frac{1}{2}\frac{n\omega p_{21}^2\tau^2}{\hbar\nu\varepsilon}\{\frac{1}{T_2}\frac{1}{1+(\omega-\omega_{21})^2\tau^2} + \frac{2}{3}(\frac{1}{T_1}-\frac{1}{T_2})[\frac{1}{1+(\omega-\omega_{21})^2\tau^2}]^2\}. \quad (34)$$

From (34) we obtain α_{eff}=0.55 cm^{-1} and α_{eff}=55 cm^{-1} for $N = 10^{21}$ m^{-3} and $N = 10^{23}$ m^{-3}, respectively. Here τ=1ps, $\omega - \omega_{21} = 10^{12}$Hz, T_1=135ps, and T_2=270ps are assumed. The figure of merit $\chi^{(3)}/\alpha\tau$ for nonlinear optical materials is 10^4esu·cm/sec, which is the largest value by far compared with that for various nonlinear optical materials reported so far.

An exciton in a 12nm thick GaAs quantum well at 2K has an absorption peak wavelength of 806nm, T_1 time constant of 180ps, T_2 time constant of 12ps, exciton density of N=2×10^{13}(m^{-2}), and oscillator strength of 15 [HONOLD]. The values are limited by the imperfections in GaAs crystal and interfaces. By solving the relations between f_0, T_1, T_2 and the coherence area A_c of the two dimensional exciton [FELDMANN][HANAMURA][YAMANISHI] self- consistently, we obtain T_1=18ps, T_2=36ps and f_0=108 for a perfect sample [HONOLD]. The $\chi^{(3)}_{\text{self}}$ and $\chi^{(3)}_{\text{mutual}}$ coefficients are about $10^{-6.5}$esu and the α_{eff} coefficient is about 55(cm^{-1}) for 1ps pulse. The figure of merit $\chi^{(3)}/\alpha\tau$ is in the same order as that for the bound exciton in CdS.

8. Squeezed State Generation and QND Measurement

Let us consider squeezed state generation and the quantum nondemolition measurement using the 2π-solitons produced by bound excitons in CdS. If we assume the same numerical parameters used in Fig. 5(a) for (25), the quantum noise is reduced by 20 dB below the standard quantum limit at a propagating distance of 400μm and 4μm for $N = 10^{21}$ m^{-3} and $N = 10^{23}$ m^{-3} and for τ=1ps. The energy loss is 2% for both cases. This is the most efficient and fastest squeezed state generation scheme proposed so far. If we assume the same numerical parameters used in Fig. 5(b) and $\omega_1 - \omega_2 = 1/\tau$, $\tau_1=\tau_2$=1ps, the measurement error $\langle\Delta\hat{n}_1^2\rangle_{\rm meas}$ calculated by (30) is about 20dB below the quantum photon number noise $< \Delta\hat{n}_1^2 >_{\rm noise} = < \hat{n}_1 >$. If $r = 0.1\mu$m, N=10^{21} m^{-3}, $z_{eff} = 400\mu$m and $< \hat{n}_2 > = 5\times 10^3$ (2mW), $< \Delta\hat{n}_1^2 >^{1/2}_{\rm meas}$ is as small as 10^2. This is the most efficient QND measurement scheme proposed so far.

Acknowledgment

The authors wish to thank for Drs. T.Takagahara, H. Ando and K. Igeta of NTT Basic Research Laboratories for their critical discussions.

References

1 H. A. Bethe, Z. Phys. **71**, 205 (1931)

2 M. Dagenais and W. F. Sharfin, J. Opt. Soc. Am. **B2**, 1179 (1985)

3 J. Feldmann, G. Peter, E. O. Göbel, P. Dawson, K. Moore, C. Foxon and R. J. Elliott, Phys. Rev. Lett. **59**, 2337(1987)

4 E. Hanamura, NATO workshop on Optical Switching in Low-Dimensional Systems, 1988, Marbella, (to be published from Plenum Press)

5 H. A. Haus, Rev. Mod. Phys. **51**, 331 (1979)

6 H. A. Haus, K. Watanabe and Y. Yamamoto, to be published in JOSA (B).

7 A. Honold, L. Schultheis, J. Kuhl and C. W. Tu, Int. Conf. Quantum Electron. (1988, Tokyo); A. Honold, to be published elsewhere.

8 N. Imoto, H. A. Haus and Y. Yamamoto, Phys. Rev. **A 32**, 2287 (1985)

9 D. Kaup, J. Math. Phys. **16**, 2036 (1975).

10 M. Kitagawa and Y. Yamamoto, Phys. Rev. **A34**, 3974 (1986)

11 G. L. Lamb, Jr., Phys. Rev. Lett. **31**, 196 (1973)

12 Y. Lai and H. A. Haus, unpublished.

13 S. L. Mc Call and E. L. Hahn, Phys. Rev. **183**, 457 (1969)

14 G. J. Milburn and D. F. Walls, Phys. Rev. **A28**,2065 (1983)

15 M. Wadachi and M. Sakagami, J. Phys. Soc. Japan **53**, 1933 (1984)

16 M. Yamanishi, NATO workshop on Optical Switching in Low-Dimensional Systems, 1988, Marbella, (to be published from Plenum Press)

17 V. E. Zakharov and A. B. Shabat, Sov. Phys. JETP **34**, 62 (1972)

The Effect of Measurement on Quantum Optical Systems

G.J. Milburn

Department of Physics, University of Queensland, St. Lucia 4067, Australia

What is the effect of a given measurement procedure on a particular quantum system? This is not 'the problem of measurement' but it is a sufficiently well posed problem to hope for a reasonably conclusive, though perhaps not very general, answer. In this talk I will consider this question in the context of two specific quantum optical measurement schemes. I will take the view that a measurement is a particular kind of physical interaction between a system (the measured system) and one or many other systems (the apparatus). The purpose is to determine how the evolution of the measured system is modified by the interaction with the apparatus.

There are two quite distinct senses in which the question at the beginning of this talk may be understood. Firstly it is the object of a measurement to yield a result. Thus one may ask 'conditional' questions of the type; what is the state of the measured system at some time given a previous history of experimental results? Such questions are relevant when one is considering the evolution of a single system subject to a sequence of measurements or when one is considering some sort of two–time correlation experiment (eg. what is the probability to obtain a result x at time $t+\tau$, given a result, y, at time t). I will refer to this kind of description as a 'selective measurement description'[1].

The standard text book description of measurement in terms of the projection postulate is a selective measurement description. A physical quantity represented by an operator $\hat{A}$, with eigenvalues $\{a\}$ and eigenstates $\{|a\rangle\}$ is measured on a system in state ρ. Given the result was a, the state after the measurement is taken as $|a\rangle\langle a|$. This is a description of an arbitrarily accurate, instantaneous measurement in the selective sense. If however we knew only that the measurement had taken place but were ignorant of the result, the state of the system would be taken as $\sum_a P(a)\,|a\rangle\langle a|$ where $P(a) = \langle a|\rho|a\rangle$ is the probability to obtain a. This would be a measurement description of another kind, here referred to as a 'nonselective measurement description'[1].

In the nonselective description, the state of the system after the measurement is diagonal, in the basis of measured quantity, the pointer basis[2]. To some the problem of measurement is to account for the evolution of the system to such a form[2,3].

In quantum optics one is primarily concerned with continual measurements, that is one in which the measurement occurs over finite periods of time, for example photon counting and homodyne detection. Such measurements do not fit the standard text book description. However they are well described by recent theories of continual measurement[4,6] and this talk will concern an application of such theories.

The experimental configuration of interest is shown in figure 1. Our object is to determine the state of the intracavity field as measurements are made on the output field

Springer Proceedings in Physics, Vol. 41 **Quantum Optics V**
Editors: J.D. Harvey and D.F. Walls

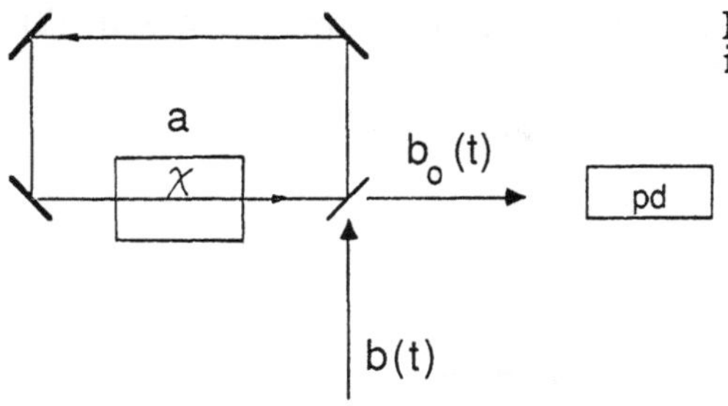

Fig. 1 Schematic representation of the intracavity measurement model

of the cavity. The measured system is the intracavity field while the apparatus is the external field, and subsequent detection devices. However, the only part of the apparatus coupled to the system is the external field. For simplicity I will assume this coupling occurs through only one output mirror. The box χ represents some internal dynamics.

The specification of any measurement procedure may be divided into, (i) specification of the system/apparatus coupling and (ii) specification of the apparatus state. In the optical system of figure 1 the coupling is determined by the rotating wave hamiltonian[7]

$$H = H_{sys} + \hbar \int_{-\infty}^{\infty} d\omega \omega b^{\dagger}(\omega) b(\omega) + i\hbar \int_{-\infty}^{\infty} d\omega \kappa(\omega) \left[b^{\dagger}(\omega) a - b(\omega) a^{\dagger} \right] \tag{1}$$

where a is the annihilation operator for the internal field and $b(\omega)$ is a positive frequency component of the external field satisfying $[b(\omega'), b^{\dagger}(\omega)] = \delta(\omega' - \omega)$. H_{sys} is the hamiltonian for the internal dynamics. Different types of measurement models may be defined by specification of different states of the incoming external fields. The simplest is direct photon counting in which the input field is the vacuum and the output field is incident on a photon counter. If, however the state of the incoming field is a highly squeezed state with large coherent amplitude a kind of homodyne measurement may be made[6], with the incoming field playing the role of a local oscillator.

The nonselective evolution of the system is determined by a master equation. In the usual approximations[6,7] the interaction picture form is

$$\begin{aligned}\frac{d\hat{s}}{dt} &= \frac{i}{\hbar}\left[\hat{H}_{sys}, \hat{\rho}\right] + A\left[ae^{-i\theta} - a^{\dagger}e^{i\theta}, \hat{\rho}\right] + \frac{\gamma}{2}(N+1)\left[2a\rho a^{\dagger} - a^{\dagger}a\rho - \rho a^{\dagger}a\right] \\ &+ \frac{\gamma}{2}N\left[2a^{\dagger}\rho a - aa^{\dagger}\rho - \rho aa^{\dagger}\right] \\ &- \frac{\gamma M^{*}}{2}\left[2a\rho a - a^{2}\rho - \rho a^{2}\right] - \frac{\gamma M}{2}\left[2a^{\dagger}\rho a^{\dagger} - a^{\dagger}a^{\dagger}\rho - \rho a^{\dagger}a^{\dagger}\right].\end{aligned} \tag{2}$$

In (2) A is determined as the product of the amplitude of the input field and a cavity enhancement factor, $\gamma/2$ is the cavity line width in a Lorentzian approximation and θ is the phase of the input field amplitude. If the input field is not squeezed $M = 0$, while if the input field is in the vacuum state $M = N = 0$.

From the point of view of homodyne measurement it is more convenient to write (2) in terms of the quadrature phase amplitude

$$\hat{X}_\phi = \frac{1}{2}\left[a\, e^{i\phi} + a^\dagger\, e^{-i\phi}\right]$$

where ϕ is the phase of M. The result is

$$\frac{d\hat{\rho}}{dt} = \frac{i}{\hbar}\left[\hat{H}_{sys},\hat{\rho}\right] - 2iA\left[\hat{X}_\phi \sin(\theta+\phi) + \hat{X}_{\phi+\pi/2}\cos(\theta+\phi), \hat{\rho}\right]$$
$$-i\frac{\gamma}{2}\left[\hat{X}_\phi, \left\{\hat{X}_{\phi+\pi/2},\hat{\rho}\right\}\right] + i\frac{\gamma}{2}\left[\hat{X}_{\phi+\pi/2}, \left\{\hat{X}_\phi,\hat{\rho}\right\}\right]$$
$$-\gamma e^{-2r}\left[\hat{X}_\phi, \left[\hat{X}_\phi,\hat{\rho}\right]\right] - \gamma e^{2r}\left[\hat{X}_{\phi+\pi/2}, \left[\hat{X}_{\phi+\pi/2},\hat{\rho}\right]\right] \qquad (3)$$

where we have used $|M| = \sinh r.\cosh r$, $N = \sinh^2 r$ assuming a minimum uncertainty squeezed input. The brackets {,} define an anticommutator.

The second commutator in (3) represents the coherent driving of the cavity mode, while the terms containing the anticommutator describe damping. The two final double commutators are the most important from the point of view of measurement. Firstly they determine a diffusion process in each of the canonically conjugate variables $\hat{X}_\phi$ and $\hat{X}_{\phi+\pi/2}$

$$\frac{d}{dt} < \hat{X}^2_{\phi+\pi/2} > = (\ldots) + \gamma e^{-2r} \qquad (4a)$$

$$\frac{d}{dt} < \hat{X}_\phi^2 > = (\ldots) + \gamma e^{2r}. \qquad (4b)$$

Secondly they tend to diagonalise the system density operator in the basis of each of the canonical variables. For example, suppose r>>1 so that the first commutator may be ignored. Let $|x>$ be an eigenstate of $\hat{X}_{\phi+\pi/2}$ then

$$\frac{d}{dt} < x|\hat{\rho}|x' > = (\ldots) - \gamma e^{2r}(x-x')^2 < x|\rho|x' >.$$

Thus there is a decay in the off–diagonal elements indicating that the eigenstates of $\hat{X}_{\phi+\pi/2}$ are the pointer basis.

Notice that under these conditions the diffusion in the conjugate variable, $\hat{X}_\phi$, is large. We conclude that for r>>1 this must describe a good measurement of $\hat{X}_{\phi+\pi/2}$, a result confirmed by further analysis[6].

As an example of a selective measurement description, I will now discuss photon counting measurements in a nondegenerate parametric amplifier.

SELECTIVE PHOTON COUNTING MEASUREMENTS in NONDEGENERATE PARAMETRIC AMPLIFICATION

Consider two intracavity modes, a and b with frequencies ω_a and ω_b, interacting via nondegenerate parametric amplification[8]. The interaction hamiltonian is

$$H_I = \hbar g \left[\varepsilon(t)\, a^\dagger b^\dagger + \varepsilon^*(t) ab\right], \qquad (5)$$

where

$$\varepsilon(t) = \varepsilon\, e^{-i(\omega_a+\omega_b)t}.$$

In this system photons are produced in pairs, one in each mode. If in time t, m photons are counted in mode b, what is the state of mode a conditioned on this result? I will assume only b–mode photons are counted and that both modes start in the vacuum state.

The projection postulate cannot be applied to this measurement; it is not instantaneous, the detection of a photon requires its destruction, thus it is not a 'first–kind' measurement and finally it is not arbitrarily accurate as all detectors have non–unit quantum efficiency. Instead I will use a theory of continual photon counting measurements[5].

The nonselective evolution is described in the interaction picture by

$$\frac{d\hat{\rho}}{dt} = -i\chi \left[a^\dagger b^\dagger + ab, \rho\right] + \gamma/2 \left[2b\rho b^\dagger - b^\dagger b\rho - \rho b^\dagger b\right] \qquad (6)$$

which represents a linear loss from mode b alone. This equation may be solved by various techniques[8,9]. From the perspective of this paper the most interesting result is that the reduced state of either mode is a thermal state with mean photon number.

$$\bar{n}_a(t) = e^{-\gamma t/2} \frac{\cosh^2(\kappa t - c)}{\cosh^2 c} - 1 \qquad (7)$$

$$\bar{n}_b(t) = \bar{n}_a(t) - g(t)\, e^{-\gamma t/2} \frac{\cosh^2(\kappa t - c)}{\cosh^2 c} \qquad (8)$$

where $\kappa = \chi\Delta$, $\Delta = \left[1 + \gamma^2/16\chi^2\right]^{1/2}$ and $\tanh c = -\gamma/4\kappa$. The function g(t) is given by

$$g(t) = 1 - f^2(t) - e^{\gamma t/2}\left[1 + \gamma/2\chi\, f(t) - f^2(t)\right] \qquad (9)$$

where

$$f(t) = \gamma/4\chi - \Delta \tanh(\kappa t - c). \qquad (10)$$

Note that as $\gamma\to 0$, $g(t)\to 0$ and thus $\bar{n}_a(t) = \bar{n}_b(t)$ as required.

If m b–mode photons are counted in time t the reduced state of mode a is

$$\hat{\rho}_a^{(m)}(t) = [P(m,t)]^{-1}\ \mathrm{tr}_b\left[\mathcal{N}^{(m)}(t)\,\hat{\rho}_a(o)\otimes\hat{\rho}_b(o)\right] \tag{11}$$

where P(m,t) is the probability to detect m b–mode photons, and

$$\mathcal{N}^{(m)}(t)\,(\,) = \int_o^t dt_m \int_o^{t_m} t_{m-1}\cdots\int_o^{t_2} dt_1\ \mathcal{S}(t-t_m)\ldots\mathcal{J}\mathcal{S}(t_1)\,(.) \tag{12}$$

$$\mathcal{S}(t)\,(.) \equiv e^{-\frac{i}{\hbar}H_I t-\gamma b^\dagger bt/2}\,(.)\,e^{\frac{i}{\hbar}H_I t-\gamma b^\dagger bt/2} \tag{13}$$

$$\mathcal{J}(.) \equiv \gamma b\rho b^\dagger. \tag{14}$$

In reference [8] equation (11) is computed using a Q–function technique. The result is

$$\hat{\rho}_a^{(m)}(t) = \sum_{n=m}^{\infty} P^{(m)}(n,t)\ |n\rangle\langle n| \tag{15}$$

where

$$P^{(m)}(n,t) = \begin{bmatrix} n \\ m \end{bmatrix}\left[1-\lambda(t)\left[1-\mu_e(t)\right]\right]^{m+1}\left[\lambda(t)\left[1-\mu_e(t)\right]\right]^{n-m} \tag{16}$$

where

$$\lambda = \frac{\bar{n}_a(t)}{1+\bar{n}_a(t)} \tag{17}$$

and

$$\mu_e(t) = \frac{\bar{n}_a(t) - \bar{n}_b(t)}{\bar{n}_a(t)} \tag{18}$$

is an effective quantum efficiency. This is identical to the result for the detection of m quanta from a thermal state with mean $\bar{n}_a(t)$ with quantum efficiency $\mu_e(t)$.

The ideal measurement limit occurs as $\mu_e(t)\to 1$. This is obtained by $\chi \ll \gamma$ with χt = constant and $\chi^2 t/\gamma \ll 1$. In this limit $\bar{n}_a(t)\to$ constant and $\bar{n}_b(t)\to 0$ i.e. all b–mode photons are counted. One has

$$\lim_{\mu_e\to 1} \hat{\rho}_a^{(m)}(t) = |m\rangle\langle m| \tag{19}$$

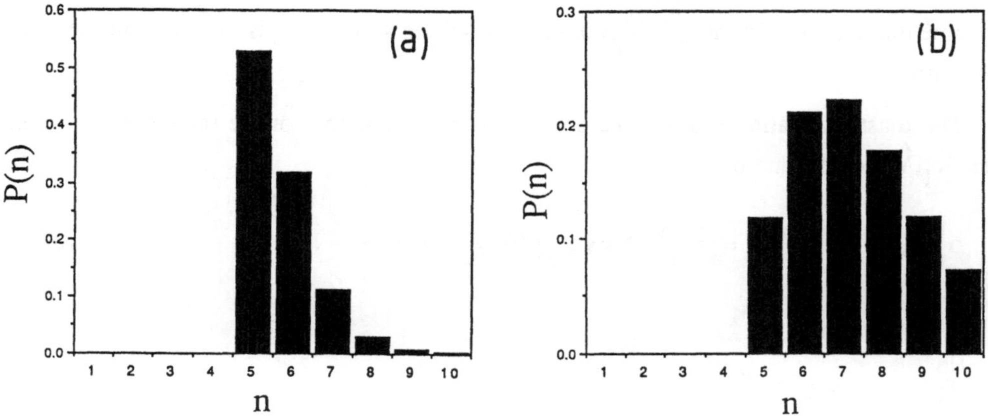

Fig. 2 A plot of the conditional post–measurement number distribution $P^{(m)}(n)$ for $m = 5$ a) $f^2 = 0.1$ b) $f^2 = 0.3$

i.e. mode a is in a pure number state with exactly m quanta. In general however $\hat{\rho}_a^{(m)}(t)$ is a mixture of number states with at least m quanta (see figure [2a,b].

As an example of a non–selective description I will now discuss homodyne detection of an a–mode quadrature phase.

NONSELECTIVE QUADRATURE MEASUREMENTS in a NONDEGENERATE PARAMETRIC AMPLIFIER

In this section I will discuss a rather special but perhaps interesting limit of the master (3).

Assume that A = 0, i.e. a squeezed vacuum input is used and that the squeezing is very large $e^{2r} >> 1$. The master equation may then be approximated by

(with an appropriate choice for ϕ)

$$\frac{d\rho}{dt} = 2i\chi \left[\hat{X}_1\hat{Y}_1 - \hat{X}_2\hat{Y}_2, \rho\right] - \Gamma \left[\hat{X}_1, \left[\hat{X}_1, \rho\right]\right] \tag{20}$$

where $\Gamma = \gamma e^{2r}$ and

$$\hat{X}_1 = \frac{1}{2}(a + a^\dagger) \; ; \; \hat{Y}_1 = \frac{1}{2}(b + b^\dagger)$$

$$\hat{X}_2 = \frac{1}{2i}(a - a^\dagger) \; ; \; \hat{Y}_2 = \frac{1}{2i}(b - b^\dagger) \tag{21}$$

under free evolution ($\Gamma = 0$), $\hat{Y}_2(t)$ is coupled to $\hat{X}_1(t)$ and thus a measurement of $\hat{X}_1(t)$ yields information on $\hat{Y}_2(t)$. If this is the case then the system should simultaneously become diagonalised in the $(\hat{X}_1,\hat{Y}_2)$ basis. As will be shown this is only true in the long time limit.

The master equation (20) can be converted to an equation for the matrix elements in the $(\hat{X}_1,\hat{Y}_2)$ eigenstate basis,

$$\rho\left[x_1,x_1',y_2,y_2',\tau\right] = {}_a\langle x_1'|\otimes_b\langle y_2'|\rho(t)|y_2\rangle_b\otimes|x_1\rangle_a. \tag{22}$$

Thus

$$\frac{\partial\rho}{\partial\tau}(x_1,x_1',y_2,y_2',\tau)$$

$$= -\left[x_1'\frac{\partial}{\partial y_2'} + x_1\frac{\partial}{\partial y_2} + y_2'\frac{\partial}{\partial x_1'} + y_2\frac{\partial}{\partial x_1}\right]\rho - \Gamma(x_1-x_1')^2\rho \tag{23}$$

where $t = \chi t$ and $\gamma = \Gamma/\chi$. This may be solved by the method of characteristics to yield

$$\rho(x_1,x_1',y_2,y_2',\tau) = \rho_o(x_1,x_1',y_2,y_2^1,\tau)\, D(x_1,x_1',y_2,y_2',\tau) \tag{24}$$

where the free evolution part is

$$\rho_o(\tau) = \frac{2}{\pi}\exp\left[-(x_1\cosh\tau - y_2\sinh\tau)^2 - (y_2\cosh\tau - x_1\sinh\tau)^2\right.$$

$$\left.-(x_1'\cosh\tau - y_2'\sinh\tau)^2 - (y_2'\cosh\tau - x_1'\sinh\tau)^2\right] \tag{25}$$

and D is a coherence decay term given by

$$D(x_1,x_1',y_2,y_2',\tau)$$

$$= \exp\left[-\frac{\gamma}{2}\left[\frac{1}{2}\sinh 2\tau + \tau\right]\left[x_1'-x_1\right]^2 - \frac{\gamma}{2}\left[\frac{1}{2}\sinh 2\tau - \tau\right]\left[y_2'-y_2\right]^2\right.$$

$$\left.+\gamma\sinh^2\tau\left[x_1'-x_1\right]\left[y_2'-y_2\right]\right]. \tag{26}$$

For very short times

$$D \simeq e^{-\gamma\tau(x_1-x_1')^2} \tag{27}$$

Thus at the beginning of the measurement coherence is destroyed in the basis of the quantity measured directly i.e $\hat{X}_1$. As time proceeds and correlations between X_1 and Y_2 develop the coherence destruction spreads to $Y_2(t)$ as well. At long times

$$D \simeq \exp\left\{-\frac{\gamma}{4}\sinh 2\tau\left[(x_1' - x_1)^2 + (y_2 - y_2')^2\right] + \gamma\sinh^2\tau\,(x_1 - x_1')(y_2 - y_2')\right\} \tag{28}$$

and the system is simultaneously diagonalised in the $(\hat{X}_1, \hat{Y}_2)$ eigenstate basis at the same rate, confirming one's expectations.

ACKNOWLEDGEMENTS

I would like to thank Cathy Holmes and Dan Walls who assisted with much of this work.

REFERENCES

1. K. Kraus: States, Effects and Operations, (Springer, Berlin, 1983)
2. W.H. Zurek: Phys. Rev. D. 24, 1516, (1981); 26, 1862, (1982).
3. M.J. Collett: Phys. Rev. A. 38, 2233, (1988)
4. A. Barchielli, L. Lanz and G.M. Prosperi: Nuovo Cimento B. 72, 79, (1982)
5. M.D. Srinivas and E.B. Davies: Opt. Acta. 28, 981, (1981)
6. G.J. Milburn: Phys. Rev. A. 37, 2970, (1988)
7. C.W. Gardiner and M.J. Collett: Phys. Rev. A. 31, 3761, (1985)
8. C.A. Holmes, G.J. Milburn and D.F. Walls: Phys. Rev. A. (to appear) (1989)
9. G.J. Milburn, A.S. Lane and D.F. Walls: Phys. Rev. A. 27, 2804, (1983)

Input-Output Formulations of Photodetection

C.W. Gardiner

Physics Department, University of Waikato, Hamilton, New Zealand

Since the original formulation of Mandel the counting formula, known as Mandel's formula [1], or in a fully quantum mechanical guise as the Kelley-Kleiner formula [2,3], has been used to analyse photodetection by a wide variety of devices, including photomultipliers and photodiodes etc., and the description given by this formula has been found completely satisfactory in accounting for observations. The formula is, however, manifestly only an approximation, obtained by assuming a narrow bandwidth detector, and by averaging over many cycles of the relevant optical frequency. Of course both of these are in almost all current situations very good approximations, but their introduction is necessary not only because they are reasonable, but also because ambiguities arise when they are not made.

I want to introduce here a different view of the problem, based on the input-output formulations introduced by Collett and myself [4]. The major new idea is to treat the photodetection process as the counting of electrons, represented by an appropriate fully quantised field, rather than merely the counting of photoionisations. This latter more traditional approach computes the probability that a detector "atom" will reach a certain energy level. When it reaches this level, it is supposed that the atom ionises irreversibly, and this amounts to a single photodetection event. In contrast, I want to measure the *electron number operator* in the output electron field.

Why do this? Firstly, because it is what happens. The assumption that the atom once ionised, releases the electron irreversibly is a very reasonable approximation, but is not entirely controllable. A second reason is that it provides a very simple and natural derivation of the counting formula, though I admit that when one develops a reasonably realistic model based on this approach, the simplicity is somewhat obscured. A third reason is related to renormalisation problems. If one wishes to avoid the rotating wave approximation, which must be necessary in high speed pulses, the problem of renormalisation of the atom's Hamiltonian arises. At its worst, this will give infinite counting rates [5]. At the very least one must face the problem that the final states to which the atom is being excited are not completely precisely defined. The input-output method avoids all of this by dealing only with asymptotic fields, either for the electron or the photon, and these are well defined.

In this paper I do not have space to discuss all the other approaches available—I shall describe my own models. The paper has three parts. I start with a very simple model, essentially of one atom and a one dimensional electromagnetic field. The electrons are treated as fully quantised *Bosons*, because this allows for great simplicity and clarity. The next section attempts to show that the results still hold, in a degree of approximation, even with Fermi electrons. This involves the adaptation of the input-out formalism to handle Fermions.

The final section reverts to Bose electrons, but treats the fully quantised propagation absorption and photoproduction problem in some detail, without making the rotating wave approximation.

Springer Proceedings in Physics, Vol. 41 **Quantum Optics V**
Editors: J.D. Harvey and D.F. Walls

1. A One Atom Model

In order to see why the photoelectron statistics are those of the incident light, let us consider the simplest possible model of a photodetector—a single atom (idealised as a harmonic oscillator), localised at some point in a one-dimensional space, and tuned to the frequency of the incoming light. In photodetection the electrons are never produced with any significant occupation of the available states so the distinction between Bosons and Fermions is immaterial; let us therefore consider Bose electrons.

We consider a simple Hamiltonian, in which a single atom interacts with the operators of both the electromagnetic field and of the electrons.

$$H = \hbar \Omega a^\dagger a + \int_{-\infty}^{\infty} d\omega \hbar \omega b^\dagger(\omega) b(\omega) + \int_{-\infty}^{\infty} d\omega \hbar \omega d^\dagger(\omega) d(\omega)$$

$$+ i\hbar \int_{-\infty}^{\infty} \bar{\gamma}(\omega) \left[b^\dagger(\omega) a - b(\omega) a^\dagger \right] d\omega$$

$$+ i\hbar \int_{-\infty}^{\infty} \bar{\kappa}(\omega) \left[d^\dagger(\omega) a - d(\omega) a^\dagger \right] d\omega \tag{1}$$

Here a is a destruction operator for the atom, $b(\omega)$ is a destruction operator for an electromagnetic field mode, $d(\omega)$ is a destruction operator for an electron field mode. The coupling between electrons, photons, and the atom is of the kind that allows an atom to be excited by the electromagnetic field, and to de-excite by emission of either an electron or a photon—the processes are all completely reversible.

Following the work of Gardiner and Collett [4] we can develop a quantum Langevin equation for a by using the following procedure

i) From the Heisenberg equations of motion for $b(\omega)$ and $d(\omega)$ we can write solutions

$$b(\omega, t) = e^{-i\omega(t-t_0)} b_0(\omega, t_0) + \bar{\gamma}(\omega) \int_{t_0}^{t} e^{-i\omega(t-t')} a(t')\, dt' \tag{2}$$

$$d(\omega, t) = e^{-i\omega(t-t_0)} d_0(\omega, t_0) + \bar{\kappa}(\omega) \int_{t_0}^{t} e^{-i\omega(t-t')} a(t')\, dt' \tag{3}$$

where t_0 is some time in the remote past.

ii) Assume that $\bar{\gamma}(\omega)$ and $\bar{\kappa}(\omega)$ are almost independent of ω, over the range of ω in which the coupling is significant, thus $\bar{\gamma}(\omega) = \sqrt{\gamma/2\pi}$, $\bar{\kappa}(\omega) = \sqrt{\kappa/2\pi}$.

iii) Define "in" fields, $b_{\text{in}}(t), d_{\text{in}}(t)$ by

$$b_{\text{in}}(t) = \frac{1}{\sqrt{2\pi}} \int d\omega e^{-i\omega(t-t_0)} b(\omega, t_0) \tag{4}$$

$$d_{\text{in}}(t) = \frac{1}{\sqrt{2\pi}} \int d\omega e^{-i\omega(t-t_0)} d(\omega, t_0). \tag{5}$$

iv) Substitute for $b(\omega,t)$ and $d(\omega,t)$ using (2) and (3) in the Heisenberg equation of motion for $a(t)$; the result is the *quantum Langevin equation*

$$\dot{a}(t) = -\mathrm{i}\Omega a(t) - \tfrac{1}{2}(\kappa+\gamma)a(t) - \sqrt{\gamma}\, b_{\mathrm{in}}(t) - \sqrt{\kappa}\, d_{\mathrm{in}}(t). \tag{6}$$

The "in" fields are essentially the inputs into the oscillator from the electromagnetic and electron fields; thus $b_{\mathrm{in}}(t)$ represents the incoming electromagnetic field, whose statistics we wish to measure. The process of measurement is to count emitted electrons, so we wish to know not the "in" field d_{in}, but a corresponding "out" field. This is supplied by writing the solution to $d(\omega,t)$ in terms of a *final condition* at time t_1 in the remote future, thus

$$d(\omega,t) = \mathrm{e}^{-\mathrm{i}\omega(t-t_1)}d(\omega,t_1) - \bar{\kappa}(\omega)\int_t^{t_1} \mathrm{e}^{-\mathrm{i}\omega(t-t')}a(t')\,dt' \tag{7}$$

and then

$$d_{\mathrm{out}}(t) = \frac{1}{\sqrt{2\pi}}\int d\omega \mathrm{e}^{-\mathrm{i}\omega(t-t_1)}d(\omega_1,t_1). \tag{8}$$

Directly from the definitions (7,8) it is easy to show that a *boundary* condition holds.

$$d_{\mathrm{out}}(t) = d_{\mathrm{in}}(t) + \sqrt{\kappa}\, a(t). \tag{9}$$

To get the relationship between $d_{\mathrm{out}}(t)$ and $b_{\mathrm{in}}(t)$ now requires us to solve the equations. We go to a frame rotating at angular frequency Ω by writing $a(t) = \mathrm{e}^{-\mathrm{i}\Omega t}A(t)$ with a similar notation for $B_{\mathrm{in}}(t)$, $D_{\mathrm{in}}(t)$, $D_{\mathrm{out}}(t)$, etc. Now assume that $\kappa+\gamma$ is sufficiently large that we can solve the equation resulting from (6) by adiabatic elimination, thus

$$A(t) = \frac{2\sqrt{\gamma}\,B_{\mathrm{in}}(t) + 2\sqrt{\kappa}\,D_{\mathrm{in}}(t)}{\kappa+\gamma} \tag{10}$$

and using (9), we get

$$D_{\mathrm{out}}(t) = \frac{\gamma-\kappa}{\gamma+\kappa}D_{\mathrm{in}}(t) - \frac{2\sqrt{\gamma\kappa}}{\gamma+\kappa}A_{\mathrm{in}}(t). \tag{11}$$

It is clear that the normally ordered moments of $D_{\mathrm{out}}(t)$are equal to those of $\frac{2\sqrt{\gamma\kappa}}{\gamma+\kappa}A_{\mathrm{in}}(t)$ provided that the state of $D_{\mathrm{in}}(t)$ corresponds to the vacuum, that is, provided we are not directing electrons at the atom, which is of course the case. If $\gamma = \kappa$, we get the ideal result that $D_{\mathrm{out}}(t) = A_{\mathrm{in}}(t)$, so that the measurement of $D_{\mathrm{out}}(t)$ is equivalent to the measurement of $A_{\mathrm{in}}(t)$. As was shown by Ghielmetti [6], this is exactly equivalent to Mandel's formula. In the case of lower efficiency, when $\gamma \neq \kappa$, the effect is similar to that of a beam splitter; that is, the output electron field is a mixture of the input photon field and the vacuum fluctuations arising from the input electron field. This is similar to a view introduced by Yuen and Shapiro [7], that such a photodetector is equivalent to an ideal photodetector preceded by a beam splitter.

Mandel's formula is not exactly equivalent to (11), which would suggest that phase information is also supplied to the electron field. The precise content is in fact that the normally ordered moments of the electron number operator

$$N(t_a, t_b) = \int_{t_a}^{t_b} D_{\text{out}}^{\dagger}(t) D_{\text{out}}(t)\, dt \tag{12}$$

are given by substituting for $D_{\text{out}}(t)$, as mentioned, thus we can say that measuring $N(t_a, t_b)$ counts the number of electrons arriving in the interval (t_a, t_b), and this is equal to

$$\frac{4\gamma\kappa}{(\gamma+\kappa)^2} \int_{t_a}^{t_b} A_{\text{in}}^{\dagger}(t) A_{\text{in}}(t)\, dt. \tag{13}$$

The equality of *normally ordered* moments, with the factor $\epsilon = \frac{4\gamma\kappa}{(\gamma+\kappa)^2} \leq 1$ as the detector efficiency is equivalent to Mandel's formula.

Notice also that the detector efficiency is strictly no greater than 100%, and reaches 100% when $\gamma = \kappa$—a fact which can be regarded as impedance matching.

This is not a relevant result to practical photodetection, which is a distributed process, in which attenuation takes place over a finite distance. A more appropriate model is presented in Sect.3.

2. A Model Using Fermi Electrons

The use of Fermi electrons is much more difficult because the harmonic oscillator can no longer be used as a model of an atom. It is simply not possible to set up a physics in which a photon raises the atom to an excited state, which emits an electron and returns to the *same* state. Instead, a three level model is the bare minimum, as in fig.1. The emitted electron leaves the atom in an intermediate state, which then makes a transition to the ground state by absorption of another electron from a bath of electrons. This is of course the essence of what happens in practice.

a) Input-Output method with Fermions: To use Fermions, an appropriate formulation of the input output methods is essential. Suppose we consider ionisation of an atom, in which a Fermion can be attached or detached. The Hamiltonian is

$$H = H_{\text{sys}} + H_{\text{int}} + H_B \tag{14}$$

where H_{sys} is the free atom Hamiltonian, and

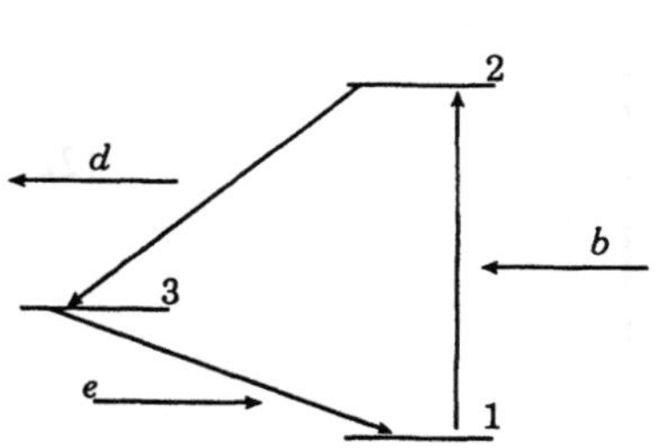

Fig 1. Three level atom model of a photodetector.

$$H_B = \int \hbar\omega\, d^\dagger(\omega) d(\omega)\, d\omega \tag{15}$$

$$H_{\rm int} = \mathrm{i}\hbar \int d\omega\, \kappa(\omega) \left\{ d^\dagger(\omega) c - c^\dagger d(\omega) \right\}, \tag{16}$$

which is exactly the same as in the Boson formulation. However, in this case, $d(\omega), d^\dagger(\omega)$, are Fermion operators so that

$$[d(\omega), d^\dagger(\omega')]_+ = \delta(\omega - \omega'). \tag{17}$$

A complication is now introduced by the concept of Fermionic and Bosonic system operators. At equal times $d(\omega), d^\dagger(\omega)$ commute with all Bosonic system operators and anticommute with all Fermionic operators. Clearly the operator c which occurs in the interaction Hamiltonian must be Fermionic.

b) Quantum Langevin Equation: The same procedure as for the Boson bath case can now be applied, to yield the Fermionic quantum Langevin equation,

$$\begin{aligned} \dot{a} = -\frac{\mathrm{i}}{\hbar}[a, H_{\rm sys}] &- \Big\{ [a, c^\dagger]_\mp \Big\{ \frac{\gamma}{2} c + \sqrt{\gamma}\, d_{\rm in}(t) \Big\} \\ &\mp \Big\{ \frac{\gamma}{2} c^\dagger + \sqrt{\gamma}\, d_{\rm in}^\dagger(t) \Big\} [a, c]_\mp \Big\} \end{aligned} \tag{18}$$

where + corresponds to a Fermionic a and − corresponds to a Bosonic a. The Bosonic equation is the same as in the usual formalism.

c) Inputs, Outputs and Causality: As previously

$$d_{\rm out}(t) = d_{\rm in}(t) + \sqrt{\gamma}\, c(t) \tag{19}$$

and

$$\left.\begin{aligned} [a(t), d_{\rm in}(t')]_\pm &= -u(t - t')\sqrt{\gamma}\,[a(t), c(t')]_\pm \\ [a(t), d_{\rm out}(t')]_\pm &= u(t' - t)\sqrt{\gamma}\,[a(t), c(t')]_\pm \end{aligned}\right\} \tag{20}$$

d) Fermionic Quantum White Noise: In this case we define

$$\langle b_0^\dagger(\omega) b_0(\omega')\rangle = \bar{N}\delta(\omega - \omega'), \qquad \langle b_0(\omega') b_0^\dagger(\omega')\rangle = (1 - \bar{N})\delta(\omega - \omega') \tag{21}$$

where $\bar{N} = \{\exp(\hbar\omega/kT) + 1\}^{-1} \le 1$

e) The Detector Model: Define

$$\left.\begin{aligned} &\left.\begin{aligned} |1\rangle\langle 1| &= n \\ |2\rangle\langle 2| &= m \\ |3\rangle\langle 3| &= l = 1 - m - n \\ |1\rangle\langle 2| &= p \end{aligned}\right\} \quad \text{Bosonic} \\ &\left.\begin{aligned} |3\rangle\langle 2| &= q \\ |1\rangle\langle 3| &= r \end{aligned}\right\} \quad \text{Fermionic} \end{aligned}\right\} \tag{22}$$

There is a field b for the photons, and two separate electron fields d and e. The first represents ejected electrons, and the second the returning electrons. The interaction Hamiltonian is

$$\begin{aligned} H_{\text{int}} = & \, \mathrm{i}\hbar \int d\omega \bar{\kappa}(\omega) \left\{ d^\dagger(\omega) q - q^\dagger d(\omega) \right\} \\ & -\mathrm{i}\hbar \int d\omega \bar{\lambda}(\omega) \left\{ e^\dagger(\omega) r^\dagger - r e(\omega) \right\} \\ & +\mathrm{i}\hbar \int d\omega \bar{\gamma}(\omega) \left\{ b^\dagger(\omega) p - p^\dagger b(\omega) \right\}, \end{aligned} \tag{23}$$

and this leads to the equations of motion

$$\left.\begin{aligned} \dot{n} &= \gamma m + \lambda l + \sqrt{\gamma} \left\{ p^\dagger b_{\text{in}} + b_{\text{in}}^\dagger p \right\} + \sqrt{\lambda} \left\{ r^\dagger e_{\text{in}}^\dagger + e_{\text{in}} r \right\} \\ \dot{m} &= -(\kappa + \gamma) m - \sqrt{\gamma} \left\{ p^\dagger b_{\text{in}} + b_{\text{in}}^\dagger p \right\} - \sqrt{\kappa} \left\{ q^\dagger d_{\text{in}} + d_{\text{in}}^\dagger q \right\} \\ \dot{l} &= -\dot{m} - \dot{n} \\ \dot{p} &= -\frac{\kappa}{2} p - \sqrt{\kappa}\, r d_{\text{in}} + \sqrt{\lambda}\, q e_{\text{in}}^\dagger \\ \dot{q} &= -\tfrac{1}{2}(\gamma + \kappa + \lambda) q - \sqrt{\gamma}\, r^\dagger b_{\text{in}} - \sqrt{\kappa}\,(1 - n) d_{\text{in}} - \sqrt{\lambda}\, e_{\text{in}} p. \\ \dot{r} &= -\frac{\lambda}{2} r + \sqrt{\gamma}\, q^\dagger b_{\text{in}} + \sqrt{\kappa}\, d_{\text{in}}^\dagger p + \sqrt{\lambda}\,(1 - m) e_{\text{in}}^\dagger \end{aligned}\right\} \tag{24}$$

f) Input-Output Boundary Conditions: We want to specify b_{in}, the incoming photon field, and to compute d_{out}, the ejected electron field. We also need to specify that

d_{in} corresponds to a vacuum—no input electrons are present in the channel to which the ejected electron couples.

e_{in} corresponds to a fully filled bath— there are numerous electrons available in the channels required to return the atom to the ground state.

We know that

$$\left.\begin{aligned} b_{\text{out}} &= b_{\text{in}} + \sqrt{\gamma}\, p \\ d_{\text{out}} &= d_{\text{in}} + \sqrt{\kappa}\, q \\ e_{\text{out}}^\dagger &= e_{\text{in}}^\dagger + \sqrt{\lambda}\, r \end{aligned}\right\} \tag{25}$$

g) Adiabatic Elimination Limit: As previously, assume that γ and κ are large and we can solve the equations for q and p by adiabatic elimination.

$$p = \frac{-\sqrt{\gamma}\,(n - m)\, b_{\text{in}} - \sqrt{\kappa}\, r d_{\text{in}} + \sqrt{\lambda}\, q e_{\text{in}}^\dagger}{\frac{1}{2}(\gamma + \kappa)} \tag{26}$$

$$q = \frac{\sqrt{\gamma}\, r^\dagger b_{\text{in}} + \sqrt{\kappa}\,(1 - n)\, d_{\text{in}} + \sqrt{\lambda}\, e_{\text{in}} p}{\frac{1}{2}(\gamma + \kappa + \lambda)} \tag{27}$$

The r equation cannot be solved by adiabatic elimination, since we do not put any fast requirement on λ. We can approximate the r equation by noting that the excitation amplitude will be small in practice, for linear operation of the detector, so that we can say $n \sim 1, m \sim l \sim r \sim p \sim q \sim 0$, so that

$$r^\dagger \approx \sqrt{\lambda} \int_{-\infty}^{t} dt' e^{-\frac{1}{2}\lambda(t-t')} e_{\text{in}}(t') \tag{28}$$

and

$$\begin{aligned} q \approx\ & \frac{2\sqrt{\gamma\lambda}}{\gamma+\kappa+\lambda} \int_{-\infty}^{t} dt' e^{-\frac{1}{2}\lambda(t-t')} e_{\text{in}}(t') b_{\text{in}}(t) + \frac{2\sqrt{\kappa}}{\gamma+\kappa+\lambda} d_{\text{in}}(t) \\ & - \frac{4\sqrt{\lambda\gamma}\, e_{\text{in}}(t) b_{\text{in}}(t)}{(\gamma+\kappa+\lambda)(\gamma+\kappa)} \end{aligned} \tag{29}$$

We will only be interested in normally ordered products, and since d_{in} is the vacuum, it will contribute nothing. It is not difficult to see that although the other two terms are similar, the first one is much bigger than the other if $\gamma, \kappa \gg \lambda$.

Thus we find

$$d_{\text{out}} \simeq \frac{\gamma-\kappa+\lambda}{\gamma+\kappa+\lambda} d_{\text{in}} + \frac{2\sqrt{\gamma\kappa\lambda}}{\gamma+\kappa+\lambda} \int_{-\infty}^{t} dt' e^{-\frac{1}{2}\lambda(t-t')} e_{\text{in}}(t') b_{\text{in}}(t). \tag{30}$$

The normally ordered correlation functions of of d_{out} are then given by those of the second term. We take, for a filled $e_{\text{in}}(t)$ that

$$\langle e_{\text{in}}^\dagger(t) e_{\text{in}}(t') \rangle = \delta(t-t') \qquad \langle e_{\text{in}}(t) e_{\text{in}}^\dagger(t') \rangle = 0 \tag{31}$$

so that, for example

$$\langle d_{\text{out}}^\dagger(t) d_{\text{out}}(t') \rangle = \left(\frac{4\gamma\kappa}{\gamma+\kappa+\lambda} \right) e^{-\frac{1}{2}\lambda|t-t'|} \langle b_{\text{in}}^\dagger(t) b_{\text{in}}(t') \rangle \tag{32}$$

provided λ is much *smaller* than the time constant of the correlation function $\langle b_{\text{in}}^\dagger(t) b_{\text{in}}(t') \rangle$ this says that the electron correlation function is that of the input photons, up to an efficiency factor of $4\gamma\kappa/(\gamma+\kappa)$, since $\gamma \ll \gamma+\kappa$. This is the same result that we found in the Bose model. The higher order normally ordered correlation functions follow similarly for *even* order correlations, but the odd order correlations will vanish, as should indeed be the case.

Thus the inclusion of Fermi electrons does yield exactly the kind of result predicted by Mandel's formula, without the spurious equality of odd order moments which happens with a Bose electron model. Nevertheless the result for even order moments in the degree of approximation used is exactly the same as that of the Bose model. The reason for this is very simple. The Bose electron model is completely linear, in that the output electron fields are linear functions of the of the input photon fields. The Fermi electron model cannot be linearised in this sense, since Fermi fields can never be equal to Bose fields, but it is linearised in the only sense that is allowable; namely there is a linear relation between the d_{out} and a bilinear function of the input Bose field and another Fermi field, which arises from the returning electrons. The equivalence of correlation functions is true only if λ, the relaxation time constant of the re-absorption of the electrons, is small compared to those of the photon absorption and electron production. This is appropriate, since it means that the virtual level to which the atom returns after emitting the photoelectron is almost stable, and correlations between photoelectrons and reabsorbed electrons cannot arise.

3. A More Realistic Detector Model

The fact that we can get essentially the same results from both Bose and Fermi electrons is encouraging, but both of these models are far too simplified. For example, the result of perfect photodetection if $\gamma = \kappa$ is not what one expects from a single atom—rather, it is the property of a lumped oscillator model of a detector. Most treatments of photodetection assume that the medium acts locally as a very weak inefficient detector, with high efficiency obtained by the eventual absorption in a sufficiently thick medium. Let us therefore model a detector of this kind, but simplify matters by assuming again Bose electrons. The model is one dimensional, as in fig 2. An appropriate Lagrangian is

$$\begin{aligned} L = \tfrac{1}{2} \int_{-\infty}^{\infty} dx \left\{ \dot{A}(x,t)^2 - c^2 \partial_x A(x,t)^2 \right\} \\ + \int_{-\infty}^{0} dx \left\{ \tfrac{1}{2}\dot{u}(x,t)^2 - \tfrac{1}{2}\omega^2 u(x,t)^2 + u(x,t) \left[2\sqrt{\kappa v}\, \dot{D}(x,t,0) + \gamma \dot{A}(x,t) \right] \right\} \\ + \tfrac{1}{2} \int_{-\infty}^{0} dx \int_{0}^{\infty} ds \left\{ \dot{D}(x,t,s)^2 - v^2 \partial_s D(x,t,s)^2 \right\} . \end{aligned} \tag{33}$$

This Lagrangian corresponds to a one dimensional electrodynamics on the range $(-\infty, \infty)$, coupled in the left half of space to a distribution of atoms represented by the operators $u(x,t)$. These atoms are coupled *locally* to a distribution of electron baths, represented by the field $D(x,t,s)$. For the electron baths and the atoms the co-ordinate x does not represent any dynamics, but merely acts as a label to distinguish the different electron baths and atoms at each point—the dynamics takes place in another variable s. Writing the bath Lagrangian in the form given in (33) has two main advantages over the more conventional Hamiltonian method.
—The input-output formalism is easily developed.
—The identification of outputs corresponding to electrons produced is straightforward.
The physics represented is something like that of a solid state photodetector, in which the electrons produced are immediately localised by thermal and other interactions in the solid, so that the photoelectric processes at two points depend on each other only to the extent enforced by

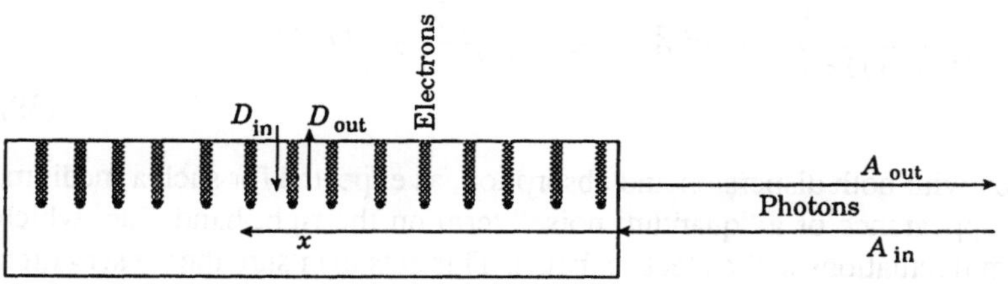

Fig 2. Spatially distributed model of a photodetector.

the coherence of the electromagnetic field. The basic concept is thus the same as in the previous part, and it is merely a matter of solving for a $D_{\rm out}$ field in terms of the electromagnetic field being input from the right.

a) Equations of Motion: For the three fields, we find, for $x < 0$

$$\left.\begin{aligned}\ddot{A} - c^2\partial_x^2 A &= -\gamma\dot{u}\\ \ddot{u} + \omega^2 u &= \gamma\dot{A} + 2\sqrt{\kappa v}\,\dot{D}(x,t,0)\\ \ddot{D} - v^2\partial_s^2 D &= -2\sqrt{\kappa v}\,\delta(s)\,\dot{u}(X)\end{aligned}\right\} \tag{34}$$

Since $D(x,t,s)$ represents the electron bath, we have to solve for it in terms of the electron inputs and outputs. The solution of the equation can be written in two equivalent forms;

$$\left.\begin{aligned}D(x,t,s) &= D_{\rm in}(x,t+s/v) - \tfrac{1}{2}\sqrt{\frac{\kappa}{v}}\,u(x,t-s/v,s)\\ &= D_{\rm out}(x,t-s/v) + \tfrac{1}{2}\sqrt{\frac{\kappa}{v}}\,u(x,t+s/v,s)\end{aligned}\right\} \tag{35}$$

We can substitute the first of equations (35) into the equation for u given in (34) to get as our basic pair of equations, for $x < 0$, (that is, inside the photodetection medium).

$$\left.\begin{aligned}\ddot{A} - c^2\partial_x^2 A &= -\gamma\dot{u}\\ \ddot{u} + \omega^2 u &= \gamma\dot{A} - \kappa\dot{u} + 2\sqrt{\kappa v}\,\dot{D}_{\rm in}(x,t)\end{aligned}\right\} \tag{36}$$

To the right, $x > 0$, where there is no medium, we simply have the free wave equation $\ddot{A} - c^2\partial_x^2 A = 0$. The other variables do not exist for $x > 0$.

b) Solution of the Equations of Motion: If we define a Fourier transform variable by

$$u(t,x) = \frac{1}{\sqrt{2\pi}}\int_{-\infty}^{\infty} d\Omega\, e^{-i\Omega t}\tilde{u}(\Omega,x) \tag{37}$$

with a similar definition for $\tilde{A}(\Omega,x)$ and $\tilde{D}(\Omega,x,s)$, we find that the u equation can be reduced to

$$\tilde{u} = \frac{(-i\Omega)\,[2\sqrt{\kappa v}\,\tilde{D}_{\rm in}(\Omega,x) + \gamma\tilde{A}]}{\omega^2 - \Omega^2 - i\kappa\Omega} \tag{38}$$

which, when substituted into the equation for $\tilde{A}$, yields

$$-\tilde{A}\left\{\Omega^2 + \frac{\gamma^2\Omega^2}{\omega^2 - \Omega^2 - i\Omega\kappa}\right\} - c^2\partial_x^2\tilde{A} = \frac{2\gamma\Omega^2\sqrt{\kappa v}}{\omega^2 - \Omega^2 - i\Omega\kappa}\tilde{D}_{\rm in}(\Omega,x) \tag{39}$$

This is a wave equation with both dispersion and absorption, as expected for such a medium. Notice, however, the appearance of a "quantum noise" term on the right-hand side, which arises from the vacuum fluctuations in the electron baths. This acts to ensure the preservation of canonical commutation relations.

Let us define a quantity ν by

$$\nu^2 = \Omega^2 + \frac{\gamma^2 \Omega^2}{\omega^2 - \Omega^2 - \mathrm{i}\,\Omega\kappa}. \tag{40}$$

In particular, we define ν so that it is the root of (40) with a *negative imaginary part*. (Notice that this means for large Ω that $\nu \sim -\Omega$). In terms of ν, (39) can be written

$$-\nu^2 \tilde{A} - c^2 \partial_x^2 \tilde{A} = (\nu^2 - \Omega^2) \frac{2\sqrt{\kappa v}}{\gamma} \tilde{D}_{\text{in}}(\Omega, x). \tag{41}$$

A general solution of this equation which satisfies the boundary condition that $\tilde{A}(\Omega, x)$ be finite as $x \to -\infty$ is (noting that $\mathbf{Im}(\nu) < 0$ by definition)

$$\tilde{A}(\Omega, x) = Q e^{i\nu x/c} + \frac{2\sqrt{\kappa v}}{\gamma} \int_{-\infty}^{\infty} dx' (\nu^2 - \Omega^2) \tilde{D}_{\text{in}}(\Omega, x', 0) \frac{e^{-\mathrm{i}\nu|x-x'|/c}}{2\mathrm{i}\,\nu c} \tag{42}$$

where Q is to be determined from boundary conditions.

c) Boundary Conditions of the Interface: The conditions of continuity of electric and magnetic fields must be met at $x = 0$, the edge of the medium. The solutions of the wave equation to the right of $x = 0$ is of the form

$$A(x,t) = a_{\text{in}}(t + x/c) + a_{\text{out}}(t - x/c), \tag{43}$$

so that the boundary conditions are

$$\dot{A}(x,t)|_{x\to 0-} = \dot{a}_{\text{in}}(t + x/c) + \dot{a}_{\text{out}}(t - x/c) \tag{44}$$

and

$$\partial_x A(x,t)|_{x\to 0-} = \frac{1}{c} \{\dot{a}_{\text{in}}(t + x/c) - \dot{a}_{\text{out}}(t - x/c)\}. \tag{45}$$

Taking the Ω frequency component of these boundary conditions and combining with (42) gives

$$Q = \frac{2\Omega \tilde{a}_{\text{in}}}{\Omega - \nu} + \frac{2\sqrt{\kappa v}}{\gamma} (\Omega + \nu)^2 \int_{-\infty}^{\infty} dx \tilde{D}_{\text{in}}(\Omega, x, 0) \frac{e^{\mathrm{i}\nu x/c}}{2\mathrm{i}\,\nu c} \tag{46}$$

$$\tilde{a}_{\text{out}} = \left(\frac{\Omega + \nu}{\Omega - \nu}\right) \tilde{a}_{\text{in}} - \frac{4\sqrt{\kappa v}}{\gamma} (\Omega + \nu)\nu \int_{-\infty}^{\infty} dx \tilde{D}_{\text{in}}(\Omega, x, 0) \frac{e^{\mathrm{i}\nu x/c}}{2\mathrm{i}\,\nu c}. \tag{47}$$

The equation for $\tilde{a}_{\text{out}}$ shows a reflected term, with reflection coefficient $\left|\frac{\Omega+\nu}{\Omega-\nu}\right|^2$, and a term which arises from the quantum fluctuations of the electron bath.

d) Attenuation Length: Since ν has an imaginary part, the factor $e^{i\nu x/c}$ in (42) represents an attenuation of the input beam with depth x. The characteristic absorption depth is $l_a = \frac{1}{2}\{\mathrm{Im}(\nu/c)\}^{-1}$. An exact expression is very complicated, but if we assume that the second term in (40) is always smaller than Ω^2, then

$$l_a \approx \frac{(\omega^2 - \Omega^2)^2 + \Omega^2\kappa^2}{\Omega^2\kappa\gamma^2} \tag{48}$$

This shows a strong resonant behaviour—the maximum absorption obviously takes place when $\omega \simeq \Omega$. The ratio of the wavelength to the minimum attenuation length is $\lambda/l_a = \gamma^2/\Omega\kappa$, and if this ratio is small, the approximations required to derive (48) are valid. This is also the regime in which one would expect to use such a detector.

e) Preservation of Commutation Relations: Although it is quite tricky, it can be shown that these solutions do preserve all canonical commutation relations.

f) Detector Efficiency and Statistics: We now want to solve for the D_{out} field, which corresponds to the ejected electrons which are counted. Notice that (35) implies that

$$D_{\mathrm{out}}(x,t) = D_{\mathrm{in}}(x,t) - \sqrt{\frac{\kappa}{v}}\,u(x,t)\,. \tag{49}$$

We use the solution for $u(x,t)$ implicit in (38), into which we substitute the solution (42) for $A(t,x)$. The solution for $\tilde{D}_{\mathrm{out}}$ can be written:

$$\begin{aligned}
\tilde{D}_{\mathrm{out}}(\Omega,x) = &-\frac{2i}{\gamma}\sqrt{\frac{\kappa}{v}}(\Omega+\nu)e^{i\nu x/c}\tilde{a}_{\mathrm{in}}(\Omega)\\
&+\left(\frac{\omega^2-\Omega^2+i\kappa\Omega}{\omega^2-\Omega^2-i\kappa\Omega}\right)\tilde{D}_{\mathrm{in}}(x,\Omega)\\
&+\frac{2i\kappa}{\gamma^2\Omega}(\nu^2-\Omega^2)(\Omega+\nu)^2\int_{-\infty}^{\infty}dx'\tilde{D}_{\mathrm{in}}(\Omega,x')\frac{e^{i\nu(x'+x)/c}}{2i\nu/c}\\
&+\frac{2i\kappa}{\gamma^2\Omega}(\nu^2-\Omega^2)^2\int_{-\infty}^{\infty}dx'\tilde{D}_{\mathrm{in}}(\Omega,x')\frac{e^{-i\nu|x-x'|/c}}{2i\nu/c}\,.
\end{aligned} \tag{50}$$

While this appears rather complicated, it does in fact represent the physics of photodetection rather naturally. There are two basic components, one arising from the input electromagnetic field, and the rest from the input electron field. The input from the latter will be assumed to consist of vacuum fluctuations only, and serves mainly to ensure the preservation of commutation relations of $\tilde{D}_{\mathrm{out}}(\Omega,x)$. As in the initial simple one atom model, the normally ordered moments of $\tilde{D}_{\mathrm{out}}(\Omega,x)$ will be given by those of the $\tilde{a}_{\mathrm{in}}(\Omega)$ term, thus ensuring that a formula like Mandel's formula is valid *locally*.

g) Output Electron Flux: If we were dealing with real electrons, rather than our present Bose electrons, we would now simply calculate the electric current operator from the output field. However, in this kind of Bose electron, the electric current is not easy to define, since a charged particle can only be represented by a non-Hermitian field, which in turn would require us to use a rather more sophisticated model of the atom than a harmonic oscillator.

The only electron flux available in this formalism is the energy flux per unit length, given by

$$N(s,t,x) = -v^2 : \frac{\partial D}{\partial t}\frac{\partial D}{\partial s} : \tag{51}$$

corresponding to a kind of Poynting vector. Thus, in a time interval (t_a, t_b), the energy flowing into the electron bath per unit length of the detector is, at the point s,

$$-v^2 \int_{t_a}^{t_b} : \frac{\partial}{\partial t} D_{\text{out}}(t - s/v, x) \frac{\partial}{\partial s} D_{\text{out}}(t - s/v, x) : dt \tag{52}$$

$$= v \int_{t_a}^{t_b} : \dot{D}_{\text{out}}(t - s/v, x)^2 : dt. \tag{53}$$

This operator quantity is normally ordered to eliminate vacuum contributions. Using (50) we can see that there will be terms arising from $\tilde{D}_{\text{in}}(\Omega, x)$, and $\tilde{a}_{\text{in}}(\Omega)$; but if there is no input electron field, the $\tilde{D}_{\text{in}}$ terms will give no contribution to the expectations of normally ordered products; so that the normally ordered moments of (51) which are what is measured, will be given by substituting into (53) the terms in (50) which involve $\tilde{a}_{\text{in}}(\Omega)$.

Suppose we assume that the input field is stationary, so that

$$\langle \tilde{a}_{\text{in}}(\Omega)\tilde{a}_{\text{in}}(\Omega') \rangle = \delta(\Omega + \Omega') S(\Omega). \tag{54}$$

Then, we can substitute (50) into (53), and take the average, to get

$$v \int_{t_a}^{t_b} \langle : \dot{D}_{\text{out}}(t - s/v, x)^2 : \rangle dt$$

$$= \frac{2\kappa}{\pi\gamma^2}(t_b - t_a) \int_{-\infty}^{\infty} d\Omega |\Omega|^2 |\Omega + \nu|^2 e^{-2\text{Im}(\nu x/c)} S(\Omega). \tag{55}$$

The total power absorbed is given by integrating this over all negative x; namely

$$\text{total absorbed power} = \int_{-\infty}^{\infty} d\Omega \frac{\kappa(t_b - t_a)}{\pi\gamma^2 \text{Im}(\nu/c)} S(\Omega)|\Omega|^2|\Omega + \nu|^2. \tag{56}$$

It is not too difficult to show that input power = total power absorbed + power reflected, i.e. from (47), using the coefficient of $\tilde{a}_{\text{in}}(\Omega)$, we can show

$$\frac{(55)}{t_b - t_a} + \frac{c}{2\pi} \int_{-\infty}^{\infty} d\Omega \left| \frac{\Omega + \nu}{\Omega - \nu} \right|^2 S(\Omega)|\Omega|^2 = \frac{c}{2\pi} \int_{-\infty}^{\infty} d\Omega S(\Omega)|\Omega|^2 \tag{57}$$

h) Detector Bandwidth: It is paradoxical that in fact the maximum efficiency is achieved by the weakest detection. If the attenuation length is very large, i.e., $\gamma^2 \ll \Omega/\kappa$, the efficiency approaches 100% at all frequencies. However this is only a crude counting efficiency. To deduce the detector bandwidth one must work out the time correlation function $\langle : N(s,t,x) N(s',t',x') : \rangle$, and appropriately integrate this over the unobserved parameters. This will be done in a later publication.

References

1 L. Mandel, *Proc. Phys. Soc. 72*, 1037, (1958).

2 P. L. Kelley and W.H. Kleiner, *Phys.Rev. 136*, 316, (1964).

3 B. R. Mollow, *Phys.Rev. 168*, 1896, (1964).
M. D. Srinivas and E.B. Davies, *Optica Acta 28*, 981, (1981).

4 C. W. Gardiner and M. J. Collett, *Phys. Rev. A31*, 3761, (1985).

5 P. D. Drummond, *Phys. Rev. A35*, 4253, (1987).

6 F. Ghielmetti, *Phys. Lett. 12*, 210, (1964).

7 H. P. Yuen and J.H. Shapiro, *IEEE Trans Inf Theory IT-26*, 78, (1980).

Einstein-Podolsky-Rosen Correlations in Nondegenerate Parametric Amplification

M.D. Reid[1] *and P.D. Drummond*[2]

[1]Physics Department, University of Waikato, Hamilton, New Zealand
[2]Physics Department, University of Queensland, Brisbane, Australia

1. INTRODUCTION

In 1935 EINSTEIN-PODOLSKY-ROSEN [1] put forward an argument which seems to suggest that a quantum mechanical description of a physical system is incomplete. The EPR paradox inspired BELL [2] to derive his famous inequalities showing that local hidden variables are incompatible with quantum mechanics. There have been recent experimental demonstrations of violation of Bell's Inequalities using correlated pairs of photons. The experiments of ASPECT et al [3] use photons produced in two-photon cascade, while the recent experiments of OU and MANDEL [4] and SHIH and ALLEY [5] have shown a violation of Bell's inequalities in the nondegenerate parametric amplifier.

In this paper, we first discuss how the EPR paradox itself may be demonstrated in nondegenerate parametric amplification by making measurements of the of quadrature phase amplitudes [6,7]. Results are given for the simple parametric amplifier and also for the parametric oscillator both below and above threshold. We point out that for a nondegenerate parametric amplifier with vacuum inputs there is a local hidden variable theory describing direct measurements of quadrature phase amplitudes [8]. Thus we cannot expect a violation of Bell's inequality for such measurements, even though the correlations are of the EPR type.

Violations of Bell's inequalities are however possible if one measures intensity correlations between signal and idler fields. There have been at least two suggested experimental schemes: one by REID and WALLS [9]; and one now experimentally realised by OU and MANDEL [4]. We summarise the result for violation of Bell's inequalities in nondegenerate parametric amplification using these schemes [9,10], pointing out that they correspond to the same transformations of signal and idler modes and hence measure the same fundamental quantum property of the field.

In summary, the nondegenerate parametric amplifier is able to demonstrate either the EPR paradox or violations of Bell's inequalities depending on the type of measurement utilised.

2. DEMONSTRATION OF THE EINSTEIN-PODOLSKY-ROSEN PARADOX

2.1 The EPR Paradox with Quadrature Phase Amplitudes

First, let us review the reasoning behind the original paradox [1]. EINSTEIN-PODOLSKY-ROSEN make three assumptions. We quote CLAUSER-SHIMONY [2].

Springer Proceedings in Physics, Vol. 41 Quantum Optics V
Editors: J.D. Harvey and D.F. Walls

Firstly, they assume quantum mechanics correctly predicts results of the experiment discussed below. Secondly, 'if without in any way disturbing the system, we can predict with certainty the value of a physical quantity, then there exists an element of physical reality corresponding to this quantity'. Thirdly, they assume no action-at-a-distance. EPR then consider two spatially separated particles which show if quantum mechanics is correct, a maximum correlation between their positions, and also their momenta. A measurement of the position of particle 2 implies with certainty the result obtained if the position of particle 1 is measured immediately. Assuming there is no action-at-a-distance, the prediction for the particle is made without disturbing the particle. The EPR concept of reality leads one to identify with particle 1 a definite predetermined value for its position. Since the momenta of the two particles are also correlated, one may use a similar argument to ascribe to particle 1 a definite predetermined value for its momentum. These arguments lead EPR to identify with particle 1 a state of definite momentum *and* position. In quantum mechanics no description of the state can simultaneously have a definite value of both momentum and position. Hence EPR proposed that quantum mechanics is an incomplete description of the state of the particle.

We note that counter-arguments exist to the EPR assumptions [11]. However, our interest is in the first assumption of the EPR argument. We wish to test the predictions of quantum mechanics for the original EPR paradox involving position and momentum variables . Most previous experiments have related to Bohm's discrete version of the paradox involving spin–$^1/_2$ particles and are troubled by poor detector efficiencies.

The paradox may be formulated in terms of the quadrature phase amplitude operators as follows [7]. Consider in the first instance two single mode fields E_A and E_B at frequencies ω_a and ω_b, with associated boson operators a and b, respectively. The quadrature phase amplitudes operators are

$$X_\theta = a\, e^{-i\theta} + a^\dagger e^{i\theta} \qquad Y_\phi = b\, e^{-i\phi} + b^\dagger e^{i\phi}. \tag{2.1}$$

We abbreviate $X_1 = X_0$, $X_2 = X_{\pi/2}$, $Y_1 = Y_0$, $Y_2 = Y_{\pi/2}$. The commutation relations and Heisenberg uncertainty principle are identical to the (normalised) commutation relations of momentum and position.

$$[X_1, X_2] = 2\,i \qquad [Y_1, Y_2] = 2i \qquad \Delta^2 X_1\, \Delta^2 X_2 \geq 1 \tag{2.2}$$

Now if the fields are prepared in such a way that the amplitude X_1 is maximally correlated with Y_2 and X_2 is maximally correlated with Y_1, the EPR reasoning will apply.

As one example we consider the output signal and idler fields of a nondegenerate parametric amplifier. In this case we have the idealised mode transformations

$$X_1(L) = X_1(0)\cosh r + Y_2(0)\sinh r \qquad Y_1(L) = Y_1(0)\cosh r + X_2(0)\sinh r$$

$$X_2(L) = X_2(0)\cosh r + Y_1(0)\sinh r \qquad Y_2(L) = Y_2(0)\cosh r + X_1(0)\sinh r \tag{2.3}$$

Here r is proportional to the length of the interaction medium and to the nonlinearity of the parametric medium. The $X_i(L)$ and $Y_i(L)$ are signal and idler outputs respectively, and the $X_i(0)$, $Y_i(0)$ the respective inputs. Calculation of the correlation coefficient gives the solution

$$C_{\theta\phi} = \frac{\langle X_\theta(L)\, Y_\phi(L)\rangle}{\sqrt{\langle X_\theta(L)^2\rangle \langle Y_\phi(L)^2\rangle}} = \sin(\theta + \phi) \tanh 2r \tag{2.4}$$

In the limit of $r \to \infty$, there is maximum correlation ($C_{\theta\phi} = 1$) for $\theta + \phi = \pi/2$. Thus X_1 and Y_2, and X_2 and Y_1, are maximally correlated and the EPR paradox applies.

2.2 The Inference Error

The quadrature phase amplitudes are measurable by homodyne detection techniques [12]. It is thus possible to determine the correlation experimentally. However, in any realistic situation r is not infinite and the correlation will not be maximum. We ask the question: can the EPR experiment be realised for less than maximum correlation, and what is the signature?

Let us suppose we measure the idler quadrature amplitude $Y_2(L)$ and obtain a result y_2. Because of the correlation between signal and idler amplitudes, we can then infer a result for the measurement of the signal amplitude $X_1(L)$. Let us choose to estimate the signal amplitude $X_1(L)$ by the scaled result $x_1{}^0 = gy_2$ where g is a constant later chosen to give greatest accuracy in the determination of $X_1(L)$. In a more general situation a measurement of $Y_\phi(L)$ might enable a better estimate of $X_1(L)$. The average error in an inference of $X_1(L)$ is given by the following variance.

$$V_1(g,\phi) = \langle (X_1(L) - gY_\phi(L))^2 \rangle \tag{2.5}$$

We choose g to minimise this error. Setting $\partial V_1(g,\phi) / \partial g = 0$ we deduce the optimal value for g to be

$$\tilde{g} = \frac{\langle X_1(L)\, Y_\phi(L)\rangle}{\langle Y_\phi(L)^2\rangle} \tag{2.6}$$

In our present example, the error is minimised for the angle $\phi = \pi/2$. The minimum error is then given by

$$V_1(\tilde{g},\tilde{\phi})_{min} = \langle X_1(L)^2\rangle - \frac{\langle X_1(L)\, Y_2(L)\rangle^2}{\langle Y_2(L)^2\rangle} . \tag{2.7}$$

Similarly, we may measure the idler amplitude $Y_1(L)$ to infer a value for the signal amplitude $X_2(L)$. Using similar arguments, the precision associated with this measurement may be determined. For the idealised parametric amplifier system with vacuum inputs, the minimum inference variances are

$$V_1(\tilde{g},\tilde{\phi})_{min} = V_2(\tilde{g},\tilde{\phi})_{min} = \frac{1}{\cosh 2r} \tag{2.8}$$

Clearly for large r, the inference errors become zero.

Now we can repeat the EPR argument. A measurement of $Y_2(L)$ of the idler beam simultaneously specifies a value for the amplitude $X_1(L)$ of the signal beam, with an average variance $V_1(g,\phi)_{min}$. Similarly, a measurement of $Y_1(L)$ of the idler beam would specify a value for $X_2(L)$ of the signal, with average variance $V_2(\bar{g},\bar{\phi})_{min}$. Provided the beams are spatially separated and assuming there is no action-at-a-distance, the EPR concept of reality would lead one to assign to the signal beam predetermined values for $X_1(L)$ (with average variance $V_1(g,\phi)_{min}$) and for $X_2(L)$ (with average variance $V_2(\bar{g},\bar{\phi})_{min}$). According to quantum mechanics, the signal amplitudes $X_1(L)$ and $X_2(L)$ are non-commuting and cannot both be specified with certainty greater than that allowed by the Heisenberg uncertainty principle $\Delta X_1 \Delta X_2 \geq 1$. Thus the assignment to the signal a description with

$$V_1(g,\phi)\, V_2(\bar{g},\bar{\phi}) < 1 \tag{2.9}$$

gives a demonstration of the apparent in completeness of quantum mechanics in terms of EPR's local realism arguments. In the simple parametric amplifier case we may have $V_1(g,\phi)\, V_2(\bar{g},\bar{\phi}) = {}^1/_{\cosh 2r}$ and hence the EPR paradox is demonstrated.

2.3 Experimental Procedure-Spectral Measurements

The quantities (2.5) and the condition (2.9) are measurable experimentally. The quadrature phase amplitudes are detectable by homodyne detection techniques. A fluctuation in the quadrature amplitudes below the standard quantum limit ('squeezing') has been observed in a number of experiments [13]. Particularly relevant to our proposed EPR experiment are the recent 'QND' experiments of LEVENSON and SHELBY [14] and SLUSHER et al [14]. Here the output signal and idler fields are spatially separated. These fields are each combined with local oscillator fields; we call them E^X_{Lo} and E^Y_{Lo} for the signal and idler respectively. The signal's local oscillator is phase shifted θ with respect to the signal field; and the idler's local oscillator is phase shifted ϕ with respect to the idler. The combined fields impinge on the surface of two spatially separated photodetectors. To describe signal and idler photocurrent fluctuations, one may introduce the photocurrent operators $i_x(t)$ and $i_y(t)$ respectively [14]. These may be shown to be proportional to the quadrature phase amplitude operators $X_\theta(L,t)$ and $Y_\phi(L,t)$ respectively. Measurements with lowest noise have made in frequency space. We thus consider the Fourier transformed quantities as follows:

$$i_x(\omega) \sim \int_{-T/2}^{T/2} e^{i\omega t}\, i_x(t)\, dt \quad \sim |E^X_{Lo}|^2\, X_\theta(L,\omega) \tag{2.10}$$

where

$$X_\theta(L,\omega) = \int_{-T/2}^{T/2} e^{i\omega t}\, X_\theta(L,t)$$

Here T is the detection time which we take to be large compared to the coherence time.

The detectable observables are the real and imaginary parts of $X_\theta(L,\omega)$ and $Y_\phi(L,\omega)$ [15]. We define

$$\overline{Re}\ X_\theta(L,\omega) = (X_\theta(L,\omega) + X_\theta(L,\omega))/\sqrt{2}, \quad \overline{Im}\ X_\theta(L,\omega) = (X_\theta(L,\omega) - X_\theta(L,\omega))/\sqrt{2i} \quad (2.11)$$

These have the following commutation relations and associated uncertainty relation,

$$[\overline{Re}\ X_1(L,\omega),\ \overline{Re}\ X_2(L,\omega)] = [\overline{Im}\ X_1(L,\omega),\ \overline{Im}\ X_2(L,\omega)] = 2i \quad (2.12)$$

$$\Delta^2 \overline{Re}\ X_1(L,\omega)\ \Delta^2 \overline{Re}\ X_2(L,\omega) \geq 1$$

Thus we consider the variances of the real spectral components.

$$V_i(g,\phi,\omega) = \langle(\overline{Re}\ X_i(L,\omega) - g\,\overline{Re}\ Y_\phi(L,\omega))^2\rangle \quad (2.13)$$

If the correlation between the signal and idler spectral components is sufficiently large that for some gains $g,\bar{g}$ and phase angles $\phi,\bar{\phi}$ we have

$$V_1(g,\phi,\omega)\ V_2(\bar{g},\bar{\phi},\omega) < 1 \quad (2.14)$$

the EPR paradox would follow. For an ideal parametric amplifier, output frequency components are related to input frequency components as given by equation (2.3) but replacing $X_i(L)$ with $\overline{Re}\ X_i(L,\omega)$, $X_i(0)$ with $\overline{Re}\ X_i(0,\omega)$, $Y_i(L)$ with $\overline{Re}\ Y_i(L,\omega)$, $Y_i(0)$ with $\overline{Re}\ Y_i(0,\omega)$. Thus for example $\overline{Re}\ X_1(L,\omega)$ and $\overline{Re}\ Y_2(L,\omega)$, and also $\overline{Re}\ X_2(L,\omega)$ and $\overline{Re}\ Y_1(L,\omega)$, are correlated. A simple check reveals as before the following solution for the minimum variance in this idealised case:

$$V_1(g,\phi)_{min} = V_2(g,\phi)_{min} = 1/\cosh 2r \quad (2.15)$$

Thus the EPR paradox will follow. Similar solutions hold for the imaginary components. The $X_\theta(\omega)$ is measured experimentally by time averaging. Provided T is sufficiently large the predicted experimental result will coincide with the result calculated by taking ensemble averages.

Perhaps the simplest way to measure the inference error (2.13) is to use the experimental set up (Fig. 1) of LEVENSON et al [14]. The output signal and idler photocurrents are individually amplified by factors g_X and g_Y respectively, and the difference current is taken. Thus we define the difference current operator

$$i_-(\omega) \sim g_X |E^X_{Lo}| (X_\theta(L,\omega) - g\,Y_\phi(L,\omega)) \quad (2.16)$$

where g is the relative amplification factor $g = g_Y|E^X_{Lo}|/g_X|E^X_{Lo}|$. Now the power spectrum of the difference current for a time stationary field, where the imaginary and real operators $\overline{Im}\ X_\theta(L,\omega)$ and $\overline{Re}\ X_\theta(L,\omega)$ etc. have identical correlation properties and are uncorrelated, may be written (we use the notation $\langle|X|^2\rangle = \langle(XX^\dagger + X^\dagger X)/2\rangle$)

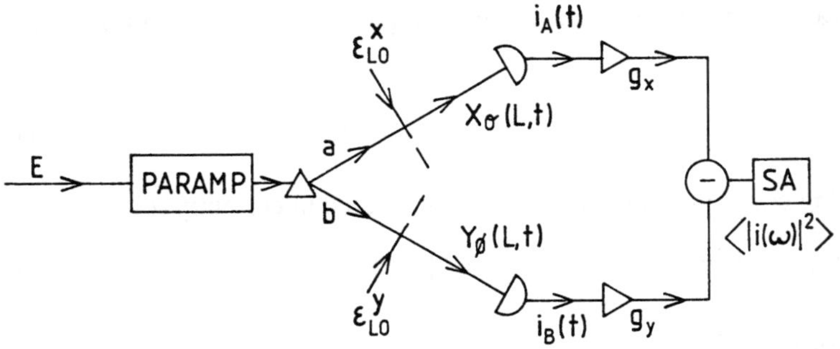

Fig. 1: Possible experimental arrangement to demonstrate the EPR paradox (for the case of a time stationary field)

$$<|i_-(\omega)|^2> \;=\; g_X^2\,|E_{Lo}^X|^2\; V_\theta(g,\phi,\omega)$$

$$\text{where} \qquad V_\theta(g,\phi,\omega) \;=\; <|X_\theta(L,\omega) - gY_\phi(L,\omega)|^2> \tag{2.17}$$

The fluctuation level (2.17) must be normalised with respect to the noise level for a minimum uncertainty state of the signal field. This can be done by replacing the signal with a coherent field in a second experiment. The noise level on the signal detector is then

$$<|i_-(\omega)|^2>_{Coh} \;=\; g_X^2\; |E_{Lo}^X| \tag{2.18}$$

This coherent or vacuum noise level has been accurately and reliably determined in squeezing experiments [14]. One may then normalise the correlated result (2.17) to achieve a value for the inference variance $V_\theta(g,\phi,\omega)$. In practice, the parameters g and ϕ are adjusted to give the minimum variance $V_\theta(g,\phi,\omega)_{min}$.

2.5 The Parametric Oscillator

One of the ways of increasing the effective interaction or amplification is to place the parametric medium inside a cavity. The system has been studied by several authors [16–18] and in particular the 'inference spectrum' as defined by equations (2.13) has been calculated both below and above threshold [19]. The model Hamiltonian for the parametric oscillator is

$$H \;=\; i\hbar g\,(a_3\, a_1^\dagger a_2^\dagger \;-\; a_3^\dagger a_1 a_2) \;+\; i\hbar\,(E\, a_3^\dagger - E\, a_3^\dagger) \;+\; \sum_{i=1}^{3} (a_1^\dagger \Gamma_i + a_1 \Gamma_i^\dagger) \tag{2.19}$$

Here g is a parametric coupling constant. The a_i are boson operators for the resonant cavity modes at frequencies ω_i respectively. We denote signal and idler modes as a_1 and a_2 respectively, and the pumped cavity mode a_3. Note $\omega_1 + \omega_2 = \omega_3$. Here E is the external driving field for the mode at frequency ω_3. The Γ_i are external reservoir operators describing loss. The cavity decay rates are denoted κ_i. We take $\kappa_1 = \kappa_2 = \kappa$ for convenience. Below threshold ($E < E_T$ where $E_T = \kappa_3\kappa/g$) the linearised equations

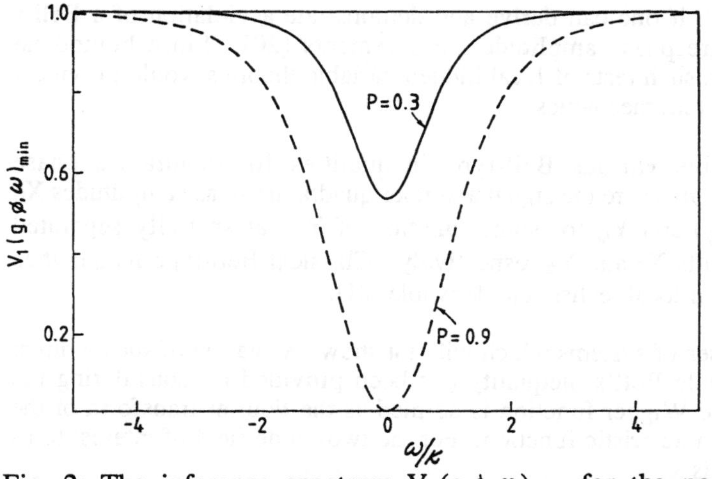

Fig. 2: The inference spectrum $V_1(g,\phi,\omega)_{min}$ for the nondegenerate parametric oscillator. Here $P = E/E_T$

for the cavity operators have been derived by REYNAUD et al [17]. These equations describe small quantum fluctuations about stable steady state classical amplitudes. Solutions for the spectral quadrature phase amplitudes $X_\theta(\omega)$ and $Y_\phi(\omega)$ external to the cavity are readily derived by Fourier transforming [17]. The solutions indicate the pairs $X_1(\omega)$ and $Y_1(\omega)$, and $X_2(\omega)$ and $Y_2(\omega)$, to be correlated. The minimum inference variance for $X_1(\omega)$ is given by

$$V_1(g,\phi,\omega)_{min} = \langle|X_1(\omega) - gY_\phi(\omega)|^2\rangle_{min} = \langle|X_1(\omega)|^2\rangle - \frac{|\langle Y_1(-\omega)X_1(\omega)\rangle|^2}{\langle|Y_1(\omega)|^2\rangle} \tag{2.20}$$

with a similar result for $V_2(g,\phi,\omega)$. For the nondegenerate parametric oscillator $V_1(g,\phi,\omega)_{min} = V_2(\bar{g},\bar{\phi},\omega)_{min}$. This inference spectrum is narrower than the usual squeezing spectrum derived by Collett and Gardiner [18] and is given explicitly in reference [19]. The solution is plotted in Fig. 2. EPR correlations are clearly evident.

The above threshold ($E > \kappa_3\kappa/g$) equations are not stable with respect to small fluctuations about a definite classical amplitude. The solutions are most readily obtained by transforming to intensity and phase variables. A recent calculation of the inference spectrum (2.13) above threshold has been presented elsewhere [12] and also indicates EPR correlations to be possible over a particular frequency bandwidth.

3. BELL'S INEQUALITIES WITH QUADRATURE PHASES?

The EPR paradox stimulated BELL [2] to derive his famous inequalities. Bell assumed the existence of a local hidden variable theory. Such a theory predicts inequalities for the results of spin $-1/2$ correlation experiments. These inequalities are then found to contradict quantum mechanical predictions.

It is logical then to ask if one can derive and demonstrate a violation of a Bell's inequality with quadrature phase amplitude measurements [20]. Since homodyne detection is very efficient, such tests of local hidden variable theories would provide a more rigorous test of quantum mechanics.

It is possible to derive various Bell-type inequalities for quadrature phase measurements. One could measure the signal and idler quadrature phase amplitudes X_θ (or some function of X_θ) and Y_ϕ (or some function of Y_ϕ) at spatially separated locations, and obtain results X_θ and Y_ϕ respectively. Classical Bell-type inequalities based on the assumptions of local realism are derivable [21].

Some insight into classes of systems which will not show a violation of such a direct quadrature phase amplitude Bell's inequality has been provided by considering the Wigner function [8]. The Wigner function is defined as the Fourier transform of the symmetrically ordered characteristic function. For the two-mode field of interest to us here the Wigner function is

$$W(\underset{\sim}{x}) = \frac{1}{(2\pi)^4} \int e^{-i\underset{\sim}{x}.\underset{\sim}{y}} \, \mathrm{Tr}\,(\rho e^{i\underset{\sim}{x}.\underset{\sim}{X}}) \, d\underset{\sim}{\lambda} \tag{3.1}$$

where $\underset{\sim}{x} = (x_1, x_2, y_1, y_2)$, $\underset{\sim}{\lambda} = (\lambda_1, \lambda_2, \mu_1, \mu_2)$, $\underset{\sim}{X} = (X_1, X_2, Y_1, Y_2)$, $d\underset{\sim}{\lambda} = d\lambda_1 d\lambda_2 d\mu_1 d\mu_2$ and ρ is the system density operator. The calculation of symmetrically ordered moments are given directly as integrals over the Wigner distribution. These symmetrically ordered moments correspond directly to quadrature phase cross correlations for the experiments using homodyne detection.

If the Wigner function is positive it provides a local hidden variable theory predicting the joint and marginal probabilities for the quadrature amplitude measurements. The hidden classical variables are the $\underset{\sim}{x} = (x_1, x_2, y_1, y_2)$. There is a well defined probability, $P(\theta,\phi,\underset{\sim}{x})$ say, of obtaining the results x_θ and y_ϕ given the variables $\underset{\sim}{x}$, and it is straightforward to show that this probability will factorise into a form $P(\theta,\phi, \underset{\sim}{x}) = P(\theta, \underset{\sim}{x})P(\phi, \underset{\sim}{x})$ thus satisfying the condition of locality. The measurable probabilities such as $P(\theta,\phi)$, the probability of obtaining results x_θ and y_ϕ for the measurements of X_θ and Y_ϕ respectively, are given by integration over all values of hidden variables $\underset{\sim}{x}$ appropriately weighted by the probability distribution $W(\underset{\sim}{x})$. The existence of theory with the above properties (locality and a positive probability distribution) are precisely the assumptions from which a (quadrature phase) Bell inequalities would follow. Of course in this case, the Wigner function is also the quantum mechanical theory and hence the correct theory. Thus we do not expect violation of such Bell inequalities where $W(\underset{\sim}{x})$ is positive.

It remains to show that the Wigner function is indeed positive for the simple nondegenerate parametric amplifier. For the ideal case where the mode transformations satisfy $\underset{\sim}{X}(L) = S\underset{\sim}{X}\,(0)$, where S is a unitary transformation as given by

(2.3), the Wigner function is

$$W(\underset{\sim}{x}) = (2/\pi)^2 \exp(-2\underset{\sim}{x}'^T \underset{\sim}{x}') \tag{3.2}$$

where $\underset{\sim}{x}' = S^{-1}\underset{\sim}{x}$. Clearly in this case a direct quadrature phase amplitude Bell inequality is not violated.

4. VIOLATION OF BELL'S INEQUALITY IN PARAMETRIC AMPLIFCATION; INTENSITY CORRELATION

4.1 The Correlation Experiment

Although a direct quadrature phase amplitude Bell's inequality is not violated in parametric amplification a Bell's inequality derived for intensity correlations is violated. This was first shown theoretically by REID and WALLS [9] who proposed the experimental scheme of Fig. 3. OU and MANDEL [4] have subsequently suggested and experimentally demonstrated violation of Bell's inequality using the scheme of Fig. 4.

Both schemes measure the joint probability $P(\theta,\phi)$ of detecting photon at the spatially separated detectors at A and B. This probability is predicted by local hidden variable theories to satisfy the following BELL-CLAUSER-HORNE inequality [2].

$$S = P(\theta,\phi) - P(\theta,\phi') + P(\theta',\phi') + P(\theta',\phi) - P(\theta',-) - P(-,\phi) \leq 0 \tag{4.1}$$

$P(\theta,-)$ and $P(-,\phi)$ are the probabilities with the beam splitter (or polariser) removed, at A and B respectively.

We analyse these schemes quantum mechanically. First the Reid-Walls scheme. The probability $P(\theta,\phi)$ of detecting photons at A and B is proportional to $\langle c_1^\dagger d_1^\dagger c\, d_1\rangle$. The operators c_i and d_i are given by

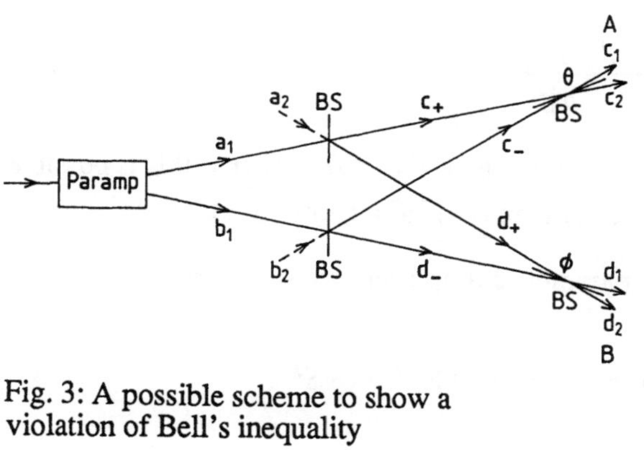

Fig. 3: A possible scheme to show a violation of Bell's inequality

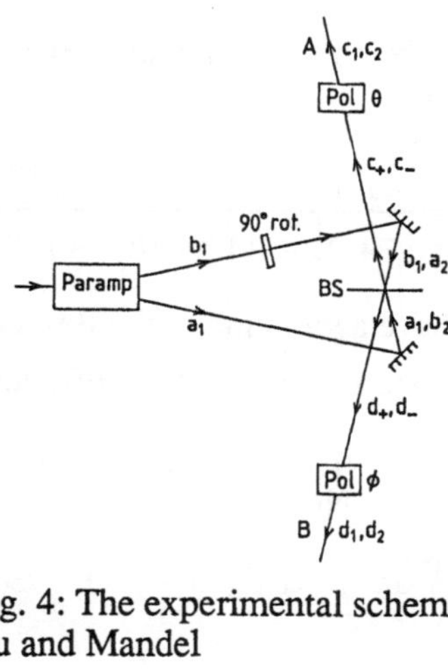

Fig. 4: The experimental scheme of Ou and Mandel

$$c_1 = c_+\cos\theta - c_-\sin\theta \quad d_1 = d_+\cos\phi - d_-\sin\phi$$
$$c_2 = c_+\sin\theta + c_-\cos\theta \quad d_2 = d_+\sin\phi + d_-\cos\phi \tag{4.2}$$

where

$$c_+ = (a_1 + a_2)/\sqrt{2} \qquad d_+ = (a_1 - a_2)/\sqrt{2}$$
$$c_- = (b_1 + b_2)/\sqrt{2} \qquad d_- = (b_1 - b_2)/\sqrt{2} \tag{4.3}$$

Here a_1 and b_1 are the output modes of the parametric amplifier, while a_2 and b_2 are vacuum inputs. Now we find

$$<c_1^\dagger d_1^\dagger c_1 d_1> = <a_1^{\dagger 2} a_1^2> \cos^2\theta\cos^2\phi + <b_1^{\dagger 2} b_1^2> \sin^2\theta\sin^2\phi + <a_1^\dagger b_1^\dagger a_1 b_1> \sin^2(\theta+\phi)$$

$$<c_1^\dagger d_2^\dagger c_1 d_2> = <a_1^{\dagger 2} a_1^2> \cos^2\theta\sin^2\phi + <b_1^{\dagger 2} b_1^2> \sin^2\theta\cos^2\phi + <a_1^\dagger b_1^\dagger a_1 b_1> \cos^2(\theta+\phi) \tag{4.4}$$

The $<c_2^\dagger d_2^\dagger c_2 d_2>$ and $<c_2^\dagger d_1^\dagger c_2 d_1>$ are given identically to $<c_1^\dagger d_1^\dagger c_1 d_1>$ and $<c_1^\dagger d_2^\dagger c_1 d_2>$ respectively, but exchanging $<a_1^{\dagger 2} a_1^2>$ and $<b_1^{\dagger 2} b_1^2>$. Now the relative probabilities are (:: denotes normal ordering)

$$P(\theta,\phi) = <:c_1^\dagger c_1 d_1^\dagger d_1:>/N, \quad P(-,\phi) = <:(c_1^\dagger c_1 + c_2^\dagger c_2)\, d_1^\dagger d_1:>/N$$

where $\quad N = <:(a_1^\dagger a_1 + a_2^\dagger a_2)(b_1^\dagger b_1 + b_2^\dagger b_2):> = <:(c_1^\dagger c_1 + c_2^\dagger c_2)(d_1^\dagger d_1 + d_2^\dagger d_2):> \qquad (4.5)$

Simplification gives

$$P(\theta,\phi) = \left(<a_1^{\dagger 2} a_1^2>\cos^2\theta\cos^2\phi + <b_1^{\dagger 2} b_1^2>\sin^2\theta\sin^2\phi + <a_1^\dagger b_1^\dagger a_1 b_1>\sin^2(\theta+\phi)\right)/N$$

$$P(\theta,-) = \left(<a_1^{\dagger 2} a_1^2>\cos^2\theta + <b_1^{\dagger 2} b_1^2>\sin^2\theta + <a_1^\dagger b_1^\dagger a_1 b_1>\right)/N \tag{4.6}$$

$$P(-,\phi) = \left(<a_1^{\dagger 2} a_1^2>\cos^2\phi + <b_1^{\dagger 2} b_1^2>\sin^2\phi + <a_1^\dagger b_1^\dagger a_1 b_1>\right)/N$$

With the choice of angles $\theta = \pi/8$, $\theta' = 3\pi/8$, $\phi = \pi/4$, $\phi' = 0$, we find $2S = -1 + \sqrt{2}C$ where

$$C = \frac{(<a_1^\dagger b_1^\dagger a_1 b_1> - 1/4(<a_1^{\dagger 2} a_1^2 + <b_1^{\dagger 2} b_1^2>))}{<a_1^\dagger b_1^\dagger a_1 b_1> + 1/2(<a_1^{\dagger 2} a_1^2> + <b_1^{\dagger 2} b_1^2>)} \tag{4.7}$$

We have a violation of the Bell inequality $S < 0$ where $C > .707$. Clearly, this is possible provided $<a_1^\dagger b_1^\dagger a_1 b_1>$ is sufficiently greater than $(<a_1^{\dagger 2} a_1^2> + <b_1^{\dagger 2} b_1^2>)/2$ [9].

Let us first consider the idealised parametric amplifier whose solutions are given by (2.3). Now

$$<a_1^{\dagger 2} a_1^2> = <b_1^{\dagger 2} b_1^2> = \sinh^4 r, \; <a_1^\dagger b_1^\dagger a_1 b_1> = \sinh^2 r\,(\sinh^2 r + \cosh^2 r). \tag{4.8}$$

We find that a violation of Bell inequality is possible for small r, as shown by REID and WALLS [9].

Now we analyse the scheme of Ou and Mandel as depicted in Fig. 4. Now a_1 and b_1 are boson operators for the polarised output fields of for example the parametric amplifier. The idler field undergoes a 90° rotation of polarisation. Thus we consider two pairs of orthogonally polarised fields incident on the beam splitter BS. The transformation giving the outputs is given by (4.3). Here the sets (a_1, a_2, c_+, d_+) and (b_1, b_2, c_-, d_-) are orthogonally polarised. In this case, b_2 and a_2 are vacuum inputs. At the spatially separated polarisers A and B we have the transformations given by (4.2), where here θ and ϕ represent the polariser angles. Calculation of the joint probabilities $\langle c_1^\dagger d_1^\dagger c_1 d_1 \rangle$, ... gives solutions identical in form to (4.4). Thus the quantities S and C are given similarly to (4.1) and (4.7), and a violation of the Bell inequality $S < 0$ is given in identical fashion. An analysis of the OU and MANDEL experiment for nondegenerate parametric amplification has been presented by TAN and WALLS [10].

REFERENCES

1. A. Einstein, B. Podolsky and N. Rosen: Phys. Rev. **47**, 777 (1935).
2. J.S. Bell: Physics **1**, 195 (1964). J.F. Clauser and A. Shimony: Rep. Prog. Phys.**41**, 1881 (1978).
3. A. Aspect, P. Grangier and G. Roger: Phys. Rev. Lett. **47**, 460 (1981), Phys. Rev. Lett. **49**, 91 (1982). A. Aspect, J. Dalibard and G. Roger: Phys. Rev. Lett. **49**, 1804 (1982).
4. Z.Y. Ou and L. Mandel: Phys. Rev. Lett. **61**, 50 (1988).
5. Y.H. Shih and C.O. Alley: Phys. Rev. Lett. **61**, 2921 (1988).
6. M.D. Reid and P.D. Drummond: Phys. Rev. Lett. **60**, 2731 (1988).
7. M.D. Reid in press Phys. Rev. A.
8. See also S. Braunstein: to be published.
9. M.D. Reid and D.F. Walls: Phys. Rev. **A34**, 1260 (1986).
10. S.M. Tan and D.F. Walls: to be published.
11. N. Bohr: Phys. Rev. **48**, 696 (1935).
12. H.P. Yuen and J.H. Shapiro: IEEE Trans. Inform Theory **IT-26**, 78 (1980).
13. See Special Issues on Squeezed States of Light, JOSA **B4** (1987), Journ. Mod. Opt. **34** (1987).
14. M.D. Levenson and R.M. Shelby: Journ. Mod. Opt. **34**, 775 (1987). M.D. Levenson, R.Shelby, M.D. Reid and D.F. Walls: Phys. Rev. Lett. **57**, 2473 (1986). R. Slusher, A. La Porte and B. Yurke: this volume.
15. B.L. Schumaker: Physics Reports **135** (1986).
16. R. Graham and H. Haken: Z. Physik **210**, 276, 319 (1968).
17. S. Reynaud, C. Fabre, E. Giacobino: JOSA **B4**, 1520 (1987).
18. M.J. Collett and C.W. Gardiner: Phys. Rev. **A30**, 1386 (1984).
19. P.D. Drummond and M. Reid: submitted to Phys. Rev. A.
20. B. Oliver and C. Stroud: (unpublished)
21. P. Grangier, M. Potasek and B. Yurke: Phys. Rev. **A38**, 3132 (1988).

Phase in Quantum Electrodynamics

D.T. Pegg[1], S.M. Barnett[2], and J.A. Vaccaro[1]

[1]School of Science, Griffith University, Nathan, Brisbane, 4111, Australia
[2]Department of Engineering Science, Oxford University, Parks Road, Oxford OX13PJ, United Kingdom

1. INTRODUCTION

The phase of a simple-harmonic vibration is a clearly defined, well understood and easily observable physical concept in classical mechanics, just as is the energy of the oscillation. One might therefore expect, following the usual procedure for quantization, that there should exist not only an Hermitian operator to represent the energy observable, but also one for the phase. The existence of an Hermitian operator to represent the phase of the electromagnetic field was indeed first postulated by DIRAC [1]. The form of the proposed operator, however, led to mathematical inconsistencies of such a nature that it is now widely accepted that phase is in a somewhat exceptional position of being a classical observable without a corresponding Hermitian operator [2 - 5]. We have shown recently [6 - 8] that the difficulties with the existence of the phase operator can be traced to an indeterminacy associated with allowing the dimension of the state space to tend to infinity prematurely. If, instead, quantum electrodynamic calculations are performed in a state space of finite but arbitrarily large dimensions, and the dimensionality of the space is allowed to tend to infinity *after* physical results, such as expectation values, are calculated, then the indeterminacy is removed and unique, well-behaved, phase properties result. A very simple mathematical analogy involves finding the limit as x tends to infinity of (x + 1)/x. Replacing x by infinity in the numerator and denominator before calculating the quotient yields an indeterminate form; performing the calculation first and then finding the limit of the result yields a well-defined unique expression. Indeed the existence of the indeterminate form is an indicator of premature limit taking.

In this paper we describe the solution to the phase problem, which involves finding a well-behaved Hermitian phase operator. We illustrate the use of the new phase operator for studying the phase properties of various states of the quantum electrodynamic field, including the squeezed vacuum.

2. EARLY PHASE OPERATORS

Dirac's recipe involving the correspondence between Poisson brackets and commutators, when applied to the Poisson bracket for the classical phase of the single-mode field, given by

$$d\hat{\phi}/dt = \{\hat{\phi},\hat{H}\} = -\omega \tag{1}$$

produces Dirac's suggested commutation relation

$$[\hat{\phi}_D,\hat{N}] = -i \tag{2}$$

with the associated uncertainty relation

$$\Delta N\Delta\phi_D \geq ½ \tag{3}$$

Springer Proceedings in Physics, Vol. 41 Quantum Optics V
Editors: J.D. Harvey and D.F. Walls

where $\hat{N}$ is the photon number operator and the subscript denotes Dirac's phase operator. We get into difficulty with (2), however, if we take the expectation value of both sides for a field in a single number state, and a second difficulty would appear to follow from (3) in that it would allow a phase uncertainty much greater than 2π. This second difficulty arises because (1) does not account for the periodic nature of $\hat{\phi}$. One would expect the time development of ϕ to be a saw-tooth function with a 2π step every period to give as the classical behaviour of the expectation value of an ensemble of oscillators or fields [8]:

$$d\langle\phi\rangle/dt = -\omega[1-2\pi P(\theta_o)] \tag{4}$$

where $P(\theta_o)\omega\delta t$ is the probability that the phase of a field is in the phase interval $\omega\delta t$ near θ_o, the value of phase where the jump occurs.

SUSSKIND and GLOGOWER [2] attempted to find an operator to represent phase by making a polar decomposition of the annihilation operator into amplitude and phase parts:

$$\hat{a} = \exp(i\hat{\phi})\hat{N}^{1/2} \tag{5}$$

This seems a very reasonable approach based directly on the corresponding classical decomposition, indeed it is so reasonable that it is difficult to accept that an operator which does not obey (5) is a genuine phase operator. For this reason operators such as the "measured" phase operator [5] are not really satisfactory for representing genuine phase. Having postulated (5) the procedure now looks simple: we know the properties of $\hat{N}$ and $\hat{a}$ so we should be able to deduce the properties of $\exp(i\hat{\phi})$. Operating on a state $|n\rangle$ with both sides of (5) gives

$$\exp(i\hat{\phi})|n\rangle = |n-1\rangle \tag{6}$$

provided $n \neq 0$. When $n = 0$, however, we obtain

$$0 \times \exp(i\hat{\phi})|0\rangle = 0$$

and the result for $\exp(i\hat{\phi})|0\rangle$ is *indeterminate*. Susskind and Glogower approached the problem of removing the indeterminacy simply by making the *additional* assumption that $\exp(i\hat{\phi})|0\rangle$ is *zero*. This additional assumption immediately destroys the unitarity of the "exponential" operator, thereby precluding the extraction of an Hermitian operator $\hat{\phi}$ from the exponent. Howbeit, this was not seen as a great problem because one could determine phase properties of functions of the cos and sin Hermitian combinations, which are defined by analogy with what *would* be the definitions if the non-unitary operator were a genuine unitary operator expressible as an exponential series.

Because the non-unitarity is a direct result of the assumption involving the vacuum state, the Susskind-Glogower operators (which henceforth we denote with a subscript s) behave well in the classical regime of high intensity fields but have peculiar properties in the quantum regime of fields for which the overlap with the vacuum is significant, and which hence require a quantum, rather than classical, description of phase. For example a calculation of $\langle 0|\hat{\cos}_s^2\phi|0\rangle$ does not yield ½, showing that the vacuum is not a state of random Susskind-Glogower phase. Further, when the definition is extended to include phase difference operators, again by analogy with what would be definitions for genuine unitary operators, we find that two independent vacuum modes are not only non-random but are actually correlated in such a manner that a measurement of Susskind-Glogower phase difference must yield a result of ±90° [2,8], i.e. in the Susskind-Glogower formalism all vacuum fields are precisely in quadrature. Other results for the Susskind-Glogower operators are that $\hat{\cos}_s\phi$ and $\hat{\sin}_s\phi$ are not compatible, and the sum of the expectation values of $\hat{\cos}_s^2\phi$ and $\hat{\sin}_s^2\phi$ is not necessarily unity.

3. THE HERMITIAN PHASE OPERATOR

If a phase operator $\hat{\phi}$ were to exist in a reasonable conjugacy relationship with $\hat{N}$, in a similar way that momentum and position are conjugate, what type of eigenvalue spectrum might we expect for $\hat{\phi}$? $\hat{N}$ has a countable infinity of eigenvalues, including zero and all the positive integers. We tend to think of $\hat{\phi}$ as having a continuous eigenvalue spectrum which, even if restricted to a 2π interval, still represents an uncountable infinity of eigenvalues. If there is to be some sort of symmetry between $\hat{\phi}$ and $\hat{N}$, so that the eigenstates of $\hat{\phi}$ span the same state space as the eigenstates of $\hat{N}$, we might expect $\hat{\phi}$ to have the "same number" of eigenstates as $\hat{N}$ or at least, if we are dealing with an infinite state space, that the eigenstates of each can be put into a one-to-one correspondence. This would suggest that the eigenvalues of $\hat{\phi}$ are perhaps not the uncountable infinity of points in the 2π interval of the linear continuum but just a dense, and perhaps equally spaced, subset of these which form a countable infinity. We do not begin our analysis with such an assumption, but we bear it in mind as a useful check to see if our approach is reasonable.

Despite the apparent absence of an Hermitian phase operator, states of well-defined phase have been known to exist for some time [9]:

$$|\theta\rangle = \lim_{s\to\infty} (s+1)^{-\frac{1}{2}} \sum_{n=0}^{s} \exp(in\theta)\,|n\rangle \tag{7}$$

where $|n\rangle$ are the s+1 number states which span the (s+1)-dimensional state space Ψ. The limiting procedure is necessary to normalize the states. The phase state is well defined in the space Ψ but care must be taken with the limiting process. As described earlier, indeterminacies can result from premature replacement of s by infinity. We therefore work in the finite but arbitrarily large space Ψ, and allow s to tend to infinity only *after* physical results such as expectation values are calculated.

We note from (7) that the state $|\theta+2\pi\rangle$ is the same as the state $|\theta\rangle$, so all different phase states can be expressed by the 2π section of the arithmetical continuum of values from θ_0 up to, but not including, $\theta_0+2\pi$. The actual choice of θ_0 is arbitrary and corresponds classically to choosing a 2π window in which to express the values of an inverse trigonometrical function. Often θ_0 is chosen as 0 or $-\pi$, but here we prefer to leave it arbitrary. The uncountable infinity of such states $|\theta\rangle$ are not mutually orthogonal, but we can pick out s+1 of these states, which include $|\theta_0\rangle$ and which do form a complete orthonormal set, which can thus form an alternative basis for Ψ. These are given [6] by $|\theta_m\rangle$ where m takes the values 0,1,2,...s and

$$\theta_m = \theta_0 + \frac{2\pi m}{(s+1)} . \tag{8}$$

When s tends to infinity these form a *countable* infinity of orthogonal states, corresponding to θ_0 plus the rational fractions of 2π, equal in number to the number states, that is, in the limit the spectrum is dense rather than continuous.

Because only one operator will have one set of s+1 states as eigenstates, there will be an uncountable infinity of different phase operators, even with s finite, corresponding to the number of different values of θ_0 which can be chosen. Thus each point on the linear continuum is the possible result of a phase measurement corresponding to some phase operator. Thus the classical idea that a result of a phase measurement can be any of the uncountable infinity of angles in a 2π interval is preserved, as each value is the eigenvalue of some phase operator; what is different from previous approaches is that we do not try to make all these values eigenvalues of the same phase operator. Instead, we have divided the uncountably infinite number of phase values into an uncountably infinite number of sets, each of which contain a countably infinite number of values corresponding to a distinct phase operator, which now has some reasonable chance of sharing a symmetric conjugacy relation with $\hat{N}$. It is thus necessary to specify which phase operator we are using. We do this by attaching a subscript to $\hat{\phi}$. To specify the operator whose eigenstates are the set $|\theta_m\rangle$ beginning with $|\theta_0\rangle$ we should attach a subscript θ_0, however, as this leads to an awkward notation, we merely use the subscript θ and define the Hermitian phase operator as

$$\hat{\phi}_\theta \equiv \sum_{m=0}^{s} \theta_m |\theta_m\rangle\langle\theta_m| \tag{9}$$

In terms of the number state basis the projector is

$$|\theta_m\rangle\langle\theta_m| = (s+1)^{-1} \sum_{n,n'} \exp[i(n'-n)\theta_m]\,|n'\rangle\langle n| \tag{10}$$

and the phase operator itself becomes

$$\hat{\phi}_\theta = \theta_0 + \frac{s\pi}{s+1} + \frac{2\pi}{s+1} \sum_{n\neq n'} \frac{\exp[i(n'-n)\theta_0]\,|n'\rangle\langle n|}{\exp[i(n'-n)2\pi/(s+1)]-1} \tag{11}$$

It is tempting to take the limit as s tends to infinity immediately giving, for s>> (n'-n),

$$(\hat{\phi}_\theta)_P = \theta_0 + \pi - i \sum_{n\neq n'} \frac{\exp[i(n'-n)\theta_0]\,|n'\rangle\langle n|}{(n'-n)} \tag{12}$$

from which we find the commutator with N, which we can reduce eventually [8] to

$$[\hat{\phi}_\theta,\hat{N}]_P = -i[1-(s+1)\,|\theta_0\rangle\langle\theta_0|\,]. \tag{13}$$

We should, however, be very cautious about taking the limit before physical results such as expectation values are calculated. In fact (12) only gives the same result as (11) when operating on states for which the overlap with $|s\rangle$ is zero or at least decreases sufficiently quickly as s approaches infinity. This is indeed true for physically preparable, or physical, states [5], and hence the subscript p, which can be prepared from the vacuum in a finite time with a bounded energy source. States with bounded (although arbitrarily large) energy are physically preparable and states with finite energy moments are such extremely good approximations to physically preparable states that they can be classed as such. Nearly all states encountered so far in quantum optics, such as single number states, superpositions of coherent states and so on, belong to this class, with the notable exception being the phase state for which even the first moment, the expectation value of energy, is infinite. Thus if we wish to make use of phase states as part of a calculation we should not use (12) or (13), but should instead use the accurate forms (11) with the resultant commutator

$$[\hat{\phi}_\theta,\hat{N}] = \frac{2\pi}{s+1} \sum_{n\neq n'} \frac{(n-n')\exp[i(n'-n)\theta_0]\,|n'\rangle\langle n|}{\exp[-i(n'-n)2\pi/(s+1)]-1} \tag{14}$$

and then take the limit at a later stage.

On the other hand, if we are dealing only with physical states $|p\rangle$ (13) correctly gives the result

$$\langle p|[\hat{\phi}_\theta,\hat{N}]|p\rangle = -i[1-2\pi P(\theta_0)] \tag{15}$$

where

$$P(\theta_0)\delta\theta = |\langle p|\theta_0\rangle|^2 \qquad (16)$$

which is the probability of finding the state with a phase within $\delta\theta$ of θ_0 where $\delta\theta$ is the interval $2\pi/(s+1)$ between eigenvalues. The first term in (15) is just the Dirac term and the second term is precisely that which, according to (4) we should add to account for the periodicity of $\hat{\phi}$. Thus the Poisson bracket-commutator correspondence is established, *but only where physical states are involved*. Because phase states are not such states, it is now clear why attempts to extrapolate, on the basis of the correspondence, from classical expressions such as (4) to find an Hermitian phase operator with well behaved eigenstates had little chance of success.

It should be noted that the trace of (14), and of (13), vanishes. This should not be unexpected as the trace of any commutator in a finite, even if arbitrarily large, state space must vanish [10]. Hence a term extra to the Dirac term should be expected, even if only to ensure a vanishing trace. We see that this trace-removing term is precisely that needed to ensure the correct periodic behaviour of phase. Furthermore, and importantly, we note that the expectation value of (13), and of (14) for a field in a single number state $|n\rangle$ vanishes, so the first difficulty mentioned with Dirac's commutator is *also* resolved.

4. CREATION AND ANNIHILATION OPERATORS

Now that we have the Hermitian phase operator, we can easily construct the corresponding unitary operators $\exp(\pm i\hat{\phi}_\theta)$ and see how they differ from the Susskind-Glogower operators. We find easily

$$\exp(i\hat{\phi}_\theta)|n\rangle = \exp[i\sum_{m=0}^{s}\theta_m|\theta_m\rangle\langle\theta_m|]|n\rangle = (s+1)^{-1/2}\sum_{m=0}^{s}\exp[-i(n-1)\theta_m]|\theta_m\rangle \qquad (17)$$

We also have, in symmetry with (7),

$$|m\rangle = \sum_{m=0}^{s}|\theta_m\rangle\langle\theta_m|n\rangle = (s+1)^{-1/2}\sum_{m=0}^{s}\exp(-in\theta_m)|\theta_m\rangle \qquad (18)$$

Comparison of (17) and (18) shows that

$$\exp(i\hat{\phi}_\theta)|n\rangle = |n-1\rangle \qquad (19)$$

provided $n\neq 0$. For $n = 0$, the resulting state is

$$(s+1)^{-1/2}\sum_m \exp(i\theta_m)|\theta_m\rangle = (s+1)^{-1/2}\exp[i(s+1)\theta_0]\sum_m \exp(-is\theta_m)|\theta_m\rangle$$
$$= \exp[i(s+1)\theta_0]|s\rangle \qquad (20)$$

Therefore, the number state representation of $\exp(i\hat{\phi}_\theta)$ is

$$\exp(i\hat{\phi}_\theta) = |0\rangle\langle 1| + |1\rangle\langle 2| + \ldots + |s-1\rangle\langle s| + \exp[i(s+1)\theta_0]|s\rangle\langle 0| \qquad (21)$$

This operator resembles the one introduced by Susskind and Glogower, but with the vital difference that the action on the vacuum state is not indeterminate and so cannot be arbitrarily set to zero. The result of acting on the vacuum with $\exp(i\hat{\phi}_\theta)$ is *uniquely and precisely determined* to be the state $\exp[i(s+1)\theta_0]|s\rangle$. The expansion (21) is fully consistent with $\exp(i\hat{\phi}_\theta)$ being a unitary operator. The conjugate operator $\exp(-i\hat{\phi}_\theta)$ is also unitary and clearly commutes with $\exp(i\hat{\phi}_\theta)$.

We can define cosine and sine operators from the unitary operators. These will correspond to cosine and sine series in the Hermitian phase operator. These operators are better behaved than their counterparts formed from the Susskind-Glogower operators. In particular we find

$$\cos^2\hat{\phi}_\theta + \sin^2\hat{\phi}_\theta = 1 \tag{22}$$

$$[\cos\hat{\phi}_\theta, \sin\hat{\phi}_\theta] = 0 \tag{23}$$

$$\langle n|\cos^2\hat{\phi}_\theta|n\rangle = \langle n|\sin^2\hat{\phi}_\theta|n\rangle = \tfrac{1}{2} \tag{24}$$

The last of these results is consistent with the phase of the vacuum being random and is in marked contrast with the result obtained using the Susskind-Glogower operators. In the Susskind-Glogower formulation the vacuum expectation values of $\cos^2\phi$ and $\sin^2\phi$ are 1/4. The major difference between our operators and those of Susskind and Glogower involves the action on the vacuum state. It should not be surprising that very different results will occur in the quantum regime involving field states with a significant vacuum component. Our operators are in *no sense* an approximation to the Susskind-Glogower operators and have similar properties to the latter only for fields with large energies where both sets of operators give results in accord with classical behaviour. Moreover, our cosine and sine operators are derived as operator functions of an Hermitian phase operator. No analogous procedure is possible in the Susskind-Glogower formulation.

The creation and annihilation operators can now be constructed by definition:

$$\hat{a} \equiv \exp(i\hat{\phi}_\theta)\hat{N}^{1/2} \tag{25}$$

$$= |0\rangle\langle 1| + 2^{1/2}|1\rangle\langle 2| + \ldots + s^{1/2}|s-1\rangle\langle s| \tag{26}$$

with $\hat{a}^\dagger$ being the Hermitian conjugate of $\hat{a}$. As s tends to infinity the action of $\hat{a}$ becomes that of the conventional annihilation operator. Note that the Hermitian amplitude operator $\hat{N}^{1/2}$ removes the $|s\rangle\langle 0|$ projector term when acting on (21) to give (26). Thus, while $\exp(i\hat{\phi}_\theta)$ uniquely determines $\hat{a}$, it is now clear why the reverse procedure (involving a decomposition of $\hat{a}$) cannot produce the vital $|s\rangle\langle 0|$ projector term in (21). Such a procedure can at best only produce an indeterminate result, requiring an *extra* assumption as used by Susskind and Glogower. The $|s\rangle\langle 0|$ projector *never* occurs in any coupling involving the electromagnetic field as these take place via $\hat{a}$ and $\hat{a}^\dagger$ [6].

We mentioned earlier that the trace of all commutators in the space Ψ must vanish, which made necessary the last term in (15). We should also anticipate an additional term in $[\hat{a},\hat{a}^\dagger]$ to ensure a vanishing trace. On calculating this commutator using (26) and its Hermitian conjugate, we indeed obtain a traceless result:

$$[\hat{a},\hat{a}^\dagger] = 1 - (s+1)|s\rangle\langle s| \tag{27}$$

The last term exactly compensates the trace obtained from the first term in the same way as the additional part to Dirac's term in $[\hat{\phi}_\theta,\hat{N}]$. However, while the addition to the $\hat{\phi}_\theta$, $\hat{N}$ commutator has direct physical consequences, the additional term in (27) has no effect when $[\hat{a},\hat{a}^\dagger]$ acts on any physical state and so has no physically observable consequences. The physical-state commutator

$$[\hat{a},\hat{a}^\dagger]_P = 1 \tag{28}$$

is sufficient when operating on physical states. This is easy to see for a physical state with an upper energy bound when s tends to infinity and is not difficult to calculate for finite moment states such as coherent states [8]. However when non-physical states such as phase states are involved, again the accurate form (27) must be used and the limit taken at a later stage.

5. PHASE PROPERTIES OF SOME FIELDS

To test if the Pegg-Barnett Hermitian phase operator is reasonable, we use it to calculate the phase properties of selected fields. The simplest is the single number state $|n\rangle$ which, when used in conjunction with (11), gives

$$\langle n|\hat{\phi}_\theta|n\rangle = \theta_0 + s\pi/(s+1) \tag{29}$$

Now we have an expectation value, we can let s tend to infinity and obtain the result $\theta_0+\pi$, i.e. a value in the centre of the chosen 2π window, as one would expect from a state with random phase. This is true for all n, including the vacuum state n = 0. It is not difficult to find also that [7]

$$\langle n|\hat{\phi}_\theta^2|n\rangle = \theta_0^2 + 2\pi\theta_0 s/(s+1) + 4\pi s(s+½)/[3(s+1)^2]. \tag{30}$$

From (29) and (30) we can find the variance of phase which, for infinite s, becomes just $\pi^2/3$ which again is precisely in accord with a random phase variable.

For a coherent state

$$|\alpha\rangle = \exp(-|\alpha|^2/2)\sum_n |\alpha|^n(n!)^{-½}\exp(in\beta)|n\rangle \tag{31}$$

we find that the mean phase is dependent on the choice of window [7], that is, on the choice of θ_0. This, however, is not surprising and has a classical origin. By calculating the phase probability distribution from $|\langle\theta_m|\alpha\rangle|^2$ we find this to be periodically and symmetrically peaked at β plus integer multiples of 2π. For an intense field the peaks are sharp, with near vanishing probability mid-way between peaks. A 2π window chosen with $\theta_0 = \beta$, therefore, will include half of the peak at β and half of the peak at $\beta+2\pi$, giving a mean of $\beta+\pi$ and maximum variance. However with a choice $\theta_0\approx\beta-\pi$, or any value of θ_0 such that the peak is almost entirely enclosed in the window, with very small values of probability at the window edge, that is, very small values of $|\langle\theta_0|\alpha\rangle|^2$, the mean phase is approximately β, with a small variance. Mathematically, the choice

$$\theta_0 = \beta - \pi s/(s+1) \tag{32}$$

gives a mean of β and a minimum variance which, for an intense coherent state becomes [7]

$$\Delta\phi^2 = (4\bar{n})^{-1} \tag{33}$$

where $\bar{n} = |\alpha|^2$ is the mean photon number. Remembering that the number variance ΔN^2 of a coherent state is $\bar{n}$, we obtain as the uncertainty relation for the coherent state

$$\Delta N\Delta\phi = ½ \tag{34}$$

which is precisely the relationship one might derive from (3), and indeed is that obtained from a phenomenological analysis [7]. Further, because with our choice of θ_0, $\langle\theta_0|\alpha\rangle$ is vanishingly small for intense fields, it follows from (13), which can be used with coherent states, that the second term in the commutator vanishes. Thus from (34) we see that an intense coherent field is a number-phase minimum uncertainty state.

States of the field which have received much recent interest are the squeezed states, i.e. two-photon coherent states [11]. The Susskind-Glogower phase properties of those states have been calculated [12-14], so it is of interest also to calculate the Pegg-Barnett phase properties for comparison [15]. These states are given by [11]

$$|\alpha,\xi\rangle = \exp(\alpha\hat{a}^{+}-\alpha^{*}\hat{a})\exp[\tfrac{1}{2}(\xi^{*}\hat{a}^{2}-\xi\hat{a}^{+2})]\,|0\rangle. \tag{35}$$

From Yuen [16]

$$\langle n|\alpha,\xi\rangle = (n!\mu)^{-\frac{1}{2}}(\nu/2\mu)^{\frac{1}{2}n}\, H_n[\beta(2\mu\nu)^{-\frac{1}{2}}]\, \exp[-\tfrac{1}{2}|\beta|^{2}+(\nu^{*}/2\mu)\beta^{2}] \tag{36}$$

where
$$\mu = \cosh r,$$
$$\nu = \exp(2i\eta)\sinh r,$$
$$\beta = \alpha\cosh r + \alpha^{*}\exp(2i\eta)\sinh r, \tag{37}$$
$$\xi = |\xi|\exp(2i\eta),$$
$$r = |\xi|,$$

and where $H_n(x)$ is the nth order Hermite polynomial.

We are interested mainly in the squeezed vacuum states, for which $\alpha = 0$, because this is where we would expect a significant difference between Susskind-Glogower and Pegg-Barnett phase properties. We use the result

$$H_{2n}(0) = \frac{(-1)^{n}(2n)!}{n!} \tag{38}$$

and the vanishing of the odd order polynomials and find that [15]

$$\langle\cos\hat{\phi}_\theta\rangle = \langle\sin\hat{\phi}_\theta\rangle = 0,$$

$$\langle\cos^{2}\hat{\phi}_\theta\rangle = \tfrac{1}{2}\left[1 - \operatorname{sech} r\cos(2\eta)\sum_{n=0}^{\infty}\frac{(2n-1)!!}{(2n)!!}\left[\frac{2n+1}{2n+2}\right]^{\frac{1}{2}}(\tanh r)^{2n+1}\right] \tag{39}$$

$$\langle\sin^{2}\hat{\phi}_\theta\rangle = \tfrac{1}{2}\left[1 + \operatorname{sech} r\cos(2\eta)\sum_{n=0}^{\infty}\frac{(2n-1)!!}{(2n)!!}\left[\frac{2n+1}{2n+2}\right]^{\frac{1}{2}}(\tanh r)^{2n+1}\right] \tag{40}$$

The difference between our result for $\langle\cos^{2}\phi\rangle$ (or $\langle\sin^{2}\phi\rangle$) and the corresponding Susskind-Glogower expression is $(\operatorname{sech} r)/4$. This difference is proportional to the probability of finding the squeezed vacuum in the vacuum state; it will be greatest for a weakly squeezed vacuum and least for a strongly squeezed vacuum, because the latter has only a small vacuum component.

For weak squeezing, r is small and we find from (39) and (40) that

$$\langle 0,\xi|\cos^{2}\hat{\phi}_\theta|0,\xi\rangle = \tfrac{1}{2}[1 - r2^{-\frac{1}{2}}\cos(2\eta) + \Delta] \tag{41}$$

$$\langle 0,\xi|\sin^{2}\hat{\phi}_\theta|0,\xi\rangle = \tfrac{1}{2}[1 + r2^{-\frac{1}{2}}\cos(2\eta) - \Delta] \tag{42}$$

where Δ is a small term of order r^{3}. It is clear that as r tends to zero both of these values approach ½, which is precisely in accord with the phase of the vacuum being random. On adding equations (41) and (42) we get

$$\langle 0,\xi|\cos^2\hat{\phi}_\theta|0,\xi\rangle + \langle 0,\xi|\sin^2\hat{\phi}_\theta|0,\xi\rangle = 1 \tag{43}$$

which is in marked contrast to the analogous calculation [12-14] using the Susskind and Glogower phase operators:

$$\langle 0,\xi|\hat{\cos}^2_{s}\phi|0,\xi\rangle + \langle 0,\xi|\hat{\sin}^2_{s}\phi|0,\xi\rangle = \tfrac{1}{2} + \tfrac{1}{4}r^2 + O(r^3).$$

The fact that the right hand side of (43) is unity is a direct consequence of using a formalism based on a unitary phase operator. Equation (43) also coincides with what would be expected from a well-defined phase probability density.

For a strongly squeezed vacuum (r large), we find [15]

$$\langle 0,\xi|\cos^2\hat{\phi}_\theta|0,\xi\rangle \simeq \tfrac{1}{2}\left[1 - \cos(2\eta)\tanh r\left[1 - \frac{\cosh r - 1}{2\sinh^2 r}\right]\right] \tag{44}$$

$$\langle 0,\xi|\sin^2\hat{\phi}_\theta|0,\xi\rangle \simeq \tfrac{1}{2}\left[1 + \cos(2\eta)\tanh r\left[1 - \frac{\cosh r - 1}{2\sinh^2 r}\right]\right]$$

These results differ from those obtained with the corresponding Susskind and Glogower operators by $\tfrac{1}{4}$sech r. As r is made larger, this difference becomes smaller because of the diminished probability of finding the strongly squeezed field in the vacuum state. In the limit of very large r the difference between the phase properties of the ideal squeezed vacuum and the superposition of phase states $2^{-1/2}(|\eta+\tfrac{1}{2}\pi\rangle + |\eta-\tfrac{1}{2}\pi\rangle)$ becomes negligible, although these states are not identical. We see that for this superposition $\langle\cos\hat{\phi}_\theta\rangle$ vanishes and $\langle\cos^2\hat{\phi}_\theta\rangle = \cos^2(\eta+\tfrac{1}{2}\pi)$, to which expression (44) tends as r becomes large.

With the Hermitian phase operator we can also calculate the mean and variance of the phase itself. From (36) and (7) we obtain for the weakly squeezed vacuum (r small) [15]

$$|\langle\theta_m|0,\xi\rangle|^2 = (s+1)^{-1}\left[1 - r2^{1/2}\cos 2(\theta_m-\eta) + r^2(3/2)^{1/2}\cos 4(\theta_m-\eta) + O(r^3)\right] \tag{45}$$

The density of phase states is $(s+1)/(2\pi)$, so in the continuum limit as s tends to infinity the phase probability density becomes

$$P(\theta) = (2\pi)^{-1}\left[1 - r2^{1/2}\cos 2(\theta-\eta) + r^2(3/2)^{1/2}\cos 4(\theta-\eta) + O(r^3)\right] \tag{46}$$

When $P(\theta)$ is represented on a polar diagram with θ as the polar angle and P the radial distance, we find that, to first order in r, $P(\theta)$ traces out an ellipse with a major axis of length $(1+2^{1/2}r)/(2\pi)$ orientated at an angle of $\eta+\tfrac{1}{2}\pi$ to the x axis and a minor axis of length $(1-2^{1/2}r)/(2\pi)$. We can compare this with the Q-function which has elliptical contours with axes of exactly the same orientation for the same state [16]. Although the polar diagram is not precisely elliptical to second order in r, it still exhibits a deviation from the circle which manifests the imposed phase information. When r becomes very large, our previous discussion of the similarity in phase properties for the greatly squeezed vacuum and an equal superposition of two phase states differing by π predicts that the polar diagram collapses to a straight line through the origin at an angle of $\eta\pm\tfrac{1}{2}\pi$ to the x axis.

The mean and variance of the phase for the weakly squeezed vacuum are now easily calculated by using (46) with

$$\langle f(\hat{\phi}_\theta)\rangle = \int_{\theta_0}^{\theta_0+2\pi} P(\theta)\, f(\theta)\, d\theta$$

The detailed results are given elsewhere [15] but we note that, as anticipated from our previous discussion of the coherent state, the mean and variance of the phase depend upon the choice of θ_o. By choosing $\theta_o = \eta-\pi$ we find that the mean phase is η and the variance takes its minimum value of

$$\Delta\phi^2 = \pi^2/3 - r/2^{1/2} + (3/2)^{1/2}r^2/8 + O(r^3) \tag{47}$$

which illustrates the "squeezing" of the phase variance as r increases from zero, its value for the unsqueezed vacuum state.

We note that the vacuum, besides being a special "unsqueezed" case of a squeezed state, is also a special zero intensity case of a coherent state. Just as a small squeezing will, to a first approximation, squeeze the vacuum probability distribution circle into a ellipse, we find also that, to a first approximation for a coherent state as defined by (31) with a small intensity, the vacuum circle will retain its shape but will be displaced [17]. The displacement vector has a magnitude $|\alpha|/\pi$ and a direction β.

6. DISCUSSION

The existence of the Hermitian phase operator resolves a number of difficulties which have hitherto been associated with quantum phase: for example, $\sin\phi$ and $\cos\phi$ are now compatible variables, the sum of their squares is unity and the vacuum has random phase. The Susskind-Glogower phase difference operator predicted that two independent modes in the vacuum state were correlated in that the spectrum of the eigenvalues of this operator for such a case contained only two values, corresponding to phase angles of $\pm\pi/2$, and the spectrum only becomes dense when the modes are in highly excited states. The spectrum for the difference of the two Hermitian phase operators for the modes is dense, however, even for number states with a low excitation number, producing a quite reasonable absence of phase correlation [8].

While immediately obvious applications of the Pegg-Barnett operator are for studying phase properties of various field states, we feel that the existence of such an Hermitian operator may have more far reaching consequences. For example another standard application of an oscillator is to measure time. We can define the unit of time in terms of an integer number of complete oscillations of, for example, the light associated with a particular atomic transition. A more precise definition of times shorter than one oscillation period would involve fractions of oscillations, that is, the phase of the oscillation. Now that we have an Hermitian phase operator $\hat{\phi}_\theta$, we can make a suitable choice of θ_o and define the Hermitian time operator as $-\hat{\phi}_\theta/\omega$ [6]. The number-phase uncertainties then translate to genuine energy-time uncertainty relations for the oscillator. An interesting consequence of this definition is that the spectrum of eigenvalues of time is dense, not continuous as it would have been if the early attempts to find an Hermitian operator $\hat{\phi}$ independent of the choice of θ_o had been successful. Of course the difference between the "instants" of time become as small as we like as we make s larger but even in the limit there is only a countable infinity of them, which in a period $2\pi/\omega$ are equal in number to the number of energy eigenstates [6].

Another application which we have not had space to explore lies in the existence now of unitary operators of the type $\exp(i\lambda\hat{\phi}_\theta)$, where λ need not be an integer. These can be used to make unitary transformations on the number states to find new bases for the state space. This might be useful for solving some problems.

7. CONCLUSION

Quantum electrodynamics has progressed without an Hermitian phase operator for over sixty years. Because of the long-standing problem of representing the phase observable by an Hermitian operator, the quantum theory of light has had to learn to live without it. For example, the amplitude and phase properties of squeezed light can all be discussed in terms of the variation with time of the uncertainty in the magnitude of the field, avoiding any mention of phase as such [18]. Because the number eigenstates span the state space, any field state can be expressed in terms of them, and finding the Hermitian phase operator whose eigenstates also span the same space merely provides an alternative way of expressing the same information contained in the state vector, so the question arises as to why we should introduce such an operator. Apart from the practical possibility that, as with all alternative formalisms, some calculations may be greatly simplified, we feel that the Hermitian phase operator removes an unsatisfactory feature that has existed since Dirac's first paper on quantum electrodynamics, namely that such a simple and useful classical concept such as phase should have to be abandoned in quantum theory because of its apparently exceptional position as a classical observable without a corresponding Hermitian operator. The number operator provides only half the picture, rather like having a momentum operator without a position operator, and the existence of the phase operator may perhaps provide a clearer understanding of the quantum nature of light.

REFERENCES

1. P.A.M. Dirac: Proc. R. Soc. Lond. A114, 243 (1927)
2. L. Susskind and J. Glogower: Physics 1, 49 (1964)
3. P. Carruthers and M.M. Nieto: Rev. Mod. Phys. 40, 411 (1968)
4. J.M. Levy-Leblond: Ann. Phys. (N.Y.) 101, 319 (1976)
5. S.M. Barnett and D.T. Pegg: J. Phys. A 19, 3849 (1986)
6. D.T. Pegg and S.M. Barnett: Europhys. Lett. 6, 483 (1988)
7. S.M. Barnett and D.T. Pegg: J. Mod. Opt. (in press)
8. D.T. Pegg and S.M. Barnett: Phys. Rev. A (in press)
9. R. Loudon: *The Quantum Theory of Light*, 1st ed. (Oxford University Press, Oxford 1973)
10. E. Merzbacher: *Quantum Mechanics*, 2nd ed. (Wiley, New York 1970) p.330
11. For reviews of squeezed states see D.F. Walls: Nature 306, 141 (1983) and R. Loudon and P.L. Knight: J. Mod. Opt. 34, 709 (1987)
12. B.C. Sanders, S.M. Barnett and P.L. Knight: Opt. Comm. 58, 290 (1986)
13. D. Yao: Phys. Lett. A 122, 77 (1987)
14. Fan Hong-Yi and H.R. Zaidi: Opt. Comm. 68, 143 (1988)
15. J.A. Vaccaro and D.T. Pegg: Opt. Comm. (in press)
16. H.P. Yuen, Phys. Rev. A 13, 2226 (1976)
17. J.A. Vaccaro (unpublished)
18. R. Loudon: Physics World 2, 16 (1989)

A Bifurcation in Squeezed State Physics: But How? or Area-of-Overlap in Phase Space as a Guide to the Phase Distribution and the Action-Angle Wigner Function of a Squeezed State

W. Schleich[1;2], *R.J. Horowicz*[2], *and S. Varro*[2]

[1]Center for Advanced Studies and Department of Physics and Astronomy, University of New Mexico, Albuquerque, NM 87131, USA
[2]Max-Planck-Institut für Quantenoptik, D-8046 Garching, Fed. Rep. of Germany

Abstract. We calculate the phase distribution of a highly squeezed state (1) with the help of the area-of-overlap principle and (2) from the definition of a phase eigenstate. This distribution undergoes a bifurcation, that is, a transition from a single to a double-peaked distribution when we decrease the product of squeeze and displacement parameter. We discuss the Wigner function of this state in action-angle variables.

1. Phase Distribution of a Squeezed State: But How?

Start from the ground state of a harmonic oscillator such as a pendulum; *suddenly* alter its frequency from ω to $s^{-1}\omega$ – change, for example, its length – while at the same time *suddenly* displace its point of suspension by an amount $\sqrt{2}\alpha$. What is the probability, $W_x[|\psi_{sq}\rangle]$ for the oscillator in such a squeezed state [1,2], $|\psi_{sq}\rangle$, to have the dimensional coordinate x? The answer is well-known: The probability curve W_x is identical to the square of the absolute value of the scalar product, $w_x[|\psi_{sq}\rangle] = \langle x|\psi_{sq}\rangle$, between a position eigenstate $|x\rangle$ and the squeezed state, that is,

$$W_x[|\psi_{sq}\rangle] \equiv |\langle x|\psi_{sq}\rangle|^2 \cong \left|[2/(\epsilon\pi)]^{1/4} \exp\left[-1/\epsilon(x-\sqrt{2}\alpha)^2\right]\right|^2 . \quad (1.1)$$

Here we have considered the limit of large squeezing, that is, $s \cong 2/\epsilon \gg 1$.

Likewise we can ask and answer the question for the corresponding probability distribution, W_p, in the momentum variable p, that is,

$$W_p[|\psi_{sq}\rangle] \equiv |\langle p|\psi_{sq}\rangle|^2 = \left|[\epsilon/(2\pi)]^{1/4} \exp[-(\epsilon/4)p^2]\right|^2 . \quad (1.2)$$

Rather than to analyze this state in terms of the conjugate variables coordinate, x, and momentum, p, as expressed by Eqs. (1.1) and (1.2) summarized in Table 1 we might want to find the squeezed state oscillator's energy distribution, $W_m[|\psi_{sq}\rangle]$. Again this request can easily be fulfilled; again the quantity W_m is the square of the absolute value of the quantum mechanical scalar product between the two states of interest, the energy eigenstate, $|m\rangle$, of the oscillator and the squeezed state, that is,

$$W_m[|\psi_{sq}\rangle] = |\langle m|\psi_{sq}\rangle|^2 . \quad (1.3)$$

The squeezed state, $|\psi_{sq}\rangle$, is not an energy eigenstate of the oscillator. Hence, it can display interesting photon count probabilities, W_m, such as super- or sub-Poissonian distributions [1,3]. Even an oscillatory behavior [4] expressed [5] in its most elementary version by

$$W_m[|\psi_{sq}\rangle] \cong \begin{cases} 0 & \text{for } m < \alpha^2 - 1/2 \\ |w_m[|\psi_{sq}\rangle]|^2 \cong |\mathcal{A}_m^{\frac{1}{2}} \exp(i\phi_m) + \mathcal{A}_m^{\frac{1}{2}} \exp(-i\phi_m)|^2 & \text{for } m > \alpha^2 - 1/2 \end{cases} \quad (1.4a)$$

Springer Proceedings in Physics, Vol. 41 Quantum Optics V
Editors: J.D. Harvey and D.F. Walls

Table 1: Probability distributions, W_x and W_p in position, x, and momentum, p, versus photon count probability, W_m, and phase distribution, W_φ, of a highly squeezed state as viewed from $x-p$ oscillator phase space.

$W_x = \lvert\langle x\vert\psi_{sq}\rangle\rvert^2 \cong [2/(\epsilon\pi)]^{1/2}\exp[-(2/\epsilon)(x-\sqrt{2}\alpha)^2]$	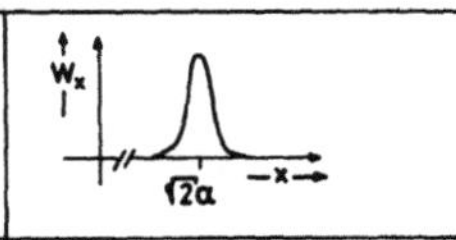	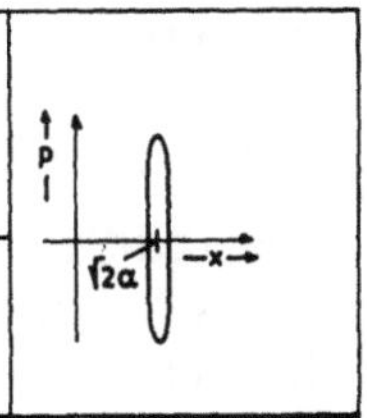
$W_p = \lvert\langle p\vert\psi_{sq}\rangle\rvert^2 \cong (\epsilon/(2\pi))^{1/2}\exp[-(\epsilon/2)p^2]$	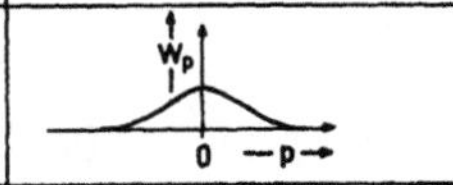	
$W_m = \lvert\langle m\vert\psi_{sq}\rangle\rvert^2 \cong \lvert A_m^{1/2} e^{i\Phi_m} + A_m^{1/2} e^{-i\Phi_m}\rvert^2$ $A_m = [\epsilon/(4\pi)]^{1/2}(m+1/2-\alpha^2)^{-1/2}\exp[-\epsilon(m+1/2-\alpha^2)]$ $\Phi_m = \int_{\sqrt{2}\alpha}^{[2(m+1/2)]^{1/2}} dx\,[2(m+1/2)-x^2]^{1/2} - \pi/4$	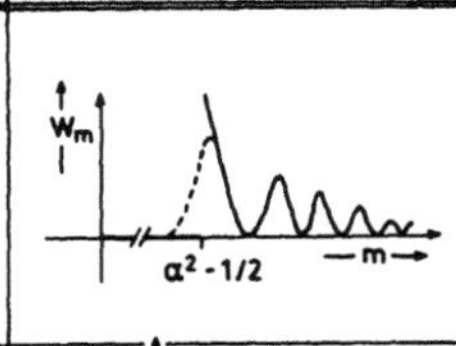	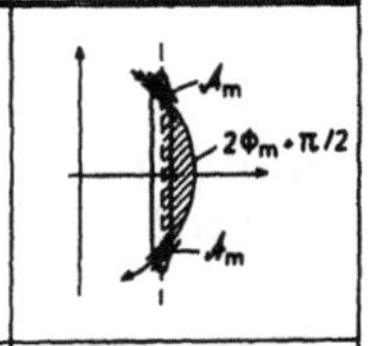
$W_\varphi = \lvert\langle\varphi\vert\psi_{sq}\rangle\rvert^2 \cong \pi^{-1/2}(\epsilon\alpha^2)^{1/2}\cos^{-2}\varphi\exp[-\epsilon\alpha^2\tan^2\varphi]$	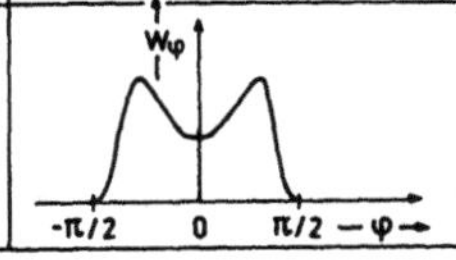	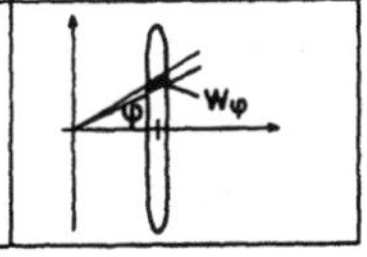

where

$$A_m \equiv [\epsilon/(4\pi)]^{1/2}\frac{\exp[-\epsilon(m+1/2-\alpha^2)]}{(m+1/2-\alpha^2)^{1/2}} \tag{1.4b}$$

and

$$\phi_m \equiv (m+1/2)\varphi_m - \alpha(m+1/2-\alpha^2)^{1/2} - \pi/4 \tag{1.4c}$$

with

$$\varphi_m \equiv \arctan[(m+1/2-\alpha^2)^{1/2}/\alpha] \tag{1.4d}$$

is possible [6,7]. This phenomenon summarized in the third row of Table 1 results [8] from the *interference* of the dark, diamond-shaped phase space zones of cross-over.

However, less straightforward to answer is an inquiry about the *phase distribution* of the squeezed state. What is the probability, $W_\varphi[|\psi_{sq}\rangle]$, for the squeezed oscillator's phase φ to lie between φ and $\varphi+d\varphi$? Guided by the above examples, Eqs. (1.1), (1.2) and (1.3), we are tempted to put forward a trail answer – an idea for an idea – of the form

$$W_\varphi[|\psi_{sq}\rangle]d\varphi \equiv |\langle\varphi|\psi_{sq}\rangle|^2 d\varphi. \tag{1.5}$$

Here the symbol $|\varphi\rangle$ denotes a phase eigenstate. But what is that state, $|\varphi\rangle$? And how to evaluate its scalar product with a squeezed state? These are the central questions of the present article. Many ways to find a clue – many ideas for an idea – offer themselves. In the present article we introduce two such clues, two such ideas.

Semiclassical quantum mechanics [9] *a la* Wentzel-Kramers-Brillouin (WKB) as an up-date of early "Atommechanik" [10]: *Area-of-overlap in phase space between two quantum states as a measure of their quantum mechanical scalar product* [8]. That is the guiding principle. To apply this very concept is the task. But how to represent in phase space a squeezed state, $|\psi_{sq}\rangle$? And how a phase state?

The Gaussian Wigner-cigar [11]

$$P_{sq}^{(W)}(x,p) = W_x[|\psi_{sq}\rangle] \cdot W_p[|\psi_{sq}\rangle] = \pi^{-1}\exp[-(2/\epsilon)(x-\sqrt{2}\alpha)^2 - (\epsilon/2)p^2] \qquad (1.6)$$

shown in Table 1 by its elliptical contour lines of exponential fall-off as a squeezed state representation is many a used method [1]. The corresponding answer for the phase state is less obvious. In its most elementary version we associate with a phase eigenstate, $|\varphi\rangle$, a state of well-defined phase, that is, a ray in $x-p$ oscillator phase space emerging from the origin and directed under an angle φ with the x-axis. More appropriate however, is the picture [12] of $|\varphi\rangle$ as a diverging beam, as shown in Fig. 1 and motivated by the analogy to polar coordinates, that is, a phase space density

$$P_{\varphi}(\varphi)rd\varphi' = r\delta(\varphi-\varphi')d\varphi'. \qquad (1.7)$$

The phase space zone of cross-over between the two distributions, Eqs. (1.6) and (1.7), shown in Fig. 2 by the shaded diamond, immediately provides the phase distribution of the squeezed state, $W_{\varphi}[|\psi_{sq}\rangle]$, shown below. When we start from a state of large coherent amplitude, $\alpha^2 \gg 1$, and gradually decrease α to the case of a squeezed vacuum, $\alpha = 0$, while keeping the squeeze, $s \cong 2/\epsilon \gg 1$, constant, the so-calculated distribution undergoes a transition from having one single maximum to two maxima shown in Fig. 3. Increasing the squeeze while keeping the displacement constant results in the same bifurcation phenomenon. In both cases the transition occurs when the height of the cigar, $(2/\epsilon)^{1/2}$, (see Eq. (1.6)) is identical to its location $x = \sqrt{2}\alpha$ on the x-axis.

The approach of Sec. 3 – an alternative to the area-of-overlap approach – makes use of the definition of a phase state [13] recently applied to propose [14] a hermitian phase operator. With the help of these states the phase probability *amplitude*, $w_{\varphi}[|\psi_{sq}\rangle] = \langle\varphi|\psi_{sq}\rangle$ of a squeezed state can be expressed by a Fourier sum of the corresponding photon number *amplitudes*, $w_m[|\psi_{sq}\rangle] \equiv \langle m|\psi_{sq}\rangle$. The distribution, W_{φ}, obtained by numerically evaluating this infinite sum, shown in the lower part of Fig. 3, agrees very well with the results of the area-of-overlap approach of Sec. 2.

In the present article we emphasize and highlight the points most relevant to the squeezed state discussion [15]. In a more extensive paper currently in preparation [12] we present the

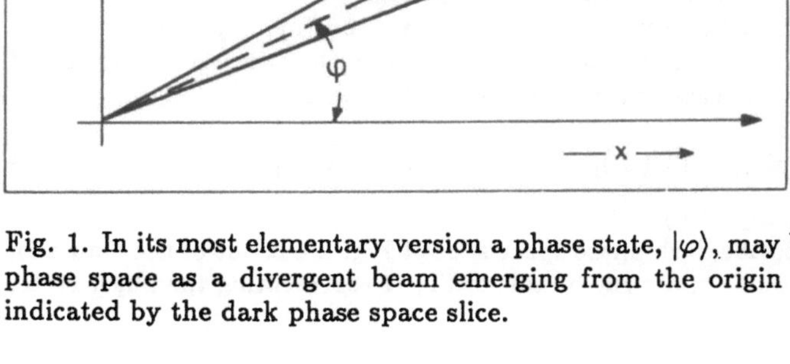

Fig. 1. In its most elementary version a phase state, $|\varphi\rangle$, may be represented in $x-p$ oscillator phase space as a divergent beam emerging from the origin and directed at an angle φ as indicated by the dark phase space slice.

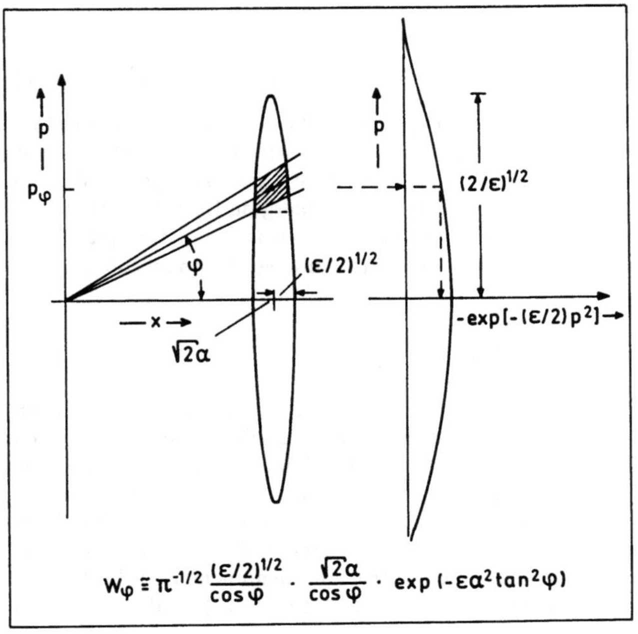

Fig. 2. The area-of-overlap principle associates with the phase distribution, W_φ, of a state highly squeezed in the x-variable, located at $x_0 = \sqrt{2}\alpha$ and represented in $x - p$ oscillator phase space by the elliptical contour line of exponential fall-off of the Gaussian cigar, Eq. (1.6), the diamond-shaped, shaded area-of-overlap between the diverging beam at angle φ and the ellipse weighted by the Gaussian distribution in p depicted on the right. The area of the diamond is governed by the product of its length, $\delta r_\varphi \sim (\epsilon/2)^{1/2}\cos^{-1}\varphi$, and its height, $r_\varphi \sim (\sqrt{2}\alpha)\cos^{-1}\varphi$, as indicated by Eq. (2.3b). Together with the weight function evaluated at the center of the shaded zone, that is, at $p_\varphi \cong \sqrt{2}\alpha\tan\varphi$, we arrive at the distribution, Eq. (2.3c), shown in the lower part of the figure.

corresponding area-of-overlap and phase state treatments of a coherent state and a number state along with the semiclassical Wigner function, $P^{(W)}_{|\psi\rangle}(m,\varphi)$, of a state, $|\psi\rangle$, in action-angle $(m - \varphi)$ variables [16,17] alluded to in the summary and outlook of Sec. 4.

2. Phase Distribution of a Squeezed State from Area-of-Overlap Principle

The guiding principle in the present search for the phase distribution, $W_\varphi[|\psi_{sq}\rangle]$, Eq. (1.5), of a squeezed state is the concept of area-of-overlap in phase space [8]: The quantum mechanical scalar product, $w_\varphi[|\psi_{sq}\rangle] \equiv \langle\varphi|\psi_{sq}\rangle$, between two quantum states such as a phase eigenstate, $|\varphi\rangle$, and a squeezed state, $|\psi_{sq}\rangle$, is identical to the area-of-overlap of the two states represented in phase space. Here we focus on the case of a highly squeezed state [2]

$$\psi_{sq}(x) \equiv \langle x|\psi_{sq}\rangle = [2/(\epsilon\pi)]^{1/4}\exp[-\epsilon^{-1}(x-\sqrt{2}\alpha)^2] \qquad (2.1)$$

expressed by the Gaussian cigar, Eq. (1.6).

In this case W_φ is identical to the dark phase space slice of Fig. 2 cut out of the Gaussian cigar, Eq. (1.6) by the diverging phase space ray bundle of Fig. 1 representing the phase state, $|\varphi\rangle$.

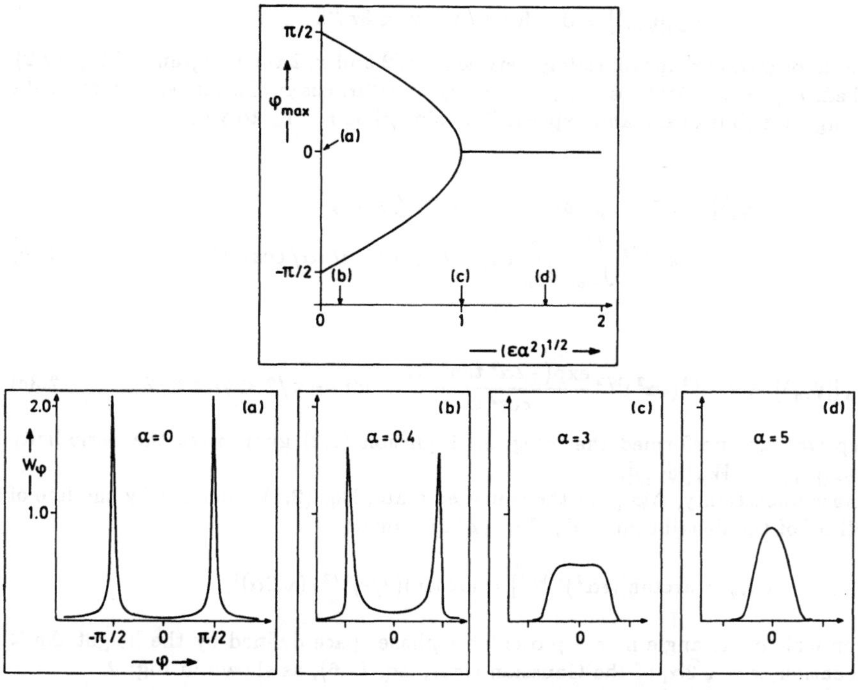

Fig. 3. The phase distribution, W_φ, of a highly squeezed state as derived from the area-of-overlap principle of Fig. 2 undergoes a bifurcation as a function of the product of the squeezed width, $\Delta x \cong (\epsilon/2)^{1/2}$ and the displacement, $\sqrt{2}\alpha$ of the Gaussian cigar, that is, as a function of $(\epsilon\alpha^2)^{1/2}$. For $(\epsilon\alpha^2)^{1/2} > 1$ the distribution, W_φ, enjoys one *single* maximum at $\varphi_{max} = 0$, whereas for $(\epsilon\alpha^2)^{1/2} < 1$ we identify *two* maxima – located symmetrically at $\varphi_{max}^{(\pm)} = \pm \arccos(\epsilon\alpha^2)^{1/2}$ – as shown in the upper part of the figure. In its lower part we display the phase distributions, W_φ, for (a) the squeezed vacuum, $\alpha = 0$, exhibiting two narrow peaks at $\varphi_{max}^{(\pm)} = \pm\pi/2$, (b) a small displacement of $\alpha = 0.4$, demonstrating two maxima, (c) the critical parameter, $(\epsilon\alpha^2) \sim 1$, showing a plateau, the transition between a double and single peaked distribution and (d) the case of large displacement, $\alpha = 5$, with a single peak. The curves in the lower part of the figure have been obtained by the method of Sec. 3, that is, by numerically evaluating Eqs. (3.3), (3.4) and (3.5). The curves in (c) and (d) are indistinguishable from the corresponding ones of the area-of-overlap principle, Eq. (2.3). Therefore we only show the numerical result.

In polar coordinates, this area reads

$$W_\varphi[|\psi_{sq}\rangle] \cong A_\varphi = \int_0^\infty dr \int_{-\pi}^{\pi} d\varphi' r\delta(\varphi - \varphi') P_{sq}^{(W)}\ (x = r\cos\varphi'; p = r\sin\varphi')$$

$$= \pi^{-1}\int_0^\infty dr\, r \exp[-(\epsilon/2)r^2\sin^2\varphi]\exp[-(2/\epsilon)\cos^2\varphi(r - \sqrt{2}\alpha/\cos\varphi)^2], \quad (2.2)$$

where in the last step we have used Eq. (1.6).

We note that for φ-values such that, $+\pi/2 < \varphi < 3\pi/2$, the integrand in Eq. (2.2) is exponentially small and hence

$$W_\varphi[|\psi_{sq}\rangle] \cong 0 \quad \text{for } \pi/2 < \varphi < 3\pi/2. \tag{2.3a}$$

For $0 < \epsilon \ll 1$ and for φ-values appropriately between $-\pi/2$ and $\pi/2$ the integrand of Eq. (2.2) is sharply peaked, $\delta r_\varphi \cong (\epsilon/2)^{1/2}\cos^{-1}\varphi$, around $r_\varphi \cong \sqrt{2}\alpha/\cos\varphi$. Hence we can evaluate the slowly varying contributions, r and $\exp[-(\epsilon/2)r^2\sin^2\varphi]$ at $r = r_\varphi$ to yield

$$W_\varphi[|\psi_{sq}\rangle] \cong \pi^{-1/2} r_\varphi \cdot \delta r_\varphi \exp[-(\epsilon/2)r_\varphi^2 \sin^2\varphi] \\ \times \pi^{-1/2}\int_{-\infty}^{\infty}\frac{dr}{\delta r_\varphi}\exp[-\delta r_\varphi^{-2}(r-\sqrt{2}\alpha/\cos\varphi)^2] \tag{2.3b}$$

or

$$W_\varphi[|\psi_{sq}\rangle] \cong \pi^{-1/2}(\epsilon\alpha^2)^{1/2}\frac{\exp(-\epsilon\alpha^2\tan^2\varphi)}{\cos^2\varphi} \quad \text{for } -\pi/2 < \varphi < \pi/2. \tag{2.3c}$$

In the last step we have performed the integral. Equation (2.2) guarantees the periodicity property $W_{\varphi+2\pi}[|\psi_{sq}\rangle] = W_\varphi[|\psi_{sq}\rangle]$.

Thus the phase uncertainty, $\Delta\varphi_{sq}$, in the squeezed state, Eq. (2.1) – defined by the line of exponential fall-off of the distribution, W_φ, Eq. (2.3c) – reads

$$\Delta\varphi_{sq} = \arctan[(\epsilon\alpha^2)^{-1/2}] = \arctan[(2/\epsilon)^{1/2}/(\sqrt{2}\alpha)],$$

and is thus identical to the angle in $x-p$ oscillator phase space defined by the height $\Delta p \cong (2/\epsilon)^{1/2}$ and location, $x = \sqrt{2}\alpha$, of the Gaussian cigar, Eq. (1.6), as shown in Fig. 2.

For $(\epsilon\alpha^2) \gg 1$, Eq. (2.3c) reduces to

$$W_\varphi[|\psi_{sq}\rangle] \cong \pi^{-1/2}(\epsilon\alpha^2)^{1/2}\exp(-\epsilon\alpha^2\varphi^2),$$

that is, to a Gaussian distribution in φ of widths $(\epsilon\alpha^2)^{-1/2} \ll 1$.

We differentiate Eq. (2.3c) with respect to φ

$$\frac{dW_\varphi}{d\varphi} = 2W_\varphi\cos^{-3}\varphi\sin\varphi[\cos^2\varphi - \epsilon\alpha^2].$$

Hence for $\epsilon\alpha^2 > 1$ the distribution, W_φ, exhibits one single maximum located at $\varphi_{max}^{(0)} = 0$, whereas for $\epsilon\alpha^2 < 1$, this maximum turns into a minimum and two new maxima located at

$$\varphi_{max}^{(\pm)}(\alpha,\epsilon) = \pm\arccos[(\epsilon\alpha^2)^{1/2}] \tag{2.4}$$

emerge as shown in the upper part of Fig. 3. This bifurcation occurs when

$$\epsilon\alpha^2 = 1 \tag{2.5}$$

that is, when the height, $\Delta p = (2/\epsilon)^{1/2}$, is identical to the displacement, $x_0 = \sqrt{2}\alpha$, of the Gaussian cigar in $x-p$ oscillator phase space.

We conclude this section by noting that the area-of-overlap procedure of finding the distribution, Eq. (2.3), that is, the integral, Eq. (2.2), is the Wigner function, $P_{sq}^{(W)}(x,p)$, Eq. (1.6) – originally given by $x-p$ variables – reexpressed in polar coordinates, $r-\varphi$, and integrated over r. This formalism – rewriting the Wigner distribution of the state of interest in polar coordinates and integrating over one variable – works only in cases when there is one single overlap in phase space, such as in this example. When there is more than one cross-over that is, *interference in phase space* great care has to be taken [18] as is also alluded to in Sec. 4.

3. Phase Distribution of a Squeezed State from Definition of Phase Eigenstate

In the preceding section we have calculated the phase distribution, $W_\varphi[|\psi_{sq}\rangle]$, of a highly squeezed state via the concept of area-of-overlap in phase space. In the present section we put forward a more mathematical treatment starting from a definition of a phase state [13]. We compare and contrast its results to those of Sec. 2.

The question of a hermitian phase operator in quantum mechanics and its eigenstates is a long-standing problem [19] originating from Dirac's original paper [20] on the quantization of the radiation field. Many a paper [21] have been written on this topic. However, only recently substantial insight into this problem has been gained [14,19] by starting from the definition of the phase eigenstates [13]

$$|\varphi\rangle \equiv \lim_{r\to\infty} |\varphi^{(r)}\rangle \tag{3.1a}$$

where

$$|\varphi^{(r)}\rangle \equiv [2\pi(r+1)]^{-1/2} \sum_{m=0}^{r} \exp(+im\varphi)|m\rangle. \tag{3.1b}$$

In this approach all computations are to be performed with the states $|\varphi^{(r)}\rangle$ for large but finite r and only in the end of the calculation one considers the limit $r\to\infty$.

With the help of Eq. (3.1b) we now calculate the phase probability amplitude

$$w_\varphi^{(r)}[|\psi_{sq}\rangle] \equiv \langle\varphi^{(r)}|\psi_{sq}\rangle = [2\pi(r+1)]^{-1/2} \sum_{m=0}^{r} \langle m|\psi_{sq}\rangle \exp(-im\varphi). \tag{3.2}$$

Therefore, the phase probability *amplitude*, $w_\varphi^{(r)}[|\psi_{sq}\rangle]$, of a squeezed state is the one-sided, finite, discrete Fourier transform of the photon number probability *amplitude* [5,8,22]

$$\begin{aligned} w_m[|\psi_{sq}\rangle] \equiv \langle m|\psi_{sq}\rangle &= \left[\frac{2\sqrt{s}}{s+1}\left(\frac{s-1}{s+1}\right)^m (2^m m!)^{-1} \exp\left(-\frac{2s}{s+1}\alpha^2\right)\right]^{1/2} \\ &\times H_m\left[\frac{s}{(s^2-1)^{1/2}}\sqrt{2}\alpha\right]. \end{aligned} \tag{3.3}$$

Here H_m denotes the m-th Hermite polynomial [23].

For $m > \alpha^2$ the quantities w_m, Eq. (3.3), decay fast [5,8]. Therefore, the sum, Eq. (3.2), converges even without the convergence factor $(r+1)^{-1/2}$, that is,

$$w_\varphi[|\psi_{sq}\rangle] = \lim_{r\to\infty}\left[(r+1)^{1/2} w_\varphi^{(r)}\right] = (2\pi)^{-1/2}\sum_{m=0}^{\infty} w_m[|\psi_{sq}\rangle]\exp(-im\varphi). \tag{3.4}$$

The phase distribution, $W_\varphi[|\psi_{sq}\rangle]$, of a squeezed state is then determined by

$$W_\varphi[|\psi_{sq}\rangle] \equiv |w_\varphi[|\psi_{sq}\rangle]|^2 \tag{3.5}$$

where w_φ follows from Eqs. (3.3) and (3.4). However, due to the term $(m!)^{-1/2}$ in $w_m[|\psi_{sq}\rangle]$, Eq. (3.3), the sum, Eq. (3.4) does not allow an *exact* analytical treatment [24]. In this section we therefore calculate the sum numerically. The lower part of Fig. 3 displays the so-calculated phase distribution, W_φ, in its dependence on the displacement parameter, for a fixed squeezing parameter $s = 21$. In agreement with the area-of-overlap results of Sec. 2 the phase distribution $W_\varphi[|\psi_{sq}\rangle]$ shows one *single* maximum at $\varphi^{(0)} = 0$ for an appropriately large displacement parameter, α, (see Figs. 3c and 3d) whereas below a critical value of α a bifurcation takes place and two maxima located at nonzero phase values approximately given by Eq. (2.4) occur. In the particular example of $\epsilon \cong 2/s \cong 0.1$ we find from Eq. (2.5), $\alpha_c \cong 3.2$ in agreement with Fig. 3c. In the limit of $\alpha = 0$ – a highly squeezed vacuum – two narrow peaks at $\varphi_{max}^{(\pm)}(\alpha = 0) = \pm\pi/2$ appear as shown in Fig. 3a, confining so the phase of a squeezed-oscillator to these phase values.

We emphasize that the phase distributions calculated numerically from Eqs. (3.3), (3.4) and (3.5) and shown in the lower part of Fig. 3 are almost indistinguishable from the approximate result, Eq. (2.3). Consequently we have depicted in the lower part of Fig. 3 only the numerical curve. For extremely small displacements, $\alpha^2 \ll 1$, Eq. (2.3) fails, as expressed most clearly from its prediction for the squeezed vacuum, $W_\varphi = 0$. For that case different approximation schemes have to be developed [12]. We conclude this section by noting that the area-of-overlap approach yields [12] up to a phase factor the result Eq. (3.4).

4. Summary and Outlook

Area-of-overlap in phase space between a diverging beam – representing a phase state – and a Gaussian cigar – depicting a squeezed state – as a measure of phase probability of a highly squeezed state? Yes! A bifurcation, that is, a transition from a single to a double peaked distribution as a function of the product of height, $(2/\epsilon)^{1/2}$ and displacement, $\sqrt{2}\alpha$, of the Gaussian cigar? Yes! The so-calculated phase probability curve almost indistinguishable from that obtained from a mathematical definition of a phase state? Yes! Three questions answered in this paper via this phase space formalism. But is there any further development possible?

The action (energy) [25] and angle distributions, W_m, Eq. (1.4) and W_φ, Eq. (2.3), – the analogs of the distributions W_x and W_p – describe the probability for one of the conjugate variables only. However, in the case of x and p there exist *joint* distributions, that is, distributions in *both* variables such as the Wigner distribution, Eq. (1.6). A joint distribution, $P_{sq}^{(W)}(m,\varphi)$, in action-angle variables – that is the new development. The semiclassical proposal of Ref. 16 when specified to the special example of a highly squeezed state

$$P_{sq}^{(W)}(m,\varphi) = \pi^{-1}\int_{-\pi/2}^{\pi/2} d\varphi' \exp(-im\varphi')\, w_{\varphi+\varphi'}[|\psi_{sq}\rangle]\, w^*_{\varphi-\varphi'}[|\psi_{sq}\rangle] \tag{4.1}$$

represents the answer. Equations (1.4) and (3.4) provide the necessary tools to evaluate [12] the above expression in terms of Airy functions – unfortunately a result of very complex mathematical nature [12]. However, the simple mock-up Wigner function

$$P_{sq}(m,\varphi) \equiv \begin{cases} 0 & \text{for } m < \alpha^2 - 1/2 \\ \mathcal{A}_m\,\delta(\varphi - \varphi_m) + \mathcal{A}_m\,\delta(\varphi + \varphi_m) + 2\mathcal{A}_m \cos 2\phi_m\,\delta(\varphi) & \text{for } m > \alpha^2 - 1/2 \end{cases} \tag{4.2}$$

where $\mathcal{A}_m, \varphi_m$ and ϕ_m are given by Eq. (1.4), contains all the essential ingredients of the semiclassical Wigner function (4.1), such as the marginals,

$$\int_{-\pi}^{\pi} d\varphi P_{sq}(m,\varphi) = 4\mathcal{A}_m \cos^2\phi_m = W_m$$

and

$$\sum_{m=0}^{\infty} P_{sq}(m,\varphi) \cong \pi^{-1/2}(\epsilon\alpha^2)^{1/2}\cos^{-2}\varphi \exp(-\epsilon\alpha^2\tan^2\varphi) = W_\varphi .$$

More insight into the distribution (4.2) springs out of the phase space considerations of the last figure of the third row in Table 1: For a fixed m-value appropriately larger than $\alpha^2 - 1/2$, the m-th circular Planck-Bohr-Sommerfeld band [8] – representing the m-th number state, $|m\rangle$

– cuts out of the Gaussian cigar – depicting the squeezed state – two symmetrically located diamond-shaped zones [5,8] each of phase space area $\mathcal{A}_m$, Eq. (1.4b), each with their centers located at an angle $+\varphi_m$ and $-\varphi_m$ with the x-axis, Eq. (1.4d). They constitute the first two positive contributions in Eq. (4.2). No other band-cigar cross-overs exist. Hence we expect no further contributions to $P_{sq}(m,\varphi)$. However, the *diamond interference* resulting in the oscillatory photon distribution, Eq. (1.4), reflects itself in the $m-\varphi$ phase space distribution (4.2) in the zero angle contribution which can change its sign. An analogous phenomenon occurs in the $x-p$ oscillator phase space Wigner function, $P^{(W)}_{|\psi\rangle}(x,p)$, of the superposition of two coherent states of identical displacement, $\sqrt{2}\alpha$, but of different phases, $+\varphi/2$ and $-\varphi/2$. Here the distribution $P^{(W)}_{|\psi\rangle}(x,p)$ consists [26] not only of two Gaussian bells both located in phase space at the radius $\sqrt{2}\alpha$ at angles $+\varphi/2$ and $-\varphi/2$ relative to the positive x-axis but also of an interference term at $\varphi=0$, which can create ditches and tongues in phase space.

The area-of-overlap in phase space approach [5,8] – the most vivid source of insight into problems such as the oscillatory photon distribution of a highly squeezed state [5,8] – provides a tool to find the phase distribution of quantum states or the Wigner function in action-angle variables.

Acknowledgments

During the course of this work we have immensely profited from discussions with S. M. Barnett, R. Chiao, C. Cohen-Tannoudji, M. Collett, T. Hänsch, R. F. O'Connell, D. Pegg, A. Schenzle, M. O. Scully, D. F. Walls and H. Walther. One of us (W.S.) is enormously grateful to J. A. Wheeler for the privilege of introducing him to, discussing and collaborating with him on semiclassical phase space physics while being at the Center for Theoretical Physics at the University of Texas at Austin and over the last years. Moreover, he would like to thank J. Harvey and D. F. Walls for organizing a most splendid conference. Many thanks go to Jeanne Williams for bringing this manuscript into the present form. Two of us (R. H. and S. V.) thank the Humboldt Foundation for a stipend. One of us (W.S.) is grateful to the Office of Naval Research for its support while being with the Center for Advanced Studies at the University of New Mexico.

Literature

1. For a review on squeezed states see D. F. Walls: Nature **306**, 141 (1983), and the special issues of J. Opt. Soc. Am. **B4**, (10) (1987), and J. Mod. Opt. **34**, (6) (1987); and, in particular, the article by G. Leuchs: Contemp. Phys. **29**, 299 (1988); and R. Loudon, P. L. Knight: J. Mod. Opt. **34**, 709 (1987).
2. In the present article we focus on quadrature squeezing in the sense of Ref. 1 rather than intensity-phase squeezing promoted most prominently by M. Kitagawa, Y. Yamamoto: Phys. Rev. **A34**, 3974 (1986); and Y. Yamamoto, S. Machida, O. Nilsson: Phys. Rev. **A34**, 4025 (1986).
3. G. J. Milburn, D. F. Walls: Phys. Rev. **A27**, 392 (1983).
4. R. S. Bondurant, B. S. Thesis, MIT, 1978 (unpublished).
5. For an extensive discussion of the photon distribution, W_m, of a highly squeezed state and various approximations see W. Schleich, J. A. Wheeler: Nature **326**, 574 (1987) and J. Opt. Soc. Am. **B4**, 1715 (1987).
6. A. Vourdas, R. M. Weiner: Phys. Rev. **A36**, 5866 (1987); Phys. Rev. **D 38**, 2209 (1988); R. Vyas, S. Singh: Phys. Rev. **A38**, 2433 (1988); V. V. Dodonov, A. B. Klimov, V. I. Man'ko: Phys. Lett. **A134**, 211 (1989); for the influence of these oscillations on the revivals of the Jaynes Cummings model see M. Venkata Satyanarayana, P. Rice, R. Vyas, H. J. Carmichael: J. Opt. Soc. Am. **B6**, 228 (1989).

7. The photon statistics of multimode squeezed light and in particular the question of oscillations in W_m has been investigated for (a) the squeezed vacuum by J. Huang, P. Kumar: In *Proceedings of the Fourth International Laser Science Conference*, Atlanta (1988), to be published; and P. Kumar: this volume and for (b) a squeezed coherent state by C. Caves: Talk at the *19th Winter Colloquium on Quantum Optics*, Snowbird, Utah (1989).
8. The area-of-overlap and interference-in-phase-space formalism is explained in J. A. Wheeler: Lett. Math. Phys. **10**, 201 (1985); W. Schleich: in *Squeezed and Nonclassical Light*, by P. Tombesi and R. Pike (Plenum, New York 1988) p. 129; W. Schleich: *Interference in Phase Space*, Habilitationsschrift, Ludwig-Maximilians-Universität, München (1988).
9. W. Pauli: in *Handbuch der Physik*, Vol. **24**, ed. by H. Geiger and K. Scheel (Springer, Berlin 1933) p. 166 and M. V. Berry, K. E. Mount: Rep. Prog. Phys. **35**, 315 (1972).
10. M. Born: in *Struktur der Materie in Einzeldarstellungen*, ed. by M. Born and J. Franck (Springer, Belin 1925).
11. M. Hillery, R. F. O'Connell, M. O. Scully, E. P. Wigner: Phys. Rep. **106**, 121 (1984).
12. W. Schleich, R. J. Horowicz, S. Varro: to be published.
13. R. Loudon, *The Quantum Theory of Light*, 1st ed., (Oxford University Press, Oxford 1973).
14. D. T. Pegg, S. M. Barnett: Europhys. Lett. **6**, 483 (1988); Phys. Rev. **A39**, 1665 (1989); S. M. Barnett, D. T. Pegg: J. Mod. Opt., to be published; see also the contribution by D. T. Pegg, S. M. Barnett and J. A. Vaccaro, this volume.
15. For a quite different approach towards phase in a squeezed state, see B. C. Sanders, S. M. Barnett, P. L. Knight: Opt. Commun. **58**, 290 (1986).
16. M. V. Berry: Phil. Trans. Roy. Soc. (London) **287**, 237 (1977). Wigner functions for any pair of operators have been developed by A. Barut: Phys. Rev. **108**, 565 (1957) and more recently by M. O. Scully, L. Cohen: In *The Physics of Phase Space*, ed. by Y. S. Kim and W. W. Zachary (Springer, Berlin 1987) p. 253. For the most recent work on Wigner functions in action-angle variables see A. Pimpale, M. Razavy: Phys. Rev. **A38**, 6046 (1988).
17. For Hamilton-Jacobi-action-angle quantum mechanics see R. A. Leacock, M. J. Padgett: Phys. Rev. **D28**, 2491 (1983).
18. For the discussion of the concept of area-of-overlap versus Wigner function and, in particular, the interference between phase space areas see W. Schleich, H. Walther, J. A. Wheeler: Found. Phys. **18**, 953 (1988); W. Schleich, D. F. Walls, J. A. Wheeler: Phys. Rev. **A38**, 1177 (1988); see also the Q-function amplitude approach by G. J. Milburn, D. F. Walls, Phys. Rev. **A38**, 1087 (1988) and G. J. Milburn: in *Squeezed and Nonclassical Light*, ed. by P. Tombesi and E. R. Pike (Plenum Press, New York 1988) p. 151.
19. For a thorough discussion of the history of the phase operator in quantum mechanics, see S. M. Barnett, D. T. Pegg: J. Phys. **A19**, 3849 (1986); see also the interesting article by S. Stenholm: Ann. Phys. (Fr.) **10**, 817 (1985).
20. P. A. M. Dirac: Proc. R. Soc. London **A114**, 243 (1927). This article is reprinted in: *Selected Papers on Quantum Electrodynamics*, edited by J. Schwinger, (Dover, New York 1958).
21. See the interesting review by P. Carruthers, M. M. Nieto: Rev. Mod. Phys. **40**, 411 (1968).
22. H. P. Yuen: Phys. Rev. **A13**, 2226 (1976).
23. G. Szegö, *Orthogonal Polynomials* (American Mathematical Society, New York 1939).
24. However, we can evaluate the sum in an approximate way by putting to use the approximation for $w_m[|\psi_{sq}\rangle]$, Eq. (1.4), and the stationary phase approximation as shown in Ref. 12. This treatment yields the phase distribution, (2.3), of the area-of-overlap approach.
25. A harmonic oscillator of frequency ω is a unique example in which the action quantization $I_m = 2\pi\hbar(m+1/2)$, is identical to the energy quantization, $E_m = \hbar\omega(m+1/2)$.
26. W. Schleich, M. Pernigo, to be published.

Part III

Atomic Spectroscopy and Cavity QED

Toward the Theoretical Limit of High Resolution in Atomic Spectroscopy

H. Takuma[1], *K. Shimizu*[1], *and F. Shimizu*[2]

[1]Institute for Laser Science, University of Electro-Communications, Chofu-shi, Tokyo 182, Japan
[2]Department of Physics, University of Tokyo, Bunkyo-ku, Tokyo 113, Japan

1. Introduction

The technique of optical spectroscopy has made revolutionary progress through the introduction of lasers as the light source, bringing tremendous improvements in both its sensitivity and resolution. The improvement in the resolution by the reduction of linewidth is a particularly powerful technique which can not be realized in the framework of pre-laser spectroscopy.

In spite of such drastic improvements which have already been accomplished, a lot of room is still left for further improvements in the resolution of laser spectroscopy, with which we will be able to make a great deal of progress in our understanding of nature. The physical causes limiting the resolution in laser spectroscopy can be attributed to the spectral width of the laser source itself and the linewidth of the spectrum to be observed. The theoretical limit of the spectral width of a laser light source, which is given by Schawlow-Townes equation, is normally extremely narrow. A great deal of improvement in the laser frequency stability has been made, the world record of sub-Herz stability being reported in [1].

Under such conditions, the spectroscopic resolution depends primarily on the linewidth of the spectral lines to be studied. Thus the reduction of the spectral linewidth has been essential in order to approach the theoretical limit of spectroscopic resolution. The line broadening mechanisms under normal conditions are known to be the first and the second order Doppler effect, pressure broadening, natural width, and finite interaction time due to thermal motion of the atoms. Methods to remove the first order Doppler broadening have already been well developed, and a great deal of Doppler-free high resolution data have been accumulated, providing important data on physics of atoms, molecules and their interaction with electromagnetic fields.

Although the linewidth is drastically narrowed by removing the first order Doppler broadening, we find various cases in which detailed structures of the atomic spectra are hidden within the residual linewidth due to natural lifetime, the second order Doppler effect and the finite time of passage of the atoms through the interaction region. We will be able to remove/reduce those line broadening effects if we have a method to reduce the natural

Springer Proceedings in Physics, Vol. 41 **Quantum Optics V**
Editors: J.D. Harvey and D.F. Walls

linewidth and also to cool the atoms down to a sufficiently low temperature. In the present paper we report our recent progress toward the limit of resolution by reducing the natural width and cooling the atoms by the use of lasers.

2. Subnatural Linewidth Spectroscopy

2.1 Background

Because the natural linewidth comes from the uncertainty in the atomic energy corresponding to the natural lifetime, narrowing may be expected by delayed observation which increases the uncertainty in the time of observation. This would certainly be possible if we could make selective observation of only such events in which the atom remained in the excited state much longer than the natural lifetime. Although such a scheme worked out quite well in the γ-ray spectroscopy[2], we observe only the statistical average of the radiation intensity emitted by an ensemble of atoms in an ordinary experimental setup for optical spectroscopy, and the statistics does not change whether we delay the observation or not. Thus we do not gain in resolution by simply carrying out delayed observation of the intensity, in spite of an enormous loss in the intensity.

Several methods to observe subnatural linewidth have already been proposed, some of them even with experimental evidence[3-10]. The earliest report came from Heidelberg University by using the rf optical double resonance method. This was essentially the rf spectroscopy of optically excited states of atoms, and line narrowing was apparently observed by delayed observation even by the use of a non-laser light source[3]. An advantage of this method is that the resolution is essentially determined by that of the rf, and high resolution can easily be obtained, as was shown in a later work on the hyperfine structure study of Li D-lines[4]. As another example in which high resolution of subnatural linewidth spectroscopy was effectively used in improving the accuracy of measurement using a similar scheme, rf Ramsey resonance on the $2^2P_{3/2}-2^2S_{1/2}$ transition of hydrogen allowed K.A. Safinya et al. to determine the fine-structure constant with an improved accuracy[7].

As the first direct observation of the optical transition, delayed observation of the quantum beat in the Hanle effect successfully gave narrowed linewidth[5,6]. However, quantum beat gives only accurate values of the sublevel separations in the excited states.

In order to observe subnatural linewidth in the direct optical transition, we proposed a novel method in which the delayed observation of the coherent transient signal due to the phase switching of the laser beam gives a linewidth corresponding to the delay time[11-13] as described in the following section.

2.2 Theory

We assume a two-level system interacting with a weak coherent laser field of nearly resonant frequency, the phase of which is shifted by ϕ at t=0:

$$E=\begin{cases} E_0 e^{i\phi} e^{i\omega t}+c.c., & t<0 \\ E_0 e^{i\omega t}+c.c., & t>0 \end{cases}$$

The Schrodinger equations in density matrix formulation for the atoms interacting with the laser beam are given as follows:

$$\frac{d}{dt}\rho_{11}=-i\frac{\mu E}{\hbar}(\rho_{12}-\rho_{21})-\gamma_{11}\rho_{11}$$

$$\frac{d}{dt}\rho_{22}=i\frac{\mu E}{\hbar}(\rho_{12}-\rho_{21})-\gamma_{22}\rho_{22}$$

$$\frac{d}{dt}\rho_{12}=(-i\omega_0-\gamma_{12})\rho_{12}-i\frac{\mu E}{\hbar}(\rho_{11}-\rho_{22}) \tag{1}$$

$$\rho_{21}=\rho_{12}^{*}$$

where i=1 corresponds to the upper state, and i=2 to the lower state.

We assume that the radiation field is weak enough for the second order perturbation method to give the solution with sufficient accuracy. The atomic state is represented by $\rho_{ij}(t;t_1)$, where t_1 is the time for the atom entering into the laser field, and the initial state is $\rho_{22}(t_1;t_1)=1$. The response of the medium is obtained by integrating the solution over various values of t_1:

$$R_{ij}(t)=\int_{-\infty}^{t}\rho_{ij}(t;t_1)\,dt_1$$

The fluorescence from the upper state is proportional to R_{11}, which is calculated as follows:

$$R_{11}(t)=\frac{\mu^2}{\hbar^2}\frac{1}{\gamma_1\gamma_2}\frac{1}{-i\Delta\omega+\gamma_{12}}|E_0|^2$$

$$+\frac{\mu^2}{\hbar^2}\frac{1}{\gamma_2}\frac{1}{-i\Delta\omega+\gamma_{12}}\frac{|E_0|^2(e^{i\phi}-1)}{i\Delta\omega-\gamma_{12}+\gamma_1}\left(e^{(i\Delta\omega-\gamma_{12})t}-e^{-\gamma_1 t}\right)+c.c. \tag{2}$$

The first term in eq.(2), which is independent of phase switching, is proportional to $E_0{}^2$ and corresponds to the steady state fluorescence without phase switching. The coherent transient response is represented by a term proportional to the second term of eq.(2), and the coherent transient term is obtained by comparing the value of R_{11} with and without phase switching. By taking the average value for $+\phi$ and $-\phi$ phase switching, we have an expression for the signal:

$$\Delta R_{11} \equiv R_{11}(\varphi,t)+R_{11}(-\varphi,t)-2R_{11}(0,t)$$
$$=4\frac{\mu^2}{\hbar^2}\frac{1}{\gamma_2}\frac{1}{(\Delta\omega)^2+\gamma_{12}^2}\frac{1}{(\Delta\omega)^2+(\gamma_1-\gamma_{12})^2}|E|^2(\cos\varphi-1)$$
$$\times\{[(\Delta\omega)^2-\gamma_{12}^2+\gamma_1\gamma_{12}][e^{-\gamma_{12}t}\cos(\Delta\omega t)-e^{-\gamma_1 t}]-\Delta\omega(\gamma_1-2\gamma_{12})e^{-\gamma_{12}t}\sin(\Delta\omega t)\} \quad (3)$$

If the lower state is the ground state of the atom, as in the case of the D-lines of alkali atoms, we have $\gamma_2=0$ and therefore $\gamma_1=2\gamma_{12}$. An example of the lineshape calculated by eq.(3) for such a case is shown on the left hand side of Fig.1. It is a wiggling signal the center component of which has a subnatural width. Such a line-shape makes the spectrum very complicated when it includes several hyperfine components. The complication can be removed by introducing a weighted average

$$\overline{\Delta R_{11}}=\gamma\int_{t_1}^{t_2}\left[R_{11}(\phi,t)+R_{11}(-\phi,t)-2R_{11}(0,t)\right]e^{\gamma t}\,dt \quad (4)$$

as shown on the right hand side of Fig.1.

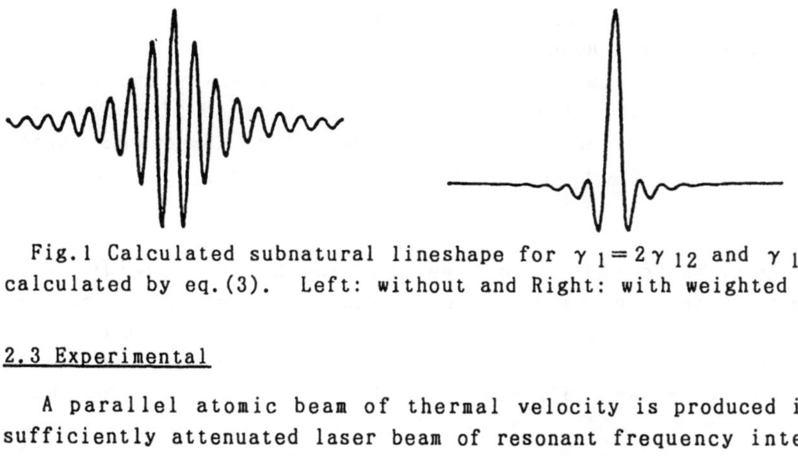

Fig.1 Calculated subnatural lineshape for $\gamma_1=2\gamma_{12}$ and $\gamma_{12}t=10$ calculated by eq.(3). Left: without and Right: with weighted average.

2.3 Experimental

A parallel atomic beam of thermal velocity is produced in vacuum, and a sufficiently attenuated laser beam of resonant frequency intersects perpendicularly with the atomic beam. The laser induced fluorescence is collected by a pair of spherical mirrors into a bundle of optical fibers and detected by a photomultiplier connected to the other end of the fiber bundle outside the vacuum chamber. The laser beam is alternately phase switched by an optoelectric modulator with a switching period T. The photoelectric pulses are counted by a multichannel scaler of total channel number 2N and a temporal resolution of T/N. The laser frequency is scanned stepwise and the count of each channel is stored in the computer memory for every step. The phase-switch signal for each channel is obtained by subtracting the count in a channel corresponding to a much longer delay time. Thus a single scan of the laser frequency is sufficient to give a set of data to calculate the

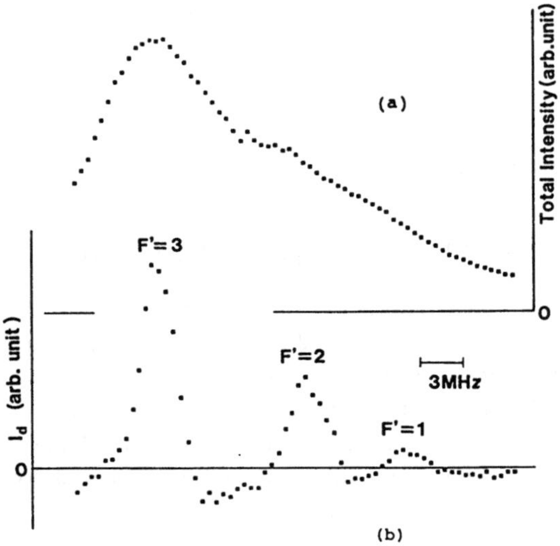

Fig.2 An example of subnatural linewidth spectrum of a part of the Li D_2 lines.

natural linewidth limited line shape and a line shape corresponding to various delay times.

An example of subnatural linewidth spectra obtained by the above method is shown in Fig.2 A part of the hyperfine structure in the Li D_2 line was observed at a delay time of 292ns, which is about 10 times the natural lifetime[13]. This result illustrates the significance of subnatural linewidth spectroscopy by showing that the three hyperfine components, which are hidden under the naturalwidth-limited line shape, are clearly observed by the present method.

Finally, one might ask to what extent we can reduce the linewidth with such a method, because the line narrowing is obtained by inevitably sacrificing signal intensity. However, the loss in the signal to noise ratio should be recovered by accumulating the signal for a long measuring time, and it is practically possible if we have sufficient absolute stability of the laser frequency.

3. Laser Cooling of Atoms

The linewidth remaining after natural linewidth reduction can be reduced by slowing down the atomic motion. The feasibility of laser cooling of atoms has been demonstrated recently by several groups with beautiful experimental successes on Na, Sc and Rb[14-18]. Especially our recent success on laser deceleration of ^{20}Ne made this beautiful technique applicable to many other kinds of atoms[19,20]. The cooling experiment has been successful only on alkali atoms simply because a tunable source to pump up the ground state atoms into the resonant states is not available at present. In our Ne deceleration experiment, the atoms make rapid round trips between the metastable state and one of the higher excited states. The energy levels related to Ne laser cooling are shown in Fig.3. The $1s_5(J=2)$ state is pumped up to the $2p_9(J=3)$ state by resonant radiation with σ^+ circular polarization. The optical pumping effect pumps up the atoms from all the magnetic sublevels to the levels of highest M in each of the electronic states, and easily brings the system into a perfect two-level dynamics, satisfying an ideal cooling condition. The change in the Doppler shift due to deceleration was compensated by adjusting the Zeeman shift of the atomic energy level.

Following the slowing down experiment, the neon cooling experiment was carried out using a setup as shown in Fig.4. The longitudinal beam is reflected back by a mirror placed near the beam source making a longitudinal cooling field, and another laser beam makes cooling fields in two independent transverse directions. The cooled atoms are trapped in a quadrupole magnetic

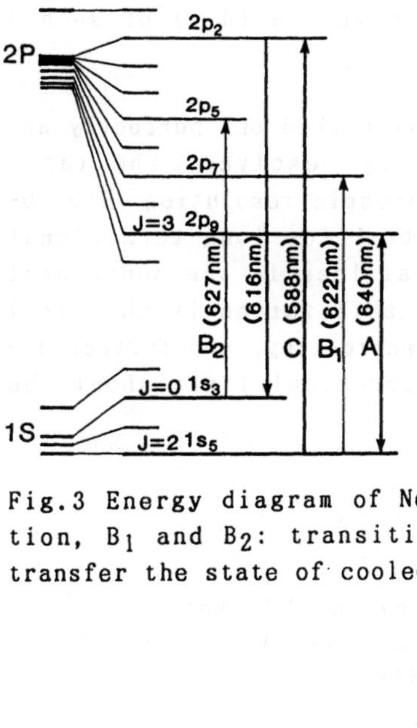

Fig.3 Energy diagram of Ne related to the laser cooling. A: cooling transition, B_1 and B_2: transition used for velocity analysis, C: transtion to transfer the state of cooled atom from $1s_5$ to $1s_3$

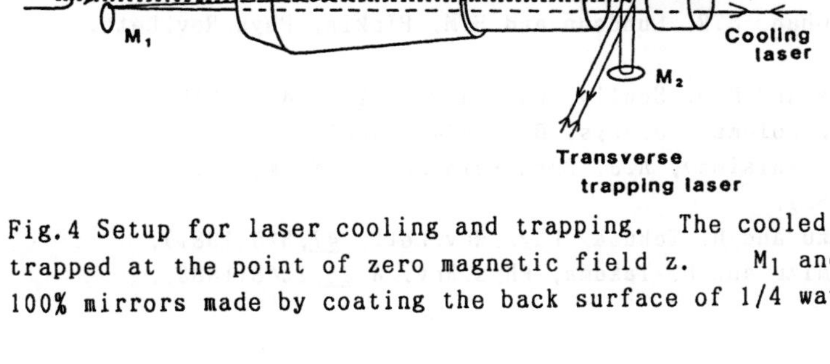

Fig.4 Setup for laser cooling and trapping. The cooled atoms are trapped at the point of zero magnetic field z. M_1 and M_2 are 100% mirrors made by coating the back surface of 1/4 wave plates.

field forming a bright sphere having a diameter of less than 1mm. In the case of ^{20}Ne, the number of trapped ions was estimated to be on the order of 10^7, the density being on the order of $10^{11}cm^{-3}$, and the temperature less than 1mK.

4. Conclusion

In the series of works described here we have shown the feasibility of two techniques which are expected to bring the spectroscopic resolution toward the theoretical limit: subnatural linewidth spectroscopy by phase switching of the laser beam allows us to reduce the natural linewidth as narrow as we wish, and the cooling scheme starting from the metastable states allows many kinds of atoms to be cooled down to submilli-Kelvin reducing the second order Doppler effect and the effect of finite time of passage. Although much further work is needed before these techniques bring us data which are extremely useful for the development of quantum physics, the experiments carried out so far seem to be sufficient in proving the validity of such a principle.

After all those line narrowing techniques have worked out perfectly and all the hyperfine components have been perfectly resolved, the laser linewidth should determine the limit of spectroscopic resolution. The absolute stability of the laser frequency is essential for long-term signal averaging, which is needed to overcome the signal loss in the subnatural linewidth spectroscopy. Thus laser frequency stabilization is the final requirement for the ultra-high resolution laser spectroscopy, and further efforts to bring the laser linewidth toward the theoretical limit should be strongly encouraged.

References

1. J.L. Hall, in Quantum Optics IV ed.by J.D. Harvey and D.F. Walls, Springer Proc. Phys.12 (Springer, Berlin, Heidelberg, New York 1986) p273.
2. K. Albrecht and W. Neuwirth, Z.Phys. 203,420(1967).
3. G. zu Putlitz, Comments At.Mol.Phys. 1,74(1969).
4. H. Orth, H. Ackerman and E.W. Otten, Z.Phys. A 273,221(1975).
5. P. Scheneck, R.C. Hilborn and H. Metcalf, Phys.Rev.Lett. 31,189(1973).
6. H. Figger and H. Walther, Z.Phys. 267,1(1974).
7. K.A. Safinya, K.K. Chan, S.R. Lundeen and F.M. Pipkin, Phys.Rev.Lett. 45,1934(1980).
8. H.W. Lee, P. Meystre and M.O. Scully, Phys.Rev. A 20, 1499(1979).
9. P.L. Knight and P.E. Coleman, J.Phys. B 13,4345(1980).
10. A.B. Doktorov, O.A. Anisimov, A.J. Burnstein and Yu.N. molin, Chem.Phys. 71,1(1982).
11. F. Shimizu, K. Umezu and H. Takuma, Phys.Rev.Lett. 47,825(1981).
12. F. Shimizu, K. Shimizu and H. Takuma, Phys.Rev. A 28,2248(1983).

13. F. Shimizu, K. Shimizu and H. Takuma, Phys.Rev. A 35,3149(1987).
14. V.I. Balykin, V.S. Letokhov and V.I. Mushin, JETP Lett. 29,560(1979).
15. W.D. Phillips and H. Metcalf, Phys.Rev.Lett. 48,596(1982).
16. W. Ertmar, R. Blatt, J.L. Hall and M. Zhu, Phys.Rev.Lett. 54,996(1985).
17. J. Prodan, W.D. Phillips, I. So, H. Metcalf and J. Dalibard, Phys.Rev. Lett. 54,992(1985).
18. S. Chu, J.E. Bjorkholm, A. Ashkin and A. Cable, Phys.Rev.Lett. 55,48(1985).
19. F. Shimizu, K. Shimizu and H. Takuma, Jpn.J.Appl.Phys. 26, L1847(1987).
20. F. Shimizu, K. Shimizu and H. Takuma, Phys.Rev. A 39, No.5(1989).

Phase Transitions of Stored Laser-Cooled Ions: A First Step Towards Crystallization in Storage Rings

R. Blümel, E. Peik, W. Quint, and H. Walther

Max-Planck-Institut für Quantenoptik and Sektion Physik, Universität München, D-8046 Garching, Fed. Rep. of Germany

Abstract. By means of laser induced fluorescence studies of two to about a hundred laser cooled Mg^+ ions stored in a Paul-trap we demonstrate that the ions occur in two stable phases: cloud and crystal. This is also verified directly by means of a photon counting imaging system. We observe transitions between these two phases at sharply defined values of the laser power, the laser detuning and the rf voltage applied to the trap. The transition points are reproducible for the cloud → crystal transition. Hysteresis is observed as a function of any one of the three parameters, keeping the other two fixed. All these experimental results are reproduced qualitatively with the help of three-dimensional molecular dynamics calculations which take photon recoil, laser polarization and saturation of the atomic transition into account. Satisfactory quantitative agreement was obtained whenever it was attempted. For a few representative cases, the dynamics of the phase transitions is presented both experimentally and theoretically. A model for the rf heating mechanism of the ions in the trap is presented along with an experimental demonstration of sympathetic cooling of $^{24}Mg^+$ by $^{26}Mg^+$ ions. Possible implications of our results for the problem of crystallization in storage rings are discussed.

1. Introduction

Historically, finite and infinite systems of charged particles in external potentials have been discussed many times in different contexts. At the turn of the century, e.g., the crystalline minimal-energy configurations of electrons embedded in a homogeneously charged sphere of positive polarity provided the first attempt at a quantitative description of atoms in the ground and excited states – J. J. Thomson's "raisin-model" [1] of an atom. In the twenties, DEBYE and HÜCKEL [2] have investigated strong electrolytical solutions. All this research culminated, when, some 50 years ago, WIGNER [3] postulated a new phase of strongly coupled Coulomb systems, the "Wigner-crystal". Crystallization is achieved when the plasma parameter Γ, which is defined as the ratio of potential and kinetic energy, exceeds a critical value, which, for the infinite 3D system, was calculated to be of the order of 170 [4, 5].

The first observation of crystallized charged particles stored in a Paul-trap was reported for aluminum micro-particles ($\approx 20\,\mu m$ diameter, 10^5 times the elementary charge) by WUERKER et al. [6]. Cooling was achieved by a light background-gas. Repeated crystallization of the particles to a regular array and melting of it were obtained by variation of the storage potential and the cooling force. Although these pioneering experiments clearly demonstrated that the Paul-trap is a marvellous instrument to gain insight into the physics of "few-body phase transitions", this aspect of the ion trap did not immediately become one of the main streams of research and for quite some time the ultimate goal in ion-trap physics was considered to utilize a single ion [7, 8] for high resolution spectroscopy [9–14] and as a frequency standard, which could easily surpass the accuracy of modern Cs atomic clocks [15]. Moreover, single ion experiments have provided a source of anti-bunched light of sub-Poissonian photon statistics [16] and have provided insight into the nature of quantum-jumps [17–19].

The first observation of the crystallization of laser cooled Mg^+-ions in a Paul-trap to an ordered structure (see Fig. 1), its melting and recrystallization, i.e., a phase transition, was then reported by DIEDRICH et al. [20–23]. Later, small ion-clusters of Hg^+-ions were investigated

Springer Proceedings in Physics, Vol. 41 **Quantum Optics V**
Editors: J.D. Harvey and D.F. Walls

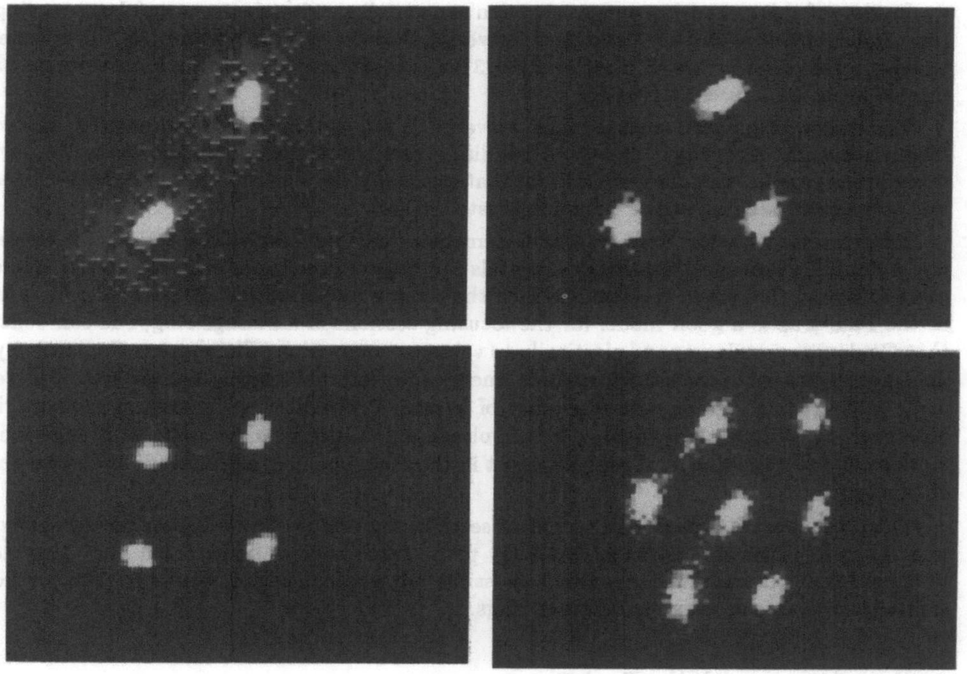

Fig. 1 : Two, three, four and seven ions confined by the dynamical potential of a Paul-trap and crystallized to an ordered structure under the influence of UV laser light tuned slightly below the resonance frequency. The average separation of the ions is 20 μm. For details see Refs. [22,23].

by WINELAND et al. [24]. The quantitative interpretation of the phase transitions observed by DIEDRICH et al. [20–23] was given by BLÜMEL et al. [25]; see also HOFFNAGLE et al. [26] for comparison.

Recently the formation of shell-structure of a large number of Be^+-ions stored in a static Penning trap was observed by GILBERT et al. [27]. That this type of experiment may be feasible was conjectured in preceding work by BOLLINGER et al. [28] and BREWER et al. [29]. The ions were cooled by radiation pressure [32–34] using laser-light. With a second laser beam the kinetic energy of the ions was investigated and it was claimed [29] that high enough plasma-parameters were attainable so that crystallization gets feasible.

Pure electron plasmas have also been produced. They bear the exciting possibility that the quantum regime may be reached at low temperatures. MALMBERG et al. [43] stored an electron plasma at cryogenic temperatures, but the plasma-parameter was not yet large enough for crystallization to occur. More successful were experiments with electron-layers on the surface of liquid helium [44] (surface state electrons). Here the formation of a two-dimensional crystal-lattice has been observed at $\Gamma_c \approx 137$ [45]. There are indications for three-dimensional Wigner-crystallization of the conduction electrons of certain alloys [46].

A completely new era of accelerator physics would begin if it were possible to produce, store and accelerate Coulomb crystals in particle accelerators and storage rings. To work with crystals instead of the usual dilute, weakly coupled particle clouds has at least one advantage: the luminosity of accelerators (storage rings) could be greatly enhanced, and (nuclear) reactions whose cross sections are too small to be investigated in currently existing accelerators would become amenable to experimentation.

In view of this promising perspective, ion storage rings with electron- and laser-cooling are now under planning and construction in several laboratories in the world [35]. It is expected that strongly coupled Coulomb-systems [36,37] can be generated resulting in three-dimensional crystal-beam structures [35,38–42].

The design of modern storage rings, however, is not particularly favourable for "growing" Wigner crystals. Focusing-, drift- and bending - sections squeeze, pull and shear the crystal. Molecular dynamics calculations indicate that especially the shear forces might lead to heating and subsequent melting of the Coulomb crystal [41,42].

Experiments with the Paul-trap have demonstrated that despite the action of a squeezing and pulling rf quadrupole field, stable crystals can be produced in the trap [20–26,47], and even more amazing, that small crystals persist if the cooling forces are switched off [48]. In as much as the Paul-trap is a good model for the focusing sections of a storage ring, one can conclude that Coulomb crystals respond elastically to volume compression. Therefore, a Coulomb crystal does not necessarily convert deformation energy into internal heating and we take this result as an encouraging sign for the feasibility of Wigner crystallization in existing storage rings. Moreover, by periodically changing the rf voltage, a Paul-trap can simulate drift and focusing sections of a storage ring, and we will report in the sequel on first calculations and experiments of this type.

The present article consists of two main sections: In Chapter 2 we discuss our recent experimental and theoretical results on few body phase transitions of laser cooled Mg^+ ions stored in a Paul-trap. In chapter 3 we discuss possible implications of our results for the physics of crystallized beams in heavy ion storage rings.

2. Experiments with the Paul-Trap

The radio frequency (rf) Paul-trap [49] grew out of the work on quadrupole mass filters [50] in the late fifties. It is well known that due to the properties of the Laplace-equation, it is impossible to trap charged particles by electrostatic potentials only. The best one can achieve is confinement in two dimensions, leaving the motion in the third dimension unstable. In order to overcome this limitation, one can either use an additional magnetic field (static Penning-trap [30,31]) or enclose the particles dynamically by an rf field applied to the electrodes of the trap. In this case, the appropriate choice of applied voltages creates a three-dimensional dynamical binding potential (dynamical Paul-trap [49,51,6]).

In the late sixties, CHURCH [52] investigated circular rf traps, which are an ideal miniature model for heavy ion storage rings. He considered primarily the storage properties of this trap design, but did not investigate crystallization. We believe that even normal Paul-traps can already give some valuable insight into the physics of crystallization in storage rings. To this end, we used a trap which is considerably larger than most of the other traps used in trapping and cooling experiments with single ions [8,12,19] and its hyperbolic electrodes are machined with high precision (tolerance: 20 μm). We believe that the large size and the mechanical precision of the trap was a prerequisite to obtain relatively large ion crystals and to observe the transitions between the crystalline and the cloud phase. Our phase transition experiments described in this chapter may well be considered as precursors for analogous experiments in ion storage rings. In this spirit we now describe our experiments in detail.

2.1 Description of the Trap

In the present experiment, a trap of ring diameter $2r_0$ = 5 mm and end-cap separation of $2z_0$ = 3.54 mm was used (see Fig. 2, and [16] further technical details). The frequency of the rf field, f, had a value of $f \approx$ 11 MHz. For the measurement of the secular and vibrational frequencies of ion-crystals, an additional AC-voltage can be applied to ring- and end-cap electrodes.

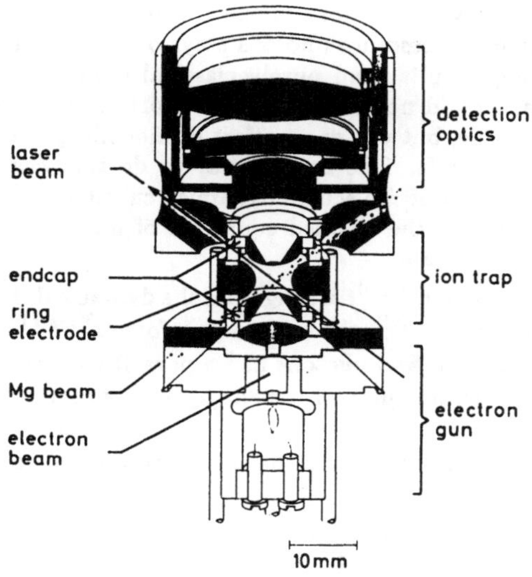

Fig. 2: Sketch of the ion-trap. The fluorescence light is observed through a hole in the upper electrode.

The trap is mounted inside a stainless-steel ultra-high vacuum chamber, the pressure of the background gas being 10^{-10} mbar. This was sufficient to store single ions for about 1 hour. The ions are loaded into the trap by means of a thermal beam of neutral Mg-atoms which are ionized close to the trap's center by an electron beam entering the trap through a small hole in the lower end-cap. In order not to distort the trap potential by the hole, it is covered by a fine molybdenum mesh (diameter of the wires: 20 μm, mesh size 200 μm). The neutral Mg-beam as well as a UV laser-beam are passed through the gaps between cap- and ring-electrodes. The UV-light is generated by a single mode dye-laser, frequency doubled in an external cavity with the help of a KDP - crystal, and has a maximum power of 1 mW. It is focussed to the center of the trap, the focal dimensions being 20 μm $\times$ 50 μm. Irradiation of the ions by the UV laser light serves two purposes. On one hand it detects the ions by laser induced fluorescence on the Mg^+ dipole transition $3S_{1/2} \rightarrow 3P_{3/2}$, on the other it cools the ions by radiation pressure. This is achieved by shifting the laser frequency by an amount Δ below the $3S_{1/2} \rightarrow 3P_{3/2}$ resonance-transition frequency. This way, a single ion can be cooled in a few seconds to a temperature below 100 mK. The fluorescence from the ions is observed through a hole in the upper end-cap (see Fig. 2), again covered by a molybdenum mesh. The large size of the trap affords a large solid angle (5.6% of 4π) for detecting the fluorescence radiation. This was done either by a photo-multiplier, or by means of a photon-counting image system (Hamamatsu, ARGUS or PIAS). For the observation of the ions, the cathode of the image system was placed in the image plane of the microscope objective attached to the trap (Fig. 2). In this way the ion pictures of Fig. 1 have been obtained.

2.2 Theoretical Background of our Molecular Dynamics Calculations

There are several reasons, why we do not want to use simplified models for the description of the phase transition experiments described below. First of all a good model description always tries to reduce the dimensionality of the problem, probably down to one dimension. Since we are dealing here with phase transition phenomena, such a simplification would be very dangerous. It is well known that critical points and the very phenomenon of a phase transition are strongly dependent on the dimensionality of the problem at hand. Actually we will see below that indeed a one-dimensional description of the trapped ions cannot reproduce the wealth of phenomena encountered for laser cooled ions in a Paul-trap, and to be specific, misses the phases transition phenomenon altogether. Second, it is not clear from the beginning, which are the relevant mechanisms responsible for, say, ion heating and the exact locations of the transition points. Here we found out that, e.g., spontaneous emission of fluorescence photons plays only a minor role in the ion heating mechanism which essentially results from purely classical dynamical effects. In our microscopic simulations, therefore, we did not put too much effort to model the spontaneous emission part as accurately as possible, but put great effort in modelling the mechanical part of the problem as realistically as possible. The natural way to do this was by means of simulating the ion dynamics with the help of molecular dynamics calculations, which in essence, consist of solving exactly (numerically) the coupled equations of motion of the interacting ions in the presence of the cooling laser beam.

The ions are subjected to five different forces: The force $\vec{F}^{(trap)}$, arising from the dynamical rf trapping field, the Coulomb interaction between the ions $\vec{F}^{(Coul)}$, the laser cooling force $\vec{F}^{(laser)}$, a random force $\vec{F}^{(rand)}$ arising from the spontaneously emitted photons and a parasitary force $\vec{F}^{(cont)}$, arising from a contact potential, which was generated unintentionally by locally coating the trap electrodes with a thin film of magnesium.

We now discuss these forces in detail. The potential generated by the rf voltage applied to the hyperbolic electrodes of the trap is given by [49,6]

$$\varphi(\vec{r},t) = g(t)\left[x^2+y^2-2z^2\right] \; ; \quad g(t) = \frac{U_0+V_0\cos(\Omega t)}{r_0^2+2z_0^2} \tag{1}$$

where U_0 and V_0 are the DC and AC components of the voltage, respectively, and $\Omega = 2\pi f$. From equation (1) we obtain the trapping force and the resulting equation of motion of a single ion in the trap

$$\vec{F}^{(trap)} = m\ddot{\vec{r}} = -e\vec{\nabla}\varphi(\vec{r},t) = -2\,e\,g(t)\,[\vec{r} - 3\vec{z}] \tag{2}$$

where $m, \vec{r}$ and $\vec{z}$ denote the mass of the ion, the position and the z-component of the position vector, respectively. The equation of motion (2) can be cast into the standard form of a Mathieu-equation [53]. The qualitative behaviour of its solutions depends crucially on the voltages U_0 and V_0. This two-dimensional parameter-space shows domains of stability and instability. Stable solutions obviously represent trapped ions, whereas unstable solutions correspond to ions escaping from the trap. Generalizing equation (2) to the case of n particles, the resulting trapping force reads:

$$\vec{F}_i^{(trap)} = -2\,e\,g(t)\,[\vec{r}_i - 3\vec{z}_i]\,. \tag{3}$$

The Coulomb interaction between the particles results in the force

$$\vec{F}_i^{(Coul)} = \frac{e^2}{4\pi\epsilon_0}\sum_{m\neq i}\frac{\vec{r}_i-\vec{r}_m}{|\vec{r}_i-\vec{r}_m|^3}. \tag{4}$$

In the process of laser cooling, every scattered photon changes the momentum of an ion on the average by an amount $\hbar\vec{k}$. An ion at position $\vec{r}(t)$ and velocity $\dot{\vec{r}}(t)$ scatters photons at a rate R, with [54]

$$R\left[\vec{r}(t);\dot{\vec{r}}(t)\right] = 2\pi\gamma\,\frac{S(\vec{r};\dot{\vec{r}})}{1+2S(\vec{r};\dot{\vec{r}})} \quad . \tag{5}$$

Here

$$S(\vec{r};\dot{\vec{r}}) = \frac{I(\vec{r})}{I_0}\cdot\frac{(\gamma/2)^2}{(\Delta-\frac{\vec{k}\cdot\dot{\vec{r}}}{2\pi})^2+(\gamma/2)^2} \tag{6}$$

is the saturation parameter and Δ and γ (=43 MHz) denote the laser detuning and the natural line width (FWHM) of the $3S_{1/2} \rightarrow 3P_{3/2}$ – transition in Mg^+, respectively. In the calculations we used $I_0 = 1.53$ W/cm^2. The beam intensity

$$I(\vec{r}) = \frac{2P}{\pi D_y D_w}\exp\left[-2(\hat{y}\cdot\vec{r}/D_y)^2\right]\,\exp\left[-2(\hat{w}\cdot\vec{r}/D_w)^2\right] \tag{7}$$

is assumed to have a Gaussian profile with beam waists D_y and D_w in polarization direction, $\hat{y}$, and $\hat{w} = \hat{k} \times \hat{y}$, respectively. Here $\hat{y}, \hat{w}$ and $\hat{k} = \vec{k}/|\vec{k}|$ denote unit vectors and P is the laser power. In the present experiments as well as in the theoretical calculations $D_y \approx 50\mu$m and $D_w \approx 20\mu$m. The ensemble average of the resulting laser cooling force acting on the i-th ion is thus

$$\vec{F}_i^{(laser)} = \hbar \vec{k} R \left[\vec{r}_i(t); \dot{\vec{r}}_i(t)\right] . \tag{8}$$

The number $N_i(l)$ of spontaneously emitted photons from the i-th ion during the l-th cycle of period $T = 1/f$ is given by

$$N_i(l) = \int_{(l-1)T}^{lT} R[\vec{r}_i(t); \dot{\vec{r}}_i(t)]\, dt . \tag{9}$$

In our simulations, we emit these N_i spontaneous photons, one by one, at the end of each rf cycle, rather than emitting them, one by one, according to the appropriate photon-statistics during the cycle. The orientation of the unit vector $\hat{u}_j$, which points into the direction of the j-th emitted photon, is chosen at random, but weighted with a $\cos^2$–distribution (dipole characteristic) with respect to the laser polarization axis. This way, only the linearly polarized spontaneously emitted photons have been simulated in our calculations. The circularly polarized ones, which occur with a $\sin^2$ - distribution, have been neglected. The random force $\vec{F}_i^{(rand)}$, is thus

$$\vec{F}_i^{(rand)}(t) = \hbar k \sum_{l=-\infty}^{\infty} \delta(t - lT) \sum_{j=1}^{N_i(l)} \hat{u}_j . \tag{10}$$

In some of our calculations we took a contact potential into account which was present in the experiment owing to the reason mentioned above. We assumed the force $\vec{F}_i^{(cont)}$ to be constant, since close to the center of the trap the contact potential gives rise to an approximately homogeneous electric field. Its magnitude was experimentally determined to be of the order of 0.1 V/cm. Altogether, the motion of the i-th particle is governed by the sum of the above forces, that is,

$$m\ddot{\vec{r}}_i = \vec{F}_i^{(trap)} + \vec{F}_i^{(Coul)} + \vec{F}_i^{(laser)} + \vec{F}_i^{(rand)} + \vec{F}_i^{(cont)} . \tag{11}$$

Given appropriate initial conditions, the equations of motion (11) are integrated forward in time by a standard Runge-Kutta [55] procedure to obtain the trajectories of the n-ion system. These trajectories provide all the information necessary to find the quantities of interest, such as the occupation probability density of the i-th ion in phase space, (i.e., the probability density of finding the i-th ion at a given position and with a given velocity), the (total) kinetic energy and the fluorescence intensity. In order to eliminate transient effects arising from initial conditions, we integrate the equations until the total kinetic energy has reached a steady state.

We note that the equations of motion (11) incorporate the micromotion which is explicitly taken into account in our molecular dynamics approach.

In the experimental situation, the imaging system is focused on the $z = 0$ – plane. Therefore, in order to allow for a comparison between the theory and the experimental results, we project the $6n$–dimensional phase space to the $z = 0$ – plane. That is, only the x_i and the y_i – components of the i-th particle are recorded as a dot in the $x - y$ – plane at the end of each rf cycle. Usually, these positions are recorded over $\approx$ 2000 cycles. We also calculate the total fluorescence, $\bar{R}$, from the n ions during ν rf cycles via

$$\bar{R} = \frac{f}{\nu} \sum_{l=1}^{\nu} \sum_{i=1}^{n} N_i(l) \tag{12}$$

where $N_i(l)$ is given by equation (9).

Depending on external parameters, such as the laser power P, the detuning Δ and the rf voltage V_0, the solutions of equation (11) can be divided into two classes: in the first the ions form an ordered structure (crystalline phase) and in the second the ions perform an erratic motion of much larger spatial extension and mean kinetic energy (cloud phase). Due to the cooling and randomizing effect of the forces $\vec{F}_i^{(laser)}$ and $\vec{F}_i^{(rand)}$, the equilibrium energy is independent of the particular choice of initial conditions. For those trap parameters, however, where both phases occur and are stable (bistability), the phase space of initial conditions is divided into two domains forming the basins of attraction for the crystalline phase, and the cloud phase, respectively. There are indications that the hyper-surface separating the two basins is actually a fractal [56].

2.3 Phase Transitions: Experiment vs. Theory

In this section we discuss our few-ion phase transition experiments in detail and compare them to the results of three-dimensional molecular dynamics simulations of equation (11). A manifestation of the existence of phase-transitions in a Paul-trap is provided by striking jumps in the fluorescence intensity (photon counts per second) of the ions as a function of the detuning Δ (see Fig. 3, where the discontinuities in the spectrum are indicated by vertical arrows). The jumps occur between two types of spectra: a broad spectrum, similar to the spectrum completed with the dashed line in Fig. 3, and a narrow spectrum, similar to the fluorescence spectrum of a single cooled ion. When we interpret the broad spectrum as a finger-print of an ion-cloud and the narrow spectrum as being characteristic for an ordered many ion situation with a "single ion signature", these jumps clearly indicate a transition from a state of erratic motion to a situation in which the ions arrange themselves in regular structures similar to the Thomson-electrons or the Wigner-crystal. The regime of detunings in which such crystals exist is depicted in Fig. 3 by the horizontal arrow. The conjecture of the existence of these two phases has been verified by direct observation (see Fig. 1) with the help of a highly sensitive imaging system, by analyzing the ion-trajectories produced by our calculations (see insets of Fig. 4) and by exploiting the different response of crystals and clouds to an additional rf voltage.

The experimental as well as the theoretical excitation spectra, and the jumps in them, have been recorded by scanning the detuning Δ from large negative values to zero. Reversing this procedure, i.e., scanning the fluorescence intensity in the backward direction, the jumps occur at different values of Δ, i.e., there appears to be bistability and hysteresis associated with transitions from the cloud-phase to the crystal phase as indicated in Figures 4 and 5. Such a hysteresis behaviour must be expected with laser-cooled ions because the cooling power of the laser strongly depends on the details of the velocity distribution of the ions (see, e.g., equation (6)). We note that depending on the detuning, either the crystal or the cloud can have a higher fluorescence intensity.

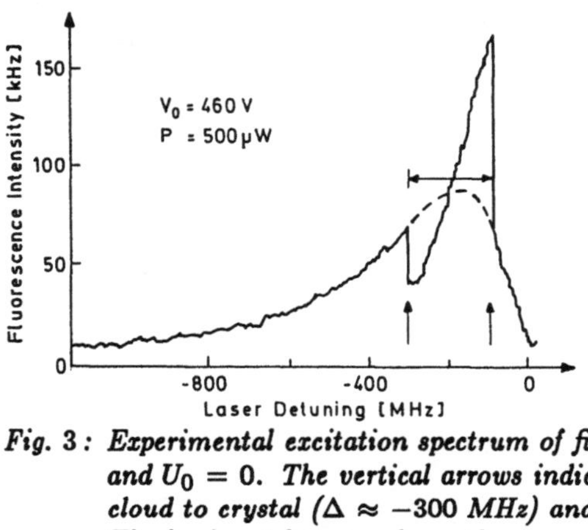

Fig. 3: Experimental excitation spectrum of five ions as a function of the laser detuning, Δ, and $U_0 = 0$. The vertical arrows indicate the detunings where phase-transitions from cloud to crystal ($\Delta \approx -300$ MHz) and from crystal to cloud ($\Delta \approx -100$ MHz) occur. The horizontal arrow shows the range of detunings in which a stable five-ion crystal is observed. The spectrum was scanned from left to right.

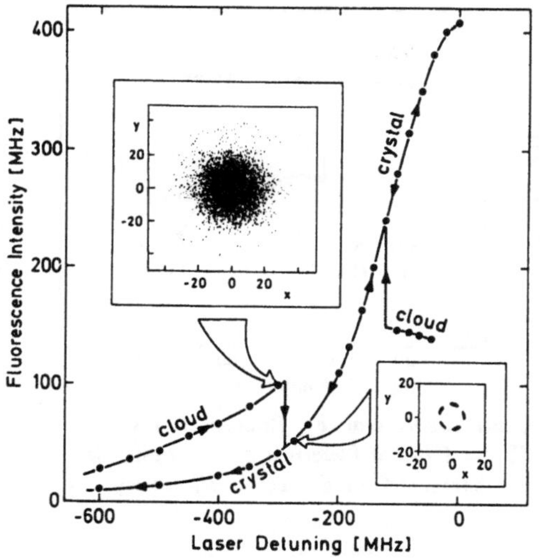

Fig. 4 : Theoretical excitation spectrum of five ions as a function of the laser detuning Δ *(*$U_0 = 0$*,* $V_0 = 780$ *V and* $P = 500\mu$ *W). The solid arrows on the fluorescence curves indicate the scanning direction. Two phase-transitions (and bistability) are apparent. The insets demonstrate directly the existence of a transition from a cloud phase to a crystalline state. They are the results of three-dimensional molecular dynamics calculations for detunings shortly before and shortly after the jump at* $\Delta = -300$ *MHz. The axes of the insets are in units of* μ*m.*

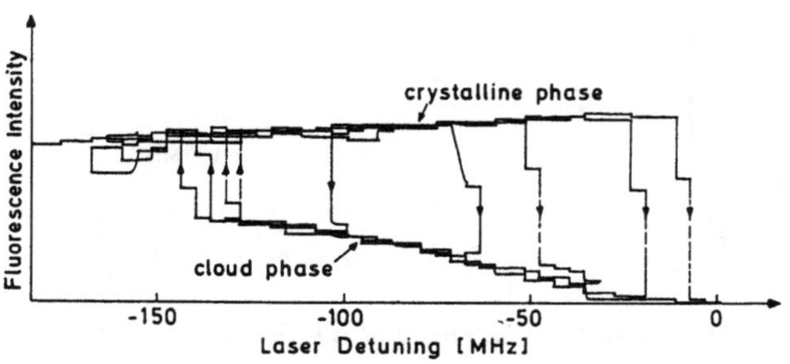

Fig. 5 : Experimental excitation spectrum of seven ions as a function of laser detuning for $U_0 = 3.5$ *V,* $V_0 = 240$ *V and* $P = 180$ μ *W.*

The behaviour of the ions in the trap is governed by essentially three parameters: the trapping voltage V_0, the laser detuning Δ and the laser power P. Hysteresis loops similar to the ones for Δ appear also whenever two of these three control parameters are kept constant and the third is swept. This is shown for the case of the rf voltage, V_0, in Fig. 6 and for the laser power, P, in Figures 7 and 8. The jumps from the cloud to the crystal are well reproducible in the experiment as well as in our simulations. The transitions from the crystal to the cloud state, i.e., the melting of the crystal occur only in the experiment (see Figures 5, 6a and 7), but not in our simulations (see Figures 4, 6b and 8). In our calculations the crystals are stable until the critical voltage corresponding to the instability condition of the Mathieu-equation is reached. For small crystals, the action of noise is necessary to bridge the energy gap between regular and cloud configurations in order to destroy the crystals. The crystal → cloud phase transition, therefore, is not a proper order → chaos transition, since in the absence of noise it

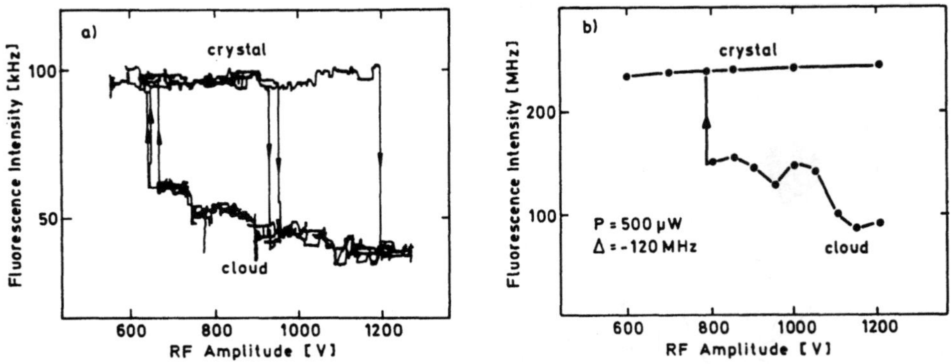

Fig. 6 : Comparison between the experimental, a), and theoretical, b), fluorescence intensity of five ions for identical trap parameters ($U_0 = 0$) as a function of rf amplitude. The experimental flourescence intensities are not corrected for detection efficiency.

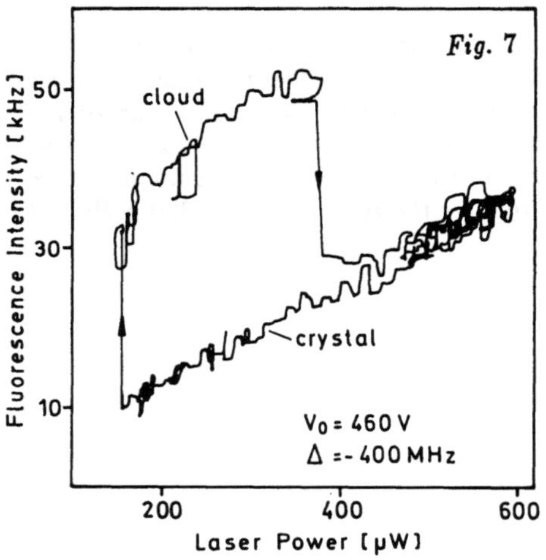

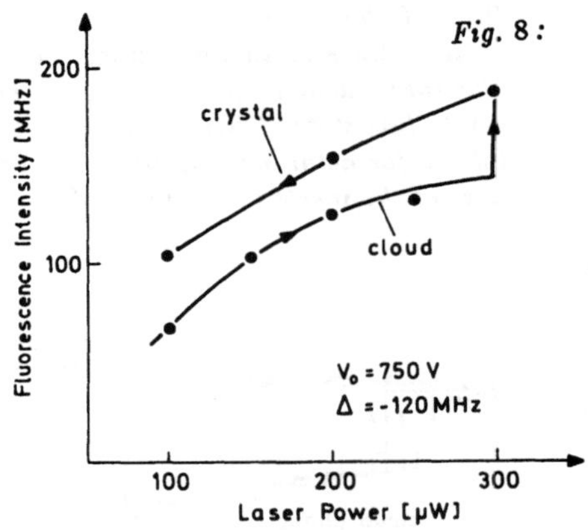

Fig. 7 : Experimental hysteresis loop in the fluorescence intensity of five ions as a function of the laser power ($U_0 = 0$).

Fig. 8 : Theoretical hysteresis in the fluorescence intensity of five ions as a function of the laser power ($U_0 = 0$). The values for V_0 and Δ differ from those of Fig. 7. Therefore the two curves cannot be compared directly.

would not occur. The experimental data show that the transition from crystal to cloud is not well defined and might be caused, e.g., by rf noise.

Motivated by our numerical simulations, we carried out a sequence of experiments in which we located very carefully the crystal → cloud transition points as a function of V_0 for several values of U_0. These experiments provide experimental evidence that ion crystals do not melt until the single particle Mathieu-instability is reached and the ions fall out of the trap [48, 56]. Therefore, both, our calculations and our experiments seem to be in conflict with the results of HOFFNAGLE et al. [26] who claim that the experimentally determined melting points of the crystals (i) are reproducible, (ii) happen well below the single particle Mathieu-instability, (iii) agree with their numerical calculations and (iv) are induced "when a bifurcation introduces a third frequency" [26]. We have not yet found any sign of such a bifurcation neither in our calculations nor in the experiments and in contrast to ref. [26] we are, in addition, led to the conclusion that it is the cloud → crystal transitions, i.e., the chaos → order transitions which are "real" and reproducible.

A possible explanation of the discrepancy might be that the authors of ref. [26] did not work with well defined initial conditions in their simulations and might not have payed enough attention to the sensitivity of ion crystals to the choice of initial phase of the rf driving field. Such a sensitivity is eliminated, if the rf field is switched on adiabatically, or V_0 is changed slowly from lower to higher values, an approach which we used in our experiments and implemented in our calculations. A further source of discrepancy concerns the way in which spontaneous emission noise was modelled in ref. [26]. The authors used displacement noise (see page 256 and 258 of ref. [26]), which generates high Fourier-components and leads to the melting of the ion crystals. Modelling spontaneous emission microscopically in the way outlined in the previous section as a stochastic *momentum* transfer, we did not observe melting of the crystals even when we enhanced the spontaneous emission rate artificially by a factor 5.

What is the time scale on which the cloud → crystal phase transitions occur? To study theoretically the dynamics of a particular jump, e.g., the one in Fig. 6b, we analyzed the fluorescence as a function of time when we suddenly switch the rf field from $V_0 = 800$ V (cloud state) to 750 V (crystal state). After a transient of about 0.5 ms, in which the fluorescence still shows large fluctuations, typical for cloud states, the fluorescence settles to its crystalline value, characterized by negligible fluctuations as depicted in Fig. 9. Experimental time-resolved phase transitions are shown in Fig. 10. Here it should be noted that Fig. 10a) and b) differ in the value of the detuning which results in quite a different dynamical behaviour.

The experimental results described above represent only a small fraction of the entire body of data accumulated. Ion-crystals, clouds and phase transitions of two to approximately 100 ions have been observed and recorded.

We used the most basic ion-crystal – consisting of only two ions – to investigate collective motions of the ions. For instance, such a two-ion crystal has two kinds of normal modes of vibration: i) three center-of-mass oscillations with secular frequencies $\omega_x = \omega_y$ and ω_z, where the distance between the ions is fixed and, ii) a breathing mode leaving the center-of-mass at rest, while the two ions vibrate against each other with an approximate frequency of $\sqrt{3}\cdot\omega_{x,y}$ or $\sqrt{3}\cdot\omega_z$, where the frequency is chosen according to the alignment of the crystal determined by U_0. When we apply an additional ac voltage of about 15 mV amplitude to the ring electrode,

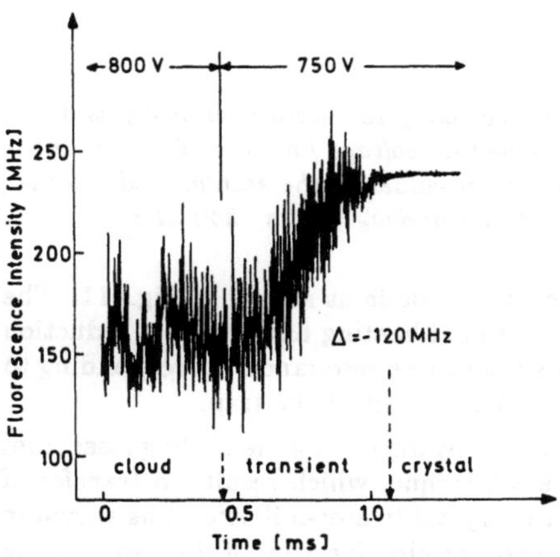

Fig. 9: Time resolved dynamics of the phase-transition of Fig. 6b) (with identical trap parameters) extracted from three-dimensional molecular dynamics calculations. Horizontal arrows indicate the time periods over which the rf voltage, V_0, has the value of 800 V and 750 V, respectively.

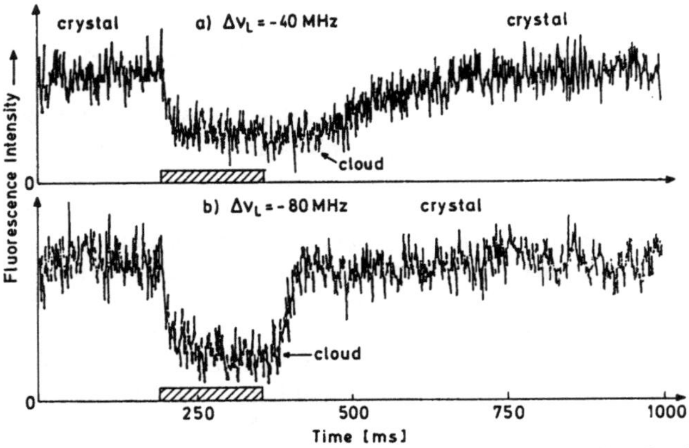

Fig. 10 : Time resolved dynamics of an experimentally obtained phase-transition of three ions. $U_0 = 0$, $P = 300\mu W$ and a): $\Delta = -40$ MHz, b): $\Delta = -80$ MHz. The rf voltage V_0 is suddenly switched from its value 140 V (crystal) to 420 V (cloud) and back to 140 V (crystal). The period over which the rf voltage of 420 V is applied is marked by the hatched rectangles.

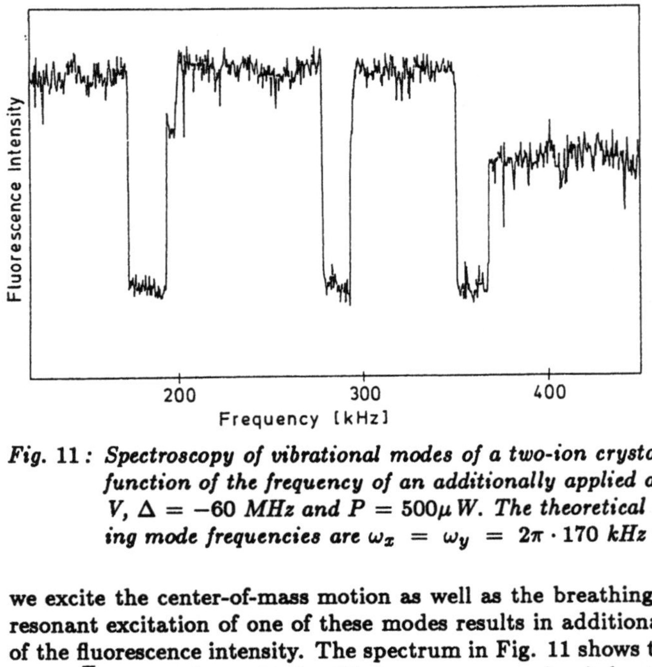

Fig. 11 : Spectroscopy of vibrational modes of a two-ion crystal: fluorescence intensity as a function of the frequency of an additionally applied ac voltage for $U_0 = 0$, $V_0 = 165$ V, $\Delta = -60$ MHz and $P = 500\mu W$. The theoretical values of the secular and breathing mode frequencies are $\omega_x = \omega_y = 2\pi \cdot 170$ kHz and $\omega_z = 2\pi \cdot 340$ kHz.

we excite the center-of-mass motion as well as the breathing mode as shown in Fig. 11. The resonant excitation of one of these modes results in additional heating and causes a reduction of the fluorescence intensity. The spectrum in Fig. 11 shows three resonances corresponding to $\omega_{x,y}$, $\sqrt{3} \cdot \omega_{x,y}$ and ω_z. At the third resonance one ion is heated out of the trap.

A collective rotation of the two-ion crystal can be induced by displacing the cooling laser from the center of the crystal. The radiation pressure causes a torque, which results in transfer of angular momentum [57] and a subsequent rotation of the crystal (water-mill effect) as shown in Fig. 12. The resulting centrifugal force increases the size of the circular orbit of the two rotating ions. The diameter of the orbit can be changed over a wide range by varying the efficiency of angular momentum transfer controlled by the scattering rate via laser displacement and detuning. The transfer of angular momentum and the corresponding increase in the diameter of

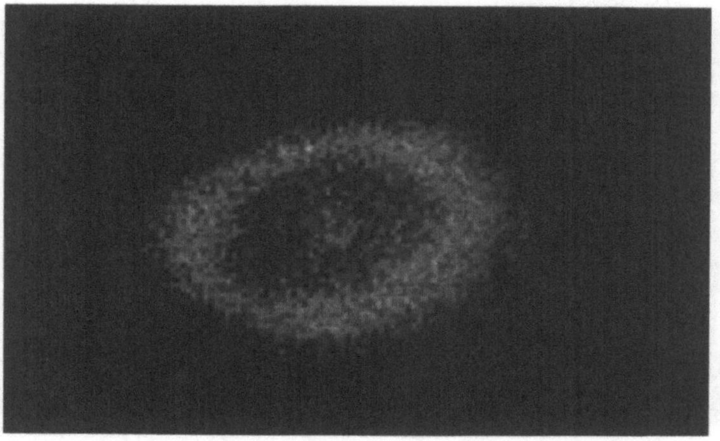

Fig. 12 : Rotating two-ion crystal for $U_0 = 5$ V, $V_0 = 210$ V, $P = 240\mu W$ and $\Delta = -120$ MHz. The integration time of the imaging system was three minutes. The semi-major axes of the ellipse are 24μm and 17μm, respectively. Towards the end of the observation period, the crystal "exploded" and one ion left the trap. The partner is cooled to the centre of the trap and gives rise to the spot in the middle of the ellipse.

the ions' orbit make phase-space domains, which lead to cloud states, accessible. Consequently, for sufficiently large diameter, the collective rotation of the two ions turns into an uncorrelated erratic motion, characteristic for a cloud.

Even the physics of a single ion still has surprises in store. In Fig. 13 we show a single ion exhibiting a stable large amplitude vibration orthogonal to the cooling laser beam. According to equation (8), there is only a cooling effect in $\vec{k}$-direction. Motion orthogonal to this direction is damped only in a minor way due to the shape of the laser focus, the presence of magnetic fields, contact potentials or trap imperfections which couple the different degrees of freedom. Probably the large amplitude motion is stable because this little cooling is counter-balanced by the stochastic heating mechanism of the spontaneously emitted fluorescence photons. When we increase the rf voltage, the amplitude of the vibration reflected in the length of the white line (integrated fluorescence signal) in Fig. 13 reduces until it suddenly collapses to a dot representing the ion sitting at the center of the trap. When we now decrease the voltage again to its original value, the large amplitude vibration reappears.

The large amplitude motion can only be observed for a single ion, since with many ions the Coulomb interaction results in a coupling of the vibrational modes and therefore in an effective cooling in all directions. Multistability of a single ion was recently reported by SAUTER et al. [58].

The cooling effect of the laser is balanced by a heating mechanism. This heating mechanism of the ions by the rf field is known [31] since the early days of rf traps. The understanding of this phenomenon, however, remained incomplete.

In order to demonstrate the occurrence of rf heating, we have loaded approximately 100 ions into the trap. At time $t = 0$ the ions form a stable cloud, in equilibrium with the cooling laser. At $t \approx 120$ ms the cooling laser is switched off; after 300 ms the laser is turned on again. The fluorescence data of 60 such experiments have been averaged and the result is shown in Fig. 14. The initial low fluorescence at 500 ms, shortly after the laser was turned on again, indicates that the ions have been heated. The cloud is cooled down again to its initial temperature during the time interval 420 ms $< t <$ 800 ms. Since the heating of the cloud occurs during the dark phase of this experiment, it can only be due to the rf field, i.e. it must be an inherent property of the Paul-trap. This heating mechanism will be discussed in detail in the next section.

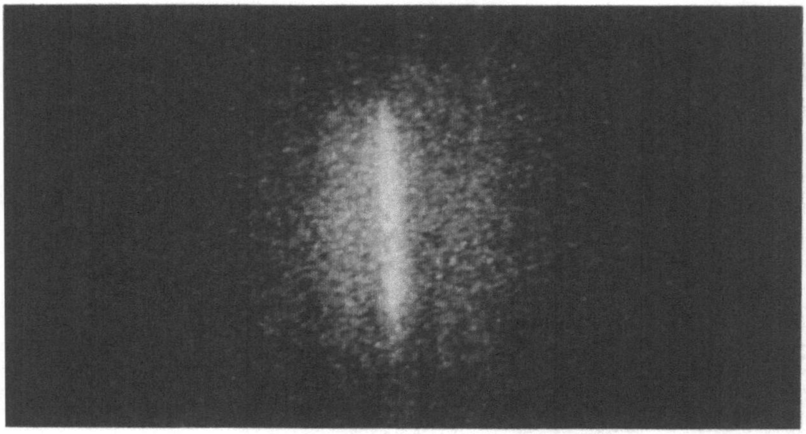

Fig. 13 : Large amplitude vibration of a single ion (maximal excursion from the trap-centre is 60μm) for $U_0 = 0$, $V_0 = 410$ V, $\Delta = -100$ MHz and $P = 220\mu W$. Increasing the rf voltage to $V_0 = 630$ V, the line collapses to a dot. Decreasing V_0 back to its original value, the large amplitude motion reappears.

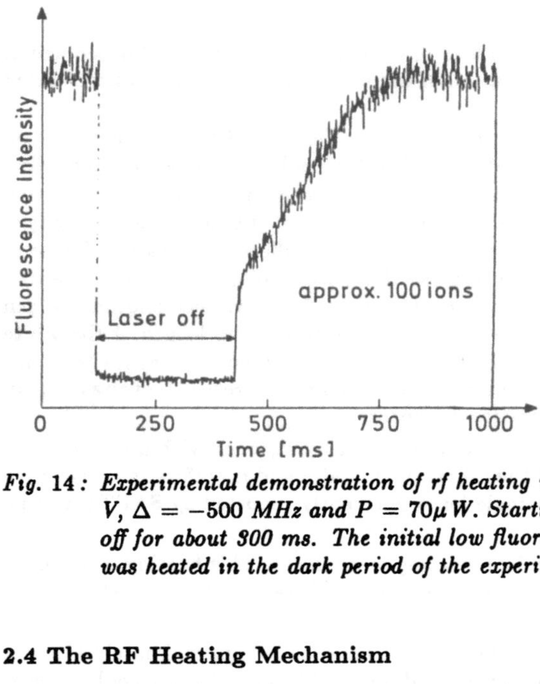

Fig. 14 : Experimental demonstration of rf heating with ≈ 100 ions for $U_0 = 3$ V, $V_0 = 180$ V, $\Delta = -500$ MHz and $P = 70\mu W$. Starting in the cloud phase the laser is switched off for about 300 ms. The initial low fluorescence at $t \approx 500$ ms proves that the cloud was heated in the dark period of the experiment.

2.4 The RF Heating Mechanism

For quite some time it was thought that an ion cloud under ideal conditions cannot absorb energy from the driving rf field. This opinion was clearly expressed by DEHMELT when in 1967 ([31], p. 65) he wrote: "... we expect that in a perfect vacuum and in a trap geometrically and electrically perfect a cloud of identical ions would not be subject to any rf heating." The premise of his statement was that certain "long-range" effects were assumed to be unimportant in an ion trap, but we will show in the sequel that the heating really results from rather close collisions of the ions and is indeed due to the rf field only.

There is an interesting analogy: Fermi's proposal for the origin of high energy cosmic radiation. In a seminal paper [59], dating from 1949, he tried to explain the large energy content of cosmic radiation by a mechanism of stochastically moving intergalactic magnetic fields. He showed that the external stochastic component of this acceleration mechanism is in fact not needed, and investigated a model for the heating of cosmic rays, which became known as "Fermi acceleration". We will show below that the mechanism of rf heating is not unlike the Fermi acceleration mechanism since both rely on the occurrence of deterministic chaos [60, 61] as the crucial condition for energy transfer and heating.

Does a one-dimensional string of ions, taking into account only $F^{(trap)}$ and $F^{(Coul)}$ in equation (11), reveal the origin of the rf heating mechanism ? To answer this question, we have monitored the average kinetic energy of two to five ions as a function of time in such a one-dimensional trap. For less than five ions we could not observe any gain in energy over several tens of milliseconds. For five ions a slow increase in energy was recorded, alas too small to compete with laser cooling. A Fast Fourier Transform of the positions $x_n(j)$ of the n-th member of the one-dimensional ion-chain taken at time $t = jT$

$$P_k^{(n)} = \frac{1}{\sqrt{N}} \sum_{j=0}^{N-1} x_n(j) \exp\left[-i\frac{2\pi}{N}jk\right] ; \quad k = 0, 1, ..., N-1 \tag{13}$$

($N = 2048$ in the present case) shows a small number of discrete frequencies dominating the spectrum and thus qualifying this model as being close to integrable and lacking a heating mechanism. Although this chain of particles cannot be used to explain the heating mechanism in a Paul-trap, it is interesting in its own right and closely related to the integrable chain of particles interacting by two-body harmonic and $\frac{1}{x^2}$ - potentials (Calogero-Moser system [62, 63]).

Two-dimensional ion-traps behave essentially like one-dimensional traps if the motion is restricted to the $x - y$ - plane. In the $x - z$ - ($y - z$ -) plane, however, strong heating occurs. In order to study this observation in more detail, we have calculated the power-spectrum $|P_k|^2$ of the positions of two ions. When we choose an initial separation of the ions which is about twice the equilibrium separation of the two-ion crystal, a set of discrete frequencies appears, like in the left power spectrum in Fig. 15. In this case, the ions perform a quasi-periodic motion. They are unable to extract energy from the rf field and due to the cooling laser they eventually end up in the crystalline state. Such a power-spectrum characterizes phase space domains which act as basins of attraction for the crystal. However, when we choose initial conditions, which correspond to typical separations in a cloud state, the spectral power of any one of the two ions shows continuous bands in frequency, similar to the power spectrum in the middle of Fig. 15. This provides evidence for the occurrence of deterministic chaos [60] in the cloud phase.

In order to investigate the heating quantitatively, we calculated the work done by the rf field per unit time (angular brackets denoting ensemble averages)

$$\begin{aligned} \kappa &= f < \sum_{i=1}^{n} \int_{1cycle} \vec{F}_i^{(trap)} \cdot d\vec{s}_i^{\,(ion)} > \\ &= f < \sum_{i=1}^{n} \int_0^T \vec{F}_i^{(trap)} \cdot \vec{v}_i^{\,(ion)}\, dt > \end{aligned} \tag{14}$$

for the case $n = 5$ ions, $V_0 = 780$ V, $U_0 = 0$, $f = 11.25$ MHz and $\Delta = -300$ MHz as a function of the rms-radius $r = \sqrt{\overline{r^2}} =< [\sum_{i=1}^{n} \vec{r_i}^{\,2}]^{1/2} >$ of the five ions. The rms-radius was controlled indirectly by the laser power P which we varied from 10 μW to 150 μW to obtain the heating curve [48, 64] displayed in Fig. 15.

For zero laser power and large r, we did not observe any net heating of the ions. This is confirmed by our experiments, in which, even in the absence of a cooling laser, large clouds of ions can be stored in a Paul-trap over several hours. The ions are far apart from each other, the Coulomb force is small and on short time scales the ions behave essentially like independent

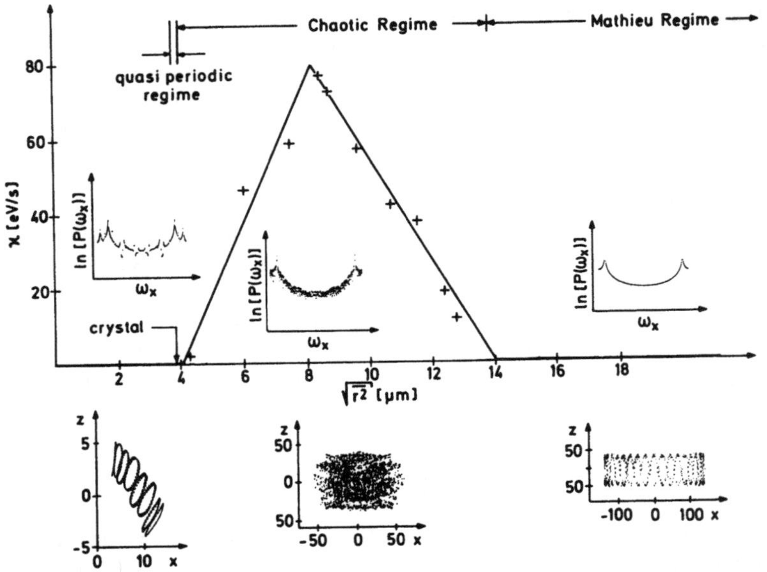

Fig. 15 : Average heating rate κ of five ions in a Paul-trap versus the rms radius of the ion configuration. The insets show the power spectrum and the corresponding stroboscopic Poincaré sections in the x − z plane of relative separation for two ions in three characteristic domains: the quasi periodic regime, the chaotic regime and the Mathieu regime. All length scales are in units of μm.

single stored ions. For this reason, we called this part of the heating diagram the "Mathieu regime". Turning on a small laser power, the rms radius r reduces drastically, but comes to a halt at about 14 μm where chaotic rf heating sets in and balances the cooling power of the laser. Increasing the laser power, results in an even smaller cloud. The smaller cloud produces more chaotic rf heating as seen clearly by the negative slope of the heating curve (Fig. 15) in the range $8\mu m < r < 14\mu m$

In the range $4\mu m < r < 8\mu m$ there is still chaotic heating but the slope of the heating curve is positive. As a consequence of the resulting triangular shape of the heating curve, at about $P = 150\mu W$, corresponding to $r \approx 8\mu m$, the chaotic heating power can no longer balance the cooling power of the laser and the cloud collapses into the "crystalline" state located at $r \approx 3.8\mu m$. A "phase transition" has occurred. Due to this collapse, the behavior of the heating rate in the range $3.8\mu m < r < 8\mu m$ cannot be studied by equilibrating laser cooling and rf heating. In this case we start out from the crystal state and slightly displace the ions to explore the "vicinity" of the crystal. We observe no heating for $3.8\mu m < r < 4\mu m$, but quasi periodic motion, and thus dubbed this regime the "quasi periodic" regime. We call the upper edge of the quasi periodic regime ($r \approx 4\mu m$) the "chaos threshold". An initial condition beyond the chaos threshold, i.e satisfying $r > 4\mu m$, leads to heating and expansion of the ion configuration and numerical data relevant for the shape of the heating curve can be taken during this "explosion phase". The laser power P is set to zero for this type of experiment. We conjecture that, apart from the trivial case of a single stored ion, the heating curve is "universal", i.e., its qualitative shape, including the existence of the chaotic regime, does not depend on the number of simultaneously trapped ions.

For the quasi periodic, the chaotic and the Mathieu-regime, repectively, we display the corresponding type of power spectrum as the insets above the abscissa of Fig. 15. The data were actually taken for the case of two ions, but they would not be very much different in the five ion case. We obtain a discrete spectrum in the quasi periodic regime and a complicated noisy spectrum in the chaotic regime. The spectrum in the Mathieu regime is again quite simple and dominated by the secular motion frequency. We also show stroboscopic pictures of the locations of the ions in the $x-y$-plane of the trap characterizing the three regions (insets below the abszissa of Fig. 15).

As a function of increasing rf voltage, the chaos threshold moves toward the radius of the crystalline configuration. However, before the chaos threshold reaches the crystal radius, the equations of motion (2) become unstable in z-direction, indicating that in this particular situation the particles would fall out of the trap. In order to achieve proper melting without loosing particles, the crystal radius has to be enlarged artificially by noise so that the size of the distorted crystal overlaps with the region of chaotic heating. Proper melting without the assistance of noise should be possible if we start in a quasi-periodic state typified in the bottom left inset of Fig. 15. Such configurations have a larger radius to begin with and the chaos threshold could be reached before the single particle Mathieu-instability sets in.

2.5 Sympathetic Cooling

Examining the excitation spectrum of the chaotic phase as a function of the laser detuning one finds, surprisingly, a finite signal for positive detunings (Fig. 16a). Naively one would expect that the fluorescence intensity has to vanish in the blue part of the spectrum because the ions in this case are heated instead of cooled by the radiation pressure force and would leave the trap. The non-zero signal at positive detunings is explained by the mechanism of sympathetic cooling. The natural isotope ratio of magnesium ^{24}Mg : ^{25}Mg : ^{26}Mg is 8:1:1 and therefore, in an ion cloud of ≈ 50 ions there are on the average ≈ 5 of the ^{26}Mg species. Due to the isotope shift, the resonance transition frequency for the $3S_{1/2} \leftrightarrow 3P_{3/2}$ dipole transition is approximately 3 GHz higher than for the ^{24}Mg ions (Fig. 16a). To understand the mechanism of laser cooling and heating in the presence of two isotopes, we distinguish three ranges of the detuning. In the first range, i.e., for negative detunings with respect to the ^{24}Mg resonance frequency, both ion species, the ^{24}Mg ions and the ^{26}Mg ions, are cooled. In the second regime, i.e., the laser frequency is shifted to the blue side of the ^{24}Mg resonance frequency, the ^{24}Mg ions are heated by the laser light as well as by the rf, but the ^{26}Mg ions are laser cooled, since for this species the laser frequency is still detuned to the red side of the transition frequency. Although fewer in number, these ions can cool the ^{24}Mg ions by exchange of energy in ^{24}Mg-^{26}Mg collisions, resulting in a non-zero fluorescence for positive detunings. In the third regime, finally, i.e., for positive detunings with respect to the ^{26}Mg resonance frequency, both isotopes are heated and no fluorescence signal is observed because the cloud size becomes much larger than the laser focus and the ions will finally be heated out of the trap (Fig. 16a). This idea is confirmed by our experiments which show that the cloud size increases rapidly as the resonance frequency is approached from the lower frequency side.

To check the picture of sympathetic cooling, the relative contributions of the two different Mg isotopes to the total fluorescence intensity is analyzed in a measurement of the intensity correlation function $G^{(2)}(\tau) \equiv < I(t) \cdot I(t+\tau) >$. Due to the Doppler effect, the secular center of mass motion of the cloud causes a temporal modulation of the fluorescence intensity which shows up in the correlation function. Note that this is a non-destructive way to measure the secular frequency as no additional force is applied to the trap to excite the secular motion of the ions. The finger prints of the two isotopes are clearly distinguished in the correlation function because the secular period is proportional to the mass of the ion. Of course this is only valid in the cloud state where the motion of the ions is chaotic and uncorrelated whereas in the crystalline phase the secular period of the center of mass motion is proportional to the average mass of the constituents of the crystal.

The results of correlation measurements for the three different values of the detuning Δ_1, Δ_2 and Δ_3 marked in the spectrum of Fig. 16a are depicted in Fig. 17a-c , respectively. With the negative detuning Δ_1 the ^{24}Mg ions are directly laser cooled and no contribution of the

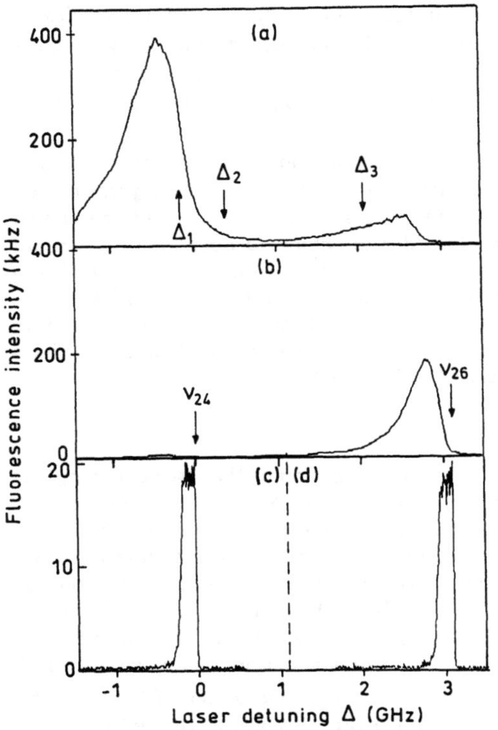

Fig. 16 : Excitation spectra of an ion cloud demonstrating sympathetic cooling and isotope separation. In (a) two peaks due to the ^{24}Mg and ^{26}Mg isotopes in their natural abundance are visible. Spectrum (b) was recorded after isotope separation by radiation pressure. Spectra (c) and (d) are of single ^{24}Mg and ^{26}Mg ions respectively for calibration of the resonance frequencies.

^{26}Mg ions is discernible in the corresponding $G^{(2)}$-measurement (Fig. 17a), because (i) the laser frequency is shifted too far from the ^{26}Mg resonance and (ii) there are much fewer ^{26}Mg ions than ^{24}Mg ions present. In this curve the third harmonic of the secular frequency is present. This is due to the nonlinearity of the excitation spectrum, which acts as the transfer function for the temporal modulation of the fluorescence intensity. Choosing a laser frequency between the resonance frequencies of the two isotopes (detuning Δ_2 in Fig. 16a), the expansion of the ^{24}Mg cloud due to laser heating is only stopped by the sympathetic cooling through collisions of ^{24}Mg ions with ^{26}Mg ions. Consequently the total fluorescence intensity is expected to be composed of the contributions of both isotopes. The corresponding correlation measurement (Fig. 17b) exhibits two different periods with a ratio of 1.08, equal to the mass ratio of the ^{24}Mg and ^{26}Mg ions in agreement with the mass dependence of the secular frequency in the pseudo potential. Finally, for the laser detuning Δ_3 (Fig. 16a), the correlation measurement (Fig. 17c) shows only the secular period of the ^{26}Mg ions. This is due to the fact that the contribution from the ^{24}Mg ions can be neglected since the detuning is too large for this isotope species.

The isotope dependent laser cooling and heating mechanism described above can also be applied to selectively increase the kinetic energy of one ion species. By fixing the laser frequency at the ^{26}Mg resonance frequency the ^{26}Mg ions are neither heated nor cooled whereas the ^{24}Mg ions are strongly heated due to the large positive detuning with respect to the ^{24}Mg resonance frequency. After a few seconds nearly all ^{24}Mg ions gain enough energy to leave the trap as it is

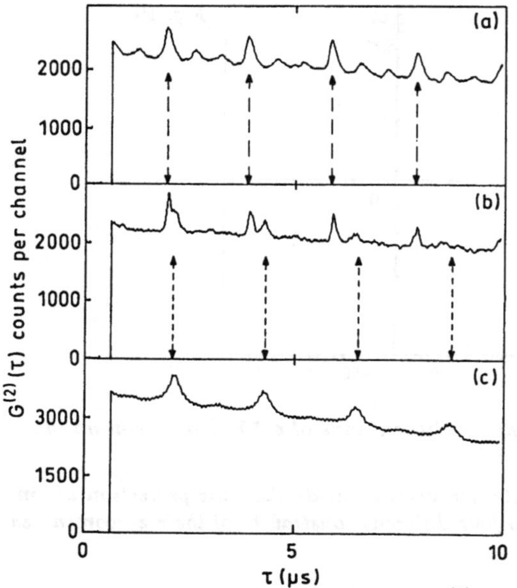

Fig. 17 : Intensity correlation functions $G^{(2)}$ of the ion cloud in Fig. 16a. for three different laser frequencies. (a): Detuning Δ_1 in Fig. 16a, (b): Δ_2, (c): Δ_3.

seen in a second recording of the excitation spectrum which was taken right after the selective heating process (Fig. 16b). Evidently there is practically no ^{24}Mg ion left in the trap and the spectral width of the ^{26}Mg ions became narrower because the ^{26}Mg ions are not heated any more by the ^{24}Mg ions. Therefore, irradiating a mixture of ^{24}Mg and ^{26}Mg ions with a UV laser tuned to the ^{26}Mg resonance transition defines a very effective procedure for laser isotope separation in the Paul-trap.

2.6 Structure and Dynamics of Many-Ion Crystals in a Paul-Trap

As the last section of chapter 2 we present theoretical results on the crystallization of up to 500 particles in a Paul-trap. We integrated the equations of motion (11) for 30, 100, 200 and 500 ions taking the dynamical trapping force $\vec{F}^{(trap)}$, the Coulomb force $\vec{F}^{(Coul)}$ and an isotropic damping force $\vec{F}^{(damp)} = -\kappa\vec{v}$ into account. The initial positions and velocities of the ions were chosen at random and calculations were performed for two different sets of initial conditions. Initially the ions formed a cloud, which was strongly heated. Therefore a large damping was necessary only achievable with a laser power of about 5 mW. As soon as a quasi crystalline configuration was reached, we slowly reduced the damping constant until we finally reached about 500 μW, a value which is well within the experimental reach. We note that the reduction of the damping constant has to be done quasi adiabatically, since otherwise the ions fall out of phase with the driving rf field and the crystal melts. For the 500 ion crystal ($U_0 = 0, V_0 = 700$ V, $f = 11.25$ MHz) and one set of initial conditions the projections of the particle positions on the $x - y$ and $x - z$ - planes are shown in Fig. 18. Although we stopped the calculations long before the ions finally settled to their equlibrium positions (as noted already in ref. [35], the crystallization process is a "painfully long and costly procedure"), the $x - z$ projection (Fig. 18) indicates that the 500 ion crystal will finally form a "layered" structure. In order to present more clearly the internal structure of these layers, Fig. 19 shows the projections of particle positions for ions satisfying $-2\mu\text{m} \leq z \leq 2\mu\text{m}$ (Fig. 19a), $2\mu\text{m} \leq z \leq 6\mu\text{m}$ (Fig. 19b), $6\mu\text{m} \leq z \leq 8.5\mu\text{m}$ (Fig. 19c), and $8.5\mu\text{m} \leq z \leq 15\mu\text{m}$ (Fig. 19d). The structure of the layers is now clearly recognizable.

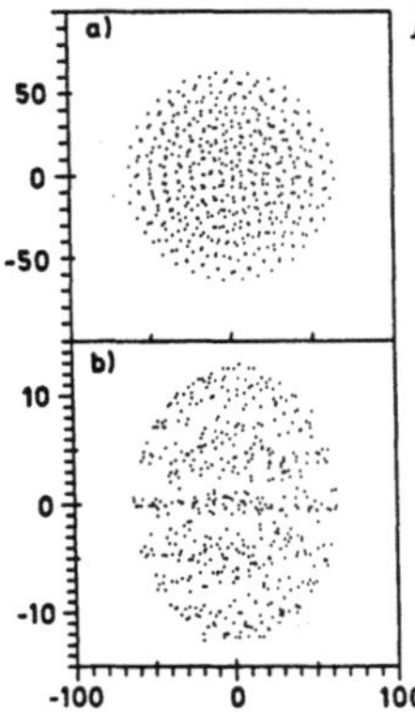

Fig. 18

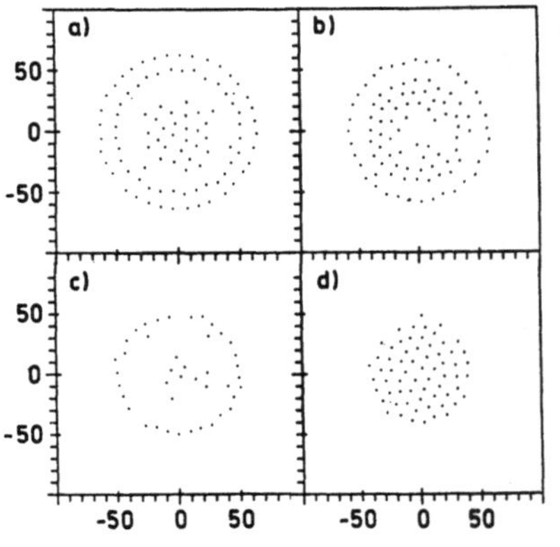

Fig. 19

Fig. 18: (Left) Projection of the positions of the constituent ions of a 500 ion crystal on the $x-y$-plane (a) and on the $x-z$-plane (b).

Fig. 19: (Right) The layered structure of the 500 ion crystal. a)-d) show the projection of ion positions to the $x-y$-plane subject to four different constraints of their z-coordinates (see text).

Crystallization in layers parallel to the $x - y$ - plane is not unreasonable, since without DC voltage ($U_0 = 0$), the Paul-trap potential is steeper in the z-direction than in the radial direction. Perhaps a shell structure [27] forms in case we apply an additional DC voltage to the trap ($U_0 \neq 0$) which results in a more spherical pseudo-oscillator.

Of course one cannot expect that an ideal crystal would form for random initial conditions. There are many possible stationary ion configurations corresponding to local minima of the total crystal energy and the probability to settle down in the absolute minimum is very small. A careful look at Fig. 19 reveals that the 500 ion crystal indeed "grew" with many imperfections. A more perfect crystal might result when growing the crystal from a small seed by injecting particles in the course of the calculation. Such a procedure is also advisable in case there is not enough cooling power available and we will come back to this point in section 3.3.

As a last numerical experiment we started in the "crystalline" configuration of Fig. 18 and switched off the cooling power altogether. We monitored the total kinetic energy of the 500 ions in the trap as a function of time for 300 rf cycles. Fig. 20 shows that the energy fluctuates wildly on the time scale of one rf cycle, but averaging the crystal energy on ± 15 cycles (interpolating line in Fig. 20) shows that there is no net gain of energy on a time scale of several hundred rf cycles. This result is surprising, since the 500 ions are still far from a stationary arrangement. We conclude that the qualitative picture of ion heating developed in the previous section still holds for such large particle numbers, and especially that even in this case there still exists a

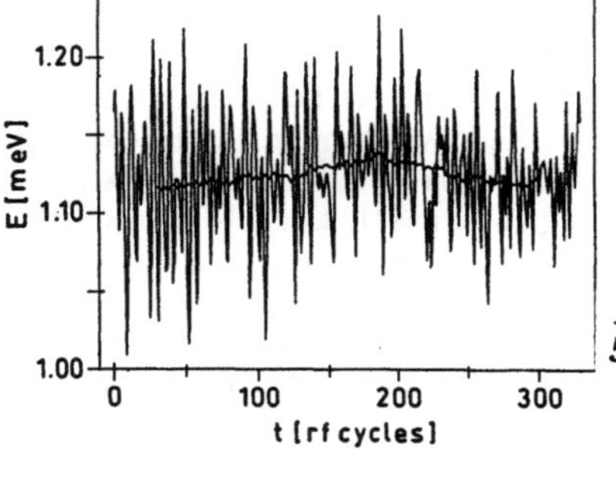

Fig. 20: Time evolution of the kinetic energy of the 500 ion crystal for zero damping.

non-heating quasi periodic dynamical phase. We think that this result is very significant since it means that even such large crystals, consisting of several "layers" do not show much heating. It is possible that this result can be generalized to the case of crystalline beams in which case a crystallized beam tube consisting of several shells also shows negligible heating.

3. Implications for Crystallization in Storage Rings

In the following sections we present some ideas on how our experience with Paul-traps might be helpful to predict the behaviour of ions in storage rings.

3.1 Focusing and Drift Sections

It has been argued (see, e.g., [35]) that the modular lattice of existing storage rings presents a severe problem for obtaining crystals in these rings. The Coulomb crystal, once generated, would be subjected to periodic squeezing and pulling in the focusing sections, to Coulomb explosion in the drift sections and shear forces in the bending sections of the storage ring. While in this paper we will not touch on the problem of shear forces [41,42], we can present some encouraging evidence that focusing and drift sections do not in principle forbid the formation of a Coulomb crystal. For this purpose, we loaded five particles into our numerical trap. In contrast to the calculations presented in previous sections, we did not apply this time a purely harmonic rf voltage, but a pulsed voltage, which was switched on for one cycle ($\approx$ 100 ns) and switched off for N cycles, where N was ranging from 1 to 100. Thus, one period of the applied voltage consists of $N+1$ rf cycles, where the first cycle is "active", and models the focusing by quadrupole lenses, and N cycles are "silent", modelling the Coulomb explosion in the drift sections. The result of the calculation(V_0 = 700 V, P = 500 μW, Δ = 300 MHz) was that stable five ion crystals exist up to $N \approx 50$, i.e., the drift sections could be up to 50 times longer than the effective length of the focusing sections. The diameter of the ion-crystals was found to grow substantially with N. We checked that even if the "silent" period was not an integer multiple of the underlying rf period $2\pi/f$, the crystals persisted, thus excluding the necessity of any sort of "phase-matching".

These results support the claim [35] that at least for "simple" Coulomb crystals in a storage ring the drift and focusing sections should be no severe obstacle for the formation of Wigner crystals in storage rings.

3.2 Beam Diagnostics

Owing to the existence of strong focusing sections in storage rings a close analogy exists between the crystallization problem in storage rings and in dynamical Paul traps. Therefore the "heating curve" for laser cooled ions in a storage ring resembles closely the one presented in section 2.4. Based on the discussion of the heating curve for Paul traps, where in the present case the beam radius has to be substituted for the radius of the ion-configuration, we propose two possible diagnostic tools. (i) As a function of the laser cooling power (or laser detuning) the typical *jumps in the fluorescence intensity* (see Figs. 3-8) should be recorded, as soon as the cooling power exceeds the top of the heating curve, whereupon the beam should show a *sudden collapse* into the crystalline state. (ii) Switching off the laser- (electron-) cooling, a non-crystalline ion-beam should show heating which starts instantaneously. If the beam was crystallized, it shoud show *delayed heating*, i.e., the crystal persists for a while and only after a certain delay time, strong heating should be observed. For a machine, which consists only of focusing and drift sections, crystallized beams composed of only a few shells are expected to live on forever. (iii) In this connection we propose *Coulomb explosion* [65] as a third diagnostic method. When turning off the focusing forces, the beam will explode due to the repulsive Coulomb forces. If the beam consists of shells, as suggested by several authors [35,38–42] and by recent experimental results in the Penning-trap [27], the different shells will hit the walls of a cylindrical electrode around the beam at different times if the ions are given enough time to reach these electrodes after their explosion. Due to the relatively long way to the electrodes, the distance between the shells can be determined in a time-of-flight measurement as is often done for investigation of exploding small molecules (see e.g. [65]). We are aware of the problem that many body effects might smear out the shell structure during the explosion phase. Model calculations addressing this point are in progress.

3.3 Loading of the Ring

It is usually argued that the number of ions which can be crystallized in a Paul-trap (or a storage ring) is limited because the laser cooling power is not sufficient to overcome the peak of the many ion heating curve (see, e.g., Fig. 15). On the other hand, the shape of the heating curve suggests to load a trap or a ring sequentially. It may be possible that a small number of ions can serve as the seed for a larger crystal, which will be grown by aggregation. This way one can produce large crystals with L particles, even if the laser power is not enough to exceed the top of the L-particle heating curve. This is so, because the seed crystal, consisting of S particles, is not heating and the top of the $(L-S)$ - particle heating curve is lower than the top of the L-particle heating curve. The question remains, to which extent the added ions perturb the already existing crystal.

3.4 The Plasma-Parameter

As mentioned in the introduction, a one-component plasma in thermal equilibrium crystallizes for $\Gamma \approx 170$. It is known, however, that the naive use of the Γ-parameter in Paul-traps and storage rings is problematic since in both devices a large fraction of the kinetic energy is not thermal, but represents the kinetic energy of a few regular degrees of freedom only, as, e.g., the secular motion or the micromotion in the trap, or betatron oscillations in storage rings. To illustrate the problem, we have calculated the plasma-parameter for the transition shown in Fig. 6 as the ratio of the time averaged potential and kinetic energies of the 5 ions. Shortly before the transition occurs ($V_0 = 800$ V), i.e., in the cloud phase, we find $\Gamma \approx 0.02$. In the crystal phase we obtained $\Gamma \approx 2$ and it is easy to show analytically that due to the micromotion this is about the upper limit for Γ. Fourier-analyzing the motion of the ions, the periodic components can be subtracted easily, which permits a calculation of the plasma-parameter in which only the kinetic energy arising from the "random" motion of the particles will enter.

Summary and Conclusions

In conclusion we have shown that ions confined in an ion-trap can be found in two modifications: in a cloud phase and in a crystalline phase. Phase transitions characterized by a discontinuous change in the fluorescence intensity can be induced by appropriately adjusting the rf voltage, the detuning and the power of the cooling laser. The transitions show hysteresis as a function of these three essential control parameters. With the help of three-dimensional molecular dynamics calculations, we reproduced qualitatively all the essential features in recent experiments, such as the size of ion clouds and crystals as well as transitions between these two phases. The dynamics of the phase transitions was explored experimentally as well as numerically via time resolved fluorescence spectra recorded for the transients of specific phase transitions. In detailed studies of one-, two- and three-dimensional simulations of ions in a Paul-trap, we have shown that depending on the radius of the ion arrangement, the dynamics of ions in a Paul-trap can be classified into four dynamical regimes: (i) The crystal, (ii) the quasi periodic regime, (iii) the chaotic regime and (iv) the Mathieu regime. The crystal, the quasi periodic regime and the Mathieu regime are the non-heating regimes. Strong heating occurs only in regime (iii) by a diffusive gain of energy due to the occurrence of deterministic chaos. The mechanism of rf heating, therefore, relies only on the deterministic features of the system, rather than on the effect of external randomness such as noise in the amplitude and frequency of the rf trapping field or collisions with the atoms of the background gas. The occurrence of chaos has the effect that in the chaotic regions of phase-space (clouds) the trajectories of molecular dynamics calculations are exponentially unstable and great care has to be taken to suppress "energy diffusion" which is generated by numerical errors which, by virtue of the exponential separation of initially close trajectories, are unavoidable in this unstable regime.

Whereas spontaneous emission of photons is the only source of heating for single ions, chaotic rf heating outweighs the heating due to spontaneous emissions by orders of magnitude in the many ion case.

The static as well as the dynamic features of laser cooled ions in a Paul-trap is a very promising subject for the study of few-body non-equilibrium phenomena. With the help of present-day powerful imaging systems, crystals, clouds and phase transitions between them can be studied as a function of particle number.

In connection with recent developments in accelerator and storage ring physics, we have shown that the Paul-trap is also a valuable tool for studying specific sections of the machine lattice. Modelling strong focusing and drift sections with a specially shaped rf voltage applied to the trap, our calculations and our experiments indicate that a periodic sequence of squeezing, pulling and Coulomb explosion does not exclude the possibility of the formation of Wigner crystals in storage rings. Finally we have proposed several methods of beam diagnostics and a procedure to load traps and storage rings in case the power of available cooling lasers is not sufficient to overcome the peak of the heating curve.

Acknowledgements

It is a pleasure to thank W. Schleich and Y. R. Shen for helpful discussions and the HLRZ in Jülich for generously providing computer time.

REFERENCES

[1] L. Föppl: Journal für die reine und angewandte Mathematik 141, 251 (1912)
[2] P. Debye, E. Hückel: Physikal. Zeitschr. 24, 185 (1923)
[3] E. P. Wigner: Trans. Faraday Soc. 34, 678 (1938)
[4] E. L. Pollock, J. P. Hansen: Phys. Rev. A8, 3110 (1973)
[5] W. L. Slattery, G. D. Doolen, H. E. DeWitt: Phys. Rev. A21, 2087 (1980)
[6] R. F. Wuerker, H. Shelton, R. V. Langmuir: Journ. Appl. Phys. 30, 342 (1959)
[7] W. Neuhauser, M. Hohenstatt, P. Toschek, H. Dehmelt: Phys. Rev. Lett. 41, 233 (1978)
[8] W. Neuhauser, M. Hohenstatt, P. Toschek, H. Dehmelt: Phys. Rev. A 22, 1137 (1980)
[9] H. G. Dehmelt: In Advances in Laser Spectroscopy, ed. by F. T. Arecchi, F. Strumia, H. Walther (Plenum, New York, 1983)
[10] D. J. Wineland, W. M. Itano, R. S. van Dyck: In Adv. Atom. Molec. Phys., Vol. 19, ed. by D. R. Bates and B. Bederson (Academic, New York, 1983)
[11] D. J. Wineland ,W. M. Itano: Phys. Lett. 82A, 75 (1981)
[12] W. Nagourney, G. Janik, H. G. Dehmelt: Proc. Natl. Acad. Sci. USA 80, 643 (1980)
[13] G. Janik, W. Nagourney, H. G. Dehmelt: J. Opt. Soc. Am. B 2, 1251 (1985)
[14] J. C. Bergquist, W. M. Itano, D. J. Wineland: Phys. Rev. A 36, 428 (1987)
[15] D. J. Wineland, W. M. Itano, J. C. Bergquist, J. J. Bollinger, H. Hemmati: Prog. Quant. Electr. 8, 139 (1984)
[16] F. Diedrich, H. Walther: Phys. Rev. Lett. 58, 203 (1987)
[17] W. Nagourney, J. Sandberg, H. G. Dehmelt: Phys. Rev. Lett. 56, 2797 (1986)
[18] Th. Sauter, W. Neuhauser, R. Blatt, P. Toschek: Phys. Rev. Lett. 57, 1696 (1986)
[19] J. C. Bergquist, R. G. Hulet, W. M. Itano, D. J. Wineland: Phys. Rev. Lett. 57, 1699 (1986)

[20] F. Diedrich, J. Krause, G. Rempe, M. O. Scully, H. Walther: In Laser Spectroscopy VIII, Springer Series in Optical Sciences, Vol. 55, 133, ed. by S. Svanberg and W. Person (Springer Verlag, Berlin, 1987)

[21] F. Diedrich, J. Krause, G. Rempe, M. O. Scully, H. Walther: In Proceedings of the Fourth Int. Conf. on Multiphoton Processes, Boulder, Colorado, 1987, ed. by S. Smith and P. Knight (Camb. Univ. Press, New York, in press)

[22] F. Diedrich, E. Peik, J. M. Chen, W. Quint, H. Walther: Phys. Rev. Lett. 59, 2931 (1987)

[23] F. Diedrich, E. Peik, J. M. Chen, W. Quint, H. Walther: Phys. Bl. 44, 12 (1988)

[24] D. J. Wineland, J. C. Bergquist, W. M. Itano, J. J. Bollinger, C. H. Manney: Phys. Rev. Lett. 59, 2935 (1987)

[25] R. Blümel, J. M. Chen, E. Peik, W. Quint, W. Schleich, Y. R. Shen, H. Walther: Nature 334, 309 (1988)

[26] J. Hoffnagle, R. G. DeVoe, L. Reyna, R. G. Brewer: Phys. Rev. Lett. 61, 255 (1988)

[27] S. Gilbert, J. J. Bollinger, D. J. Wineland: Phys. Rev. Lett. 60, 2022 (1988)

[28] J. J. Bollinger, D. J. Wineland: Phys. Rev. Lett. 53, 348 (1984)

[29] L. R. Brewer, J. D. Prestage, J. J. Bollinger, D. J. Wineland: In Strongly Coupled Plasma Physics, ed. by F. J. Rogers and H. E. DeWitt (Plenum, New York, 1987)

[30] F. M. Penning: Physica 3, 873 (1936)

[31] H. G. Dehmelt: In Adv. Atom. Molec. Phys., Vol. 3, 53, ed. by D. R. Bates, I. Estermann (Academic Press, New York, 1967)

[32] D. J. Wineland, W. M. Itano: Phys. Rev. A 20, 1521 (1979)

[33] J. Javanainen: J. Appl. Phys. 23, 175 (1980)

[34] S. Stenholm: Rev. Mod. Phys. 58, 699 (1986)

[35] D. Habs: In Frontiers of Particle Beams (Springer, New York, 1988), and D. Habs: MPI Heidelberg, MPI H-1987-V10

[36] S. Ichimaru: Rev. Mod. Phys. 54, 1017 (1982)

[37] H. Totsuji: In Strongly Coupled Plasma Physics, ed. by F. J. Rogers and H. E. DeWitt (Plenum, New York, 1987)

[38] J. P. Schiffer, P. Kienle: Z. Phys. A 321, 181 (1985)

[39] A. Rahman, J. P. Schiffer: Phys. Rev. Lett. 57, 1133 (1986)

[40] J. P. Schiffer, O. Poulsen: Europhys. Lett. 1, 55 (1986)

[41] J. P. Schiffer: preprint, Argonne Nat. Lab. PHY-5094-HI-88

[42] J. P. Schiffer, A. Rahman: preprint, Argonne Nat. Lab. PHY-5121-ME-88

[43] J. H. Malmberg, T. M. O'Neil: Phys. Rev. Lett. 39, 1333 (1977)

[44] C. C. Grimes, G. Adams: Phys. Rev. Lett. 42, 795 (1979)

[45] G. Deville, A. Valdes, E. Andrei, F. I. B. Williams: Phys. Rev. Lett. 53, 588 (1984)

[46] R. Gerhardts: Phys. Bl. 42, 23 (1986)

[47] Th. Sauter, H. Gilhaus, I. Siemers, R. Blatt, W. Neuhauser, P. E. Toschek: Z. Phys. D 10, 153 (1988)

[48] R. Blümel, J. M. Chen, F. Diedrich, E. Peik, W. Quint, W. Schleich, Y. R. Shen, H. Walther: In Proceedings of the Eleventh International Conference on Atomic Physics (ELICAP), Paris, July 4-8, 1988, (World Scientific, 1989)

[49] W. Paul, O. Osberghaus, E. Fischer: Ein Ionenkäfig, Forschungsberichte des Wirtschafts- und Verkehrsministeriums Nordrhein-Westfalen 415 (1958)

[50] W. Paul, M. Raether: Z. Physik 140, 262 (1955)

[51] E. Fischer: Z. Physik 156, 1 (1959)

[52] D.A. Church: J. Appl. Phys. 40, 3127 (1969)
[53] E. T. Whittaker, G. N. Watson: A Course of Modern Analysis (Camb. Univ. Press, Cambridge, 1927)
[54] R. Loudon: The Quantum Theory of Light (Clarendon Press, Oxford, 1986)
[55] J. Stoer, R. Bulirsch: Einführung in die Numerische Mathematik, Vol. 2 (Springer, Berlin, 1978)
[56] R. Blümel, C. Kappler, W. Quint, H. Walther: Phys. Rev. A, in print
[57] D. J. Wineland, J. J. Bollinger, W. M. Itano, J. D. Prestage: J. Opt. Soc. Am. B 2, 1721 (1985)
[58] Th. Sauter, H. Gilhaus, W. Neuhauser, R. Blatt, P.E. Toschek: Europhys. Lett. 7, 317 (1988)
[59] E. Fermi: Phys. Rev. 75, 1169 (1949)
[60] H.G. Schuster: Deterministic Chaos (Physik-Verlag, Weinheim, 1984)
[61] A.J. Lichtenberg, M.A. Lieberman: In Regular and Stochastic Motion, Appl. Math. Sci., Vol. 38 (Springer, New York, 1983)
[62] F.J. Calogero: Math. Phys. 12, 419 (1971)
[63] J. Moser: Advanc. Math. 16, 1 (1975)
[64] R. Blümel, J. M. Chen, F. Diedrich, E. Peik, W. Quint, H. Walther: In Proceedings of the Workshop on Crystalline Ion Beams, Wertheim, Main, October 4-7, 1988, organized by GSI, Darmstadt
[65] The Structure of small Molecules and Ions, ed. by R. Naaman and Z. Vager (Plenum Press, New York, 1988)

Modulation Spectroscopy and Cavity QED

M.G. Raizen, R.J. Thompson, R.J. Brecha*, H.J. Kimble*, and H.J. Carmichael**

Department of Physics, University of Texas at Austin, Austin, TX78712, USA

Although discussed long ago by Purcell[1], the study of the radiative processes of atoms in cavities has been the subject of great interest in recent years. The presence of the cavity introduces two new rates not present for atoms in free space: the cavity decay rate κ and the coupling rate g of the atoms to the cavity field, which is given by

$$g = \frac{\mu}{\hbar} \sqrt{\frac{\hbar\omega}{2\varepsilon_o V}} \qquad [1]$$

where μ is the dipole moment of the atomic transition and V is the mode volume of the cavity. In addition, the cavity mode subtends a solid angle $\Delta\Omega$ relative to the 4π solid angle of free space. Modification of the free space rate of spontaneous emission γ was achieved in previous work by having a cavity with large $\Delta\Omega$ in a regime in which γ is small compared to κ, g. For $\gamma << g^2/\kappa << \kappa$ (the so-called overdamped regime), enhancement of spontaneous emission was observed by tuning the cavity for coincident atomic and cavity resonance and inhibition of spontaneous emission observed by detuning the cavity[2-6]. These effects can be explained by using perturbation theory. The altered rate is obtained from Fermi's golden rule using the (cavity) modified density of states.

For N atoms in a cavity and for $g\sqrt{N} >> \kappa >> \gamma$, a ringing regime for intracavity emission and absorption has been investigated[7-8]. For a single atom and $g >> (\kappa, \gamma)$ the system approaches the Jaynes-Cummings model which has no dissipation. Experiments conducted in this regime demonstrated oscillatory behavior as well as related phenomena such as collapse and revival[9].

In this paper we present an experimental study of two-level atoms in a cavity in the optical domain and find that large modifications of linewidths and resonant frequencies can occur even if the cavity subtends a very small solid angle. We

* Present address: Division of Physics, Mathematics, and Astronomy, California Institute of Technology, Pasadena, CA.

**Permanent address: Department of Physics, University of Arkansas, Fayetteville, AK.

Springer Proceedings in Physics, Vol. 41 **Quantum Optics V**
Editors: J.D. Harvey and D.F. Walls

work in a regime for which $g\sqrt{N} > \gamma > \kappa$ ($g\sqrt{N}$ is the effective coupling rate of N atoms in a cavity) and with a very weak intracavity field. Here, dissipation of both the atoms and the cavity plays an important role and the coupling is so strong that an emitted photon is likely to be reabsorbed by the collective atomic polarization before it can escape. In such a case perturbation theory cannot be used. A complete analysis of the normal modes of the coupled atom-cavity system has been previously derived[10-11]. The eigenvalues in weak field and for the (uncoupled) cavity resonance frequency ω_c equal to the atomic transition frequency ω_A are

$$\lambda_{\pm} = -\frac{\kappa + \frac{\gamma}{2}}{2} \pm \sqrt{\frac{(2\kappa - \gamma)^2}{16} - g^2 N} \quad , \qquad [2]$$

where we assume a rotating frame at the common frequency $\omega_c = \omega_A$. In the strongly coupled regime the square root term is imaginary and provides the normal mode frequency known as the vacuum-field Rabi splitting[12]. The real term in Eq. [2] is an average of cavity and atomic polarization decay rates. When $\kappa << \gamma$ this real part gives a decay rate which is 1/2 the natural free-space rate. We refer to this phenomenon as subnatural linewidth averaging. It arises from the dynamical processes in the coupling of the atoms to the cavity mode and is discussed in detail elsewhere[13].

The experiments that we report constitute a spectroscopic study of the structure of the coupled system. A detailed description is given in Ref. (14). Two-state sodium atoms are prepared in an atomic beam by optical pumping (3 $S_{1/2}$, F = 2, m_F = 2 → 3 $P_{3/2}$, F = 3, m_F = 3, transition at 589nm). They intersect at 90° the axis of a high finesse optical cavity. Several cavities have been employed, the most recent having a finesse F = 26,000 and length 3.2mm, giving $\kappa/\gamma \sim 0.095 \pm 0.01$. An input beam is mode matched to the TEM_{00} mode of the cavity and the cavity is servo-locked to the empty cavity (no sodium atoms) resonance. An electrooptic modulator generates weak AM sidebands on the input beam and the transmitted output of the cavity is detected by optical heterodyne with an intense local oscillator in a balanced detection scheme such as was used for squeezed state detection[15]. The resulting photocurrent is sent to an electronic spectrum analyzer whose analysis frequency corresponds to the frequency of the heterodyne signal between the local oscillator and one of the swept AM sidebands (through the use of a tracking generator and RF mixers). By such a method the response function of the composite atom-field system was studied for probe frequencies ± 60MHz around the signal beam carrier frequency. Figure [1] shows the output of the empty cavity obtained with this method together with a Lorentzian fit. With atoms present the AM sidebands are swept symmetrically about ω_A, while the cavity is again locked to the empty cavity resonance at ω_c.

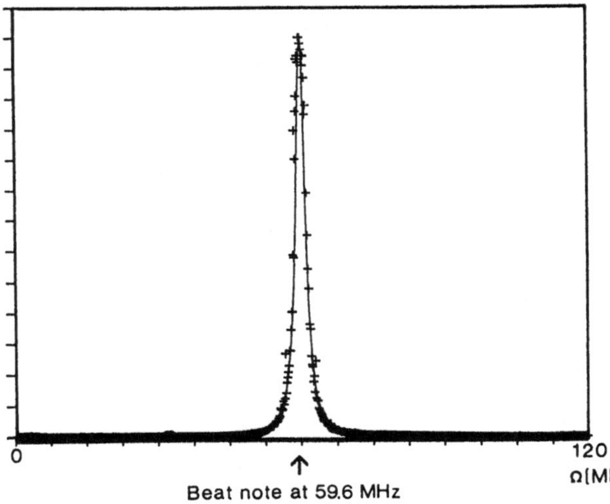

Figure 1 - Transmitted spectrum of empty cavity locked on resonance using heterodyne signal at 59.6MHz. Data points give the scaled power of the beat note between the local oscillator and a transmitted AM sideband in a range ± 60MHz around a signal carrier frequency. The solid line is a Lorentzian fit with FWHM = 2.3MHz.

To determine the normal mode frequency from our measurements, the effective coupling rate $g\sqrt{N}$ is required. This frequency is related to the cooperativity parameter C of optical bistability by $g\sqrt{N} = \sqrt{\kappa\gamma C}$. Measurements of the input/output characteristics are used to determine C as well as the required weak field input [16].

Figure [2] shows a transmission spectrum of the swept AM sidebands for the coupled atom-cavity system. The LO and the signal carrier are offset 59.6MHz and the spectrum analyzer and analysis frequency is scanned from 0 to 120MHz. We see spectral features offset from the empty cavity response of Fig. [1]. The asymmetry evident between the peaks is accounted for by introducing a small atomic detuning. The solid line is a theoretical fit, derived from the linearized Maxwell Bloch equations. To obtain the fit we used the measured parameters κ/γ, C and allowed for less than ± 1.5MHz atomic detuning, as well as an overall amplitude scaling factor. From traces such as Fig. [2] we obtain a frequency splitting and a linewidth for each member of the doublet. By changing the number of atoms (hence C) we have obtained an absolute comparison of the predicted splitting frequency Ω_{th} vs. the measured splitting frequency Ω_{exp} and have found very good agreement over a large range of N (~50 times change). From linewidth measurements we have found widths for each of the split lines of 7.75 ± .75MHz (FWHM) which is significantly below the 10MHz natural linewidth. The theoretical width for this comparison is given by the real part of the eigenvalue $(2\kappa + \gamma/\Gamma)/4$, where Γ accounts approximately for transit broadening ($\Gamma = 1$ is for no transit broadening). In our experiments $\kappa/\gamma = 0.095 \pm 0.01$, $\Gamma \sim 0.8$, so that a full width of 7.2MHz is the expected linewidth.

The spectrum of a single atom in a cavity in the strongly coupled regime is predicted by a completely quantum mechanical calculation[13,17] to have analogous spectral features to those that we have described above. It is of great interest to observe these effects experimentally. In our current experiments we have followed the spectral splitting to low values of N, where no bistability is present in the input-output scan and so the measurement of C is based upon a higher atomic density calibration of C relative to a downstream fluorescent signal which can be detected even when there are too few atoms present for the system to be bistable. From our calibration we find that the lowest atomic number for which the peaks of the normal mode splitting were

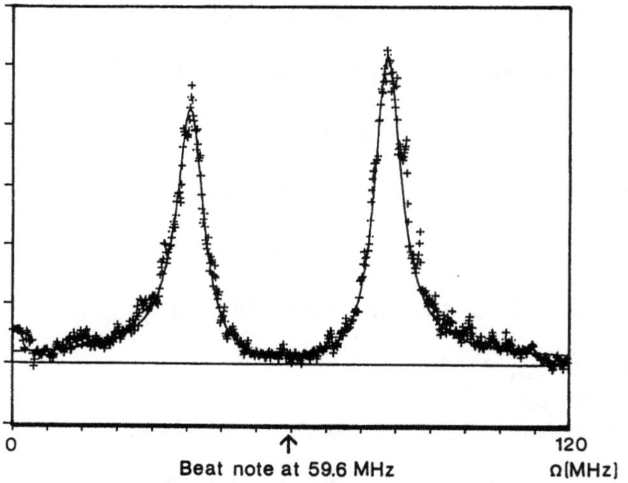

Figure 2 - Transmitted spectrum of cavity with atoms. Cavity is locked on (uncoupled cavity) resonance to the weak field heterodyne signal at 59.6MHz, and atomic resonance is within 1.5MHz of this. Data points give the scaled power of the beat note between the local oscillator and a transmitted AM sideband in a range ±60MHz around signal carrier frequency. The empty cavity FWHM is 1.7MHz, and corresponds to a different cavity than the one shown in Fig. 1. The solid line is a theoretical fit for the transmitted spectrum using the measured values of μ = 0.085 and C = 41. For this fit we used Γ = 0.8, an atomic detuning of 1MHz, as well as an overall scaling factor.

resolved was N ~ 40 atoms. It appears from these measurements that this method has the potential sensitivity to observe the splitting for a single atom in a cavity, although the cavity needs to be considerably shorter and the finesse appropriately higher to resolve the spectral peaks.

In summary, we have conducted a spectroscopic study of a system of atoms coupled to a single cavity mode. We observe directly the normal mode splitting and subnatural linewidth averaging. This technique appears to have the potential for studying the splitting induced by a single atom in a cavity.

This work was supported in part by the Office of Naval Research (Contract No. N00014-87-K-0156), the National Science Foundation (Grant No. PHY8351074), and the Venture Research Unit of British Petroleum, Inc. M. G. Raizen acknowledges support from IBM.

References

1. E. M. Purcell: Phys. Rev. 69, 681 (1946)
2. D. Kleppner: Phys. Rev. Lett. 47, 233 (1981)
3. P. Goy, J. M. Raimond, M. Gross and S. Haroche: Phys. Rev. Lett. 50, 1903 (1983)
4. W. Jhe, A. Anderson, E. A. Hinds, D. Meschede, L. Moi, S. Haroche: Phys. Rev. Lett. 58, 666 (1987)
5. D. J. Heinzen, J. J. Childs, J. E. Thomas, M. S. Feld: Phys. Rev. Lett. 58, 1320 (1987)
6. F. DeMartini, G. Innocenti, G. R. Jacobovitz and P. Mataloni: Phys. Rev. Lett. 59, 2955 (1987)
7. Y. Kaluzny, P. Goy, M. Gross, J. M. Raimond, S. Haroche: Phys. Rev. Lett. 51, 1175 (1983)
8. R. J. Brecha, L. A. Orozco, M. G. Raizen, Min Xiao, H. J. Kimble: J. Opt. Soc. Am. B 3, 238 (1986)
9. Gerhard Rempe, Herbert Walther, Norbert Klein: Phys. Rev. Lett. 58, 353 (1987)
10. H. J. Carmichael: Phys. Rev. A 33, 3262 (1986)
11. G. S. Agarwal: Phys. Rev. Lett. 53, 1732 (1984)
12. J. J. Sanchez-Mondragon, N, B. Narozhny, J. H. Eberly: Phys. Rev. Lett. 51, 550 (1983)
13. H. J. Carmichael, R. J. Brecha, M. G. Raizen, H. J. Kimble, P. R. Rice: submitted to Phys. Rev. Lett.
14. M. G. Raizen, R. J. Thompson, R. J. Brecha, H. J. Kimble, H. J. Carmichael: in preparation
15. M. G. Raizen, L. A. Orozco, Min Xiao, T. Boyd, H. J. Kimble: Phys. Rev. Lett. 59, 198 (1987)
16. A. T. Rosenberger, L. A. Orozco, H. J. Kimble: Phys. Rev. A 28, 2569 (1983)
17. P. R. Rice, H. J. Carmichael: J. Opt. Soc. Amer. B 5, 1661 (1988)

“Taming” Sodium

W.J. Sandle, D.E. McClelland,[†] *R. Holzner,*[‡] *and D.M. Warrington*

Physics Department, University of Otago, Dunedin, New Zealand

Some of the difficulties faced by experimentalists when they wish to compare quantum optics behaviour in an atomic vapour with model predictions are reviewed, and a recent experiment in nonlinear birefringence, in which the behaviour of the sodium D_1 line has been simplified by employing the Back-Goudsmit effect, is summarised.

1. Introduction

My title emphasises 'sodium' so perhaps it may be appropriate at the outset to remind you very briefly why sodium has a special place in the laser experimentalist's armoury. Atomic sodium has its strongly allowed 3^2S — 3^2P resonance transition at a very convenient wavelength for interaction with dye laser light, and the metal vapour itself is easy to handle since the metal sublimes in the accessible temperature range 100-300°C.

However, using sodium in a quantum optics experiment is a far from simple matter. As is shown in Fig. 1, the resonance line is considerably more complicated than the idealised two-level transition of theoretical quantum optics. The line comprises the well known fine structure doublet 589.6 nm (D_1) and 589.0 nm (D_2), each component of which has considerable further hyperfine structure. In addition, when the sodium is used in the form of a thermal vapour, inhomogeneous broadening due to the Doppler effect has to be taken into account and collisional broadening of the transitions and collisional redistribution among sublevels occur.

Finally, other nonlinear physical mechanisms beyond simple 'two-level' saturation behaviour operate. In general the laser connects many sub-states in the lower 'level' to many sub-states in the upper 'level'. The interaction with the laser field therefore gives rise to redistribution of population among the substates of single 'levels' (that is to say it causes "optical pumping") as well as the overall redistribution between lower and upper 'levels' (i.e. "two-level saturation"). The optical pumping, furthermore, is not simply of one single type: in "Zeeman optical pumping" it is confined to redistribution among the Zeeman substates of each of the well-defined hyperfine states; but in "hyperfine optical pumping" it occurs as a population adjustment between different hyperfine levels.

The complexity of sodium and the range of nonlinear mechanisms provides quite a challenge to an experimentalist. On the one hand, advantage can be taken of the complexity to introduce new physical techniques: for example, laser cooling below the one-photon recoil energy (Doppler) limit [1,2]. On the other hand, one can attempt to 'tame' sodium so that its behaviour closely approximates one of the simple atomic models used in theoretical quantum optics.

My aim here is to give a simple introduction to the complexity of sodium's D-lines, and discuss a couple of ways—one carried out at Otago—in which their use can be related simply, and quantitatively, to theory.

---

† Now at the Department of Physics and Theoretical Physics (Faculties), Australian National University, Canberra, ACT, Australia

‡ Now at Sektion Physik, Universität Zürich, Zürich, Switzerland

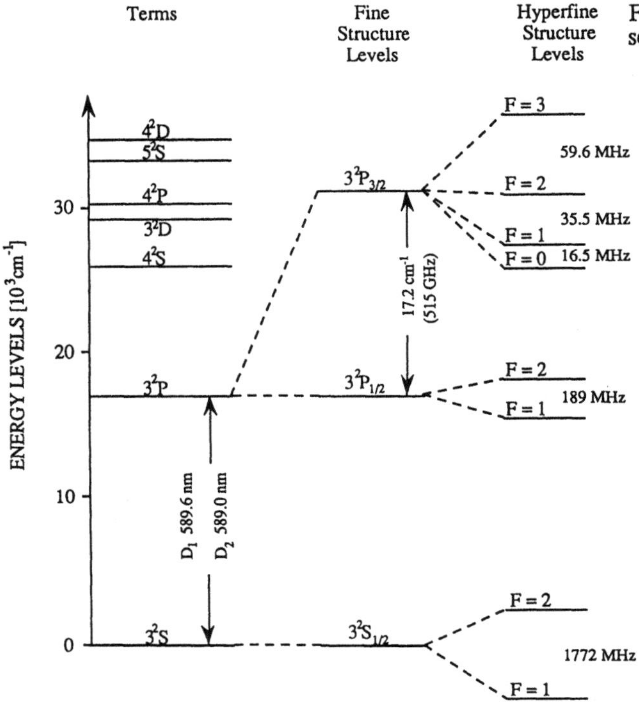

Figure 1. Level structure of the atomic sodium resonance D lines

2. Theory

Quantum opticians like to think about 'two–state' or (as they are called in the field) 'two-level' systems. Unfortunately, there are no atoms which behave strictly in this way. The basic reason is that single–photon, optical–dipole interactions connect atomic states of opposite parity, and so there must be a change in single electron angular momentum quantum number l in such an interaction. Thus at least one of the l values must be different from zero. When we remember that a non-zero l value implies Zeeman substate degeneracy, we understand why the simple two-state transition must be an idealisation. In the case of sodium the ground configuration 3s is characterised by $L=l=0$ for the active electron outside the closed–shell core, forming the "term" 3^2S, and the excited configuration of interest, 3p, is characterised by $L=l=1$, forming the term 3^2P [3].

Fine and Hyperfine–Structure

You will remember I said earlier that two separate "fine–structure" transitions were involved. These come from the splitting of the 3^2P term into two states by the spin–orbit interaction $\zeta\hat{L}\cdot\hat{S}$ [3]. [A simple picture is that this splitting arises from the alignment energy of the intrinsic magnetic dipole of the outer (active) electron in the effective magnetic field due to orbital motion in the electric field of the nucleus and core electrons.] For our present purposes, the important point is that these states are labelled by values $\frac{1}{2}$ and $\frac{3}{2}$ for the total electronic angular momentum quantum number J, and are described as $3^2P_{1/2}$ and $3^2P_{3/2}$. The ground state becomes $3^2S_{1/2}$ under the effect of the spin–orbit interaction, and remains unsplit.

The hyperfine levels, arising from the formally similar (but physically distinct) interaction $A\,\hat{I}\cdot\hat{J}$ [3] due to coupling between the *nuclear* angular momentum operator, $\hat{I}$, and electronic angular momentum operator, $\hat{J}$, are, in their turn, labelled by values of the total angular momentum quantum number F. The $J=\frac{1}{2}$, S and P, states are each split into two levels $F = 1$ or 2 (however it is important to note that the value of A differs between the S and P states). The splitting of the $3^2P_{3/2}$ state is a little more complicated: $J=\frac{3}{2}$ couples to $I=\frac{3}{2}$ to produce four levels with $F = 0,1,2$ or 3. The intervals between these levels can be calculated under the approximation that J remains a good quantum number [3] to be A, $2A$, and $3A$. As can be seen from Fig. 1, the relative sizes of the experimental intervals are quite closely in accord with these simple predictions.

Zeeman Structure

Finally I need to consider the effect of a magnetic field, $\boldsymbol{B}$. The combined hyperfine—Zeeman interaction now becomes

$$A\,\hat{\boldsymbol{I}}\cdot\hat{\boldsymbol{J}} + g_J\,\mu_B\,\hat{\boldsymbol{J}}\cdot\boldsymbol{B} - g_I\,\mu_N\,\hat{\boldsymbol{I}}\cdot\boldsymbol{B}.$$

The last term (direct interaction of the nuclear magnetic moment with $\boldsymbol{B}$) is small enough to be ignored, but the other two terms must be considered together, since they will be of comparable magnitude for the fields to be considered later (fractions of a tesla) and do not commute. The detailed effect of the magnetic field strength B on the energy sub–levels, given by the "Breit–Rabi" formula [3] and shown plotted in Fig. 2 for a $J=\frac{1}{2}$ state, then changes character as the strength of B changes. It is important to note that F is no longer a good quantum number over the whole range of B; the associated "decoupling" of I and J for large B is termed the "Back–Goudsmit" effect.

For extremely large values of B, it will in fact be the case that L and S also become decoupled, and precess independently about the magnetic field direction (the "Paschen–Back" effect), but these field values (several tens of tesla) are beyond those considered here.

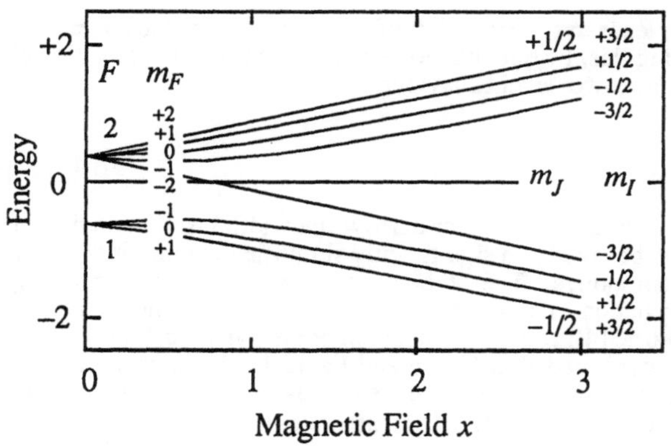

Figure 2. Breit-Rabi diagram for Zeeman and hyperfine–structure splitting in a $J=\frac{1}{2}$, $I=\frac{3}{2}$ state. Energy is plotted in units of the zero–field hyperfine separation; and x is $\mu_B\,(g_J+g_I)B$, also expressed in the same units. The experiment described in Section 5 was carried out at $x=6$

Selection Rules

The "selection rules", which govern how sub–levels are connected by laser radiation, are the key to the degree of practical complication of the laser—atom interaction.

First, one must imagine a whole diagram (such as Fig. 2) replacing the single lower 'level' of idealised quantum optics. Next one imagines a separate, perhaps even more complicated diagram (e.g. $J'=\frac{3}{2}$ coupling with $I=\frac{3}{2}$ to make $F'=0,1,2$ or 3 with a total of 16 sublevels), replacing the single idealised upper 'level'.

The question now is—How are the sub–levels in the lower 'level' are connected to the sub–levels in the upper 'level'?

The answer is—It depends on the strength of the magnetic field!

In the weak field limit, when the magnitude (F) of $\hat{\boldsymbol{F}}$ and its projection (m_F) along $\boldsymbol{B}$ are 'good' quantum numbers, the selection rules are

$$\Delta F \equiv F' - F = 0, +1, -1, \tag{1}$$

and

$$\Delta m_F \equiv m_F' - m_F = 0, +1, -1; \tag{2}$$

the latter [selection rule (2)] corresponding respectively to interaction with π (linear), σ^+ (one sense of circular), or σ^- (other sense of circular) polarised radiation.

In the strong field limit, on the other hand, when not only the magnitudes (I and J) of $\hat{I}$ and $\hat{J}$, but also their projections (m_I and m_J) along $\boldsymbol{B}$ are 'good' quantum numbers, the selection rules are

$$\Delta m_I \equiv m_I' - m_I = 0 \text{ (always)}, \tag{3}$$

$$\Delta J \equiv J' - J = 0, +1, -1, \tag{4}$$

and

$$\Delta m_J \equiv m_J' - m_J = 0, +1, -1; \tag{5}$$

the last, in this case, corresponding respectively to interaction with π, σ^+, or σ^- polarised radiation. [The condition on ΔJ is automatically satisfied in the examples $J = \frac{1}{2} \leftrightarrow J' = \frac{1}{2}$, or $J = \frac{1}{2} \leftrightarrow J' = \frac{3}{2}$.]

It is the selection rule (3), $\Delta m_I = 0$, which leads to the high–B field simplification of the sodium D_1 line to be discussed later, but before describing this, I would like to explain an important application of the $B = 0$ *optically pumped* D_2 line to obtain true two–level behaviour.

3. Two–Level Behaviour

A number of researchers in the early 1970's realised that under narrow linewidth laser excitation of sodium atoms in an atomic beam, the D_2 line behaved in a significantly simplified manner if the laser was tuned to the $F = 2$ —$F' = 3$ hyperfine component. Then the atoms must, by selection rule (1), return to their initial $F = 2$ state following excitation. This method was (for example) applied to atomic beam deflection by Schieder, Walther and Wöste [4], and to investigation of the dynamical Stark effect by Schuda, Stroud and Hercher [5], Walther [6], and Wu, Grove and Ezekiel [7]. In each of these cases, linearly polarised coupling was used; because of the different strengths of the $m_F = 0$—$m_F' = 0$, $m_F = 1$—$m_F' = 1$ and $m_F = 2$—$m_F' = 2$ transitions, which all participated simultaneously in the laser interaction, there was only *approximate* two–state character.

It appears that Abate, in 1974 [8], was the first to put forward the idea that under *circularly* polarised $F = 2$ —$F' = 3$ excitation (say σ^+ with $\Delta m_F = +1$) *pure* two–state ($m_F = 2$—$m_F' = 3$) behaviour was possible provided all the atoms were initially assembled in an $F = 2$ extremum ($m_F = 2$) substate. The atoms then remain in these two states because the lower state is the only one accessible from the upper by radiative decay. He described how the initial preparation could be achieved by optical pre-pumping in an atomic beam. This method was applied to careful measurements of the fluorescent spectrum from a laser–driven atom by Hartig, Rasmussen, Schieder and Walther [9], and Grove, Wu and Ezekiel [10], to photon antibunching by Kimble, Dagenais and Mandel [11], and to optical bistability by Grant and Kimble [12].

4. $J = 1/2 \leftrightarrow J' = 1/2$ Behaviour

The simplification treated in the previous section does not work for laser excitation of general polarisation, since the atoms 'unpump' from the extremum m_F state. For such general excitation the transition which has simpler character is D_1, and the underlying $J = \frac{1}{2}$ — $J' = \frac{1}{2}$ structure becomes qualitatively manifest under conditions of strong collisional relaxation.

This structure leads to quite different forms of nonlinear behaviour from the two–state case. For example, Mitschke, Deserno, Lange and Mlynek [13] describe in terms of this structure sustained lower–level spin precession in a weak applied magnetic field under linearly polarised, continuous–wave D_1 excitation in an optical cavity. So, even with steady excitation of *pure* polarisation, a physically realisable nonlinear system may exhibit interesting features arising from mode–mode competition.

The $J = \frac{1}{2} — J' = \frac{1}{2}$ behaviour is interesting as a paradigm for the simplest system exhibiting polarisation mode—mode competition effects. The question then is—how can quantitatively pure $J = \frac{1}{2} — J' = \frac{1}{2}$ behaviour be experimentally realised? The key to answering this question lies with the discussion of Section 2. Under application of a magnetic field of sufficient strength to decouple I and J, the D_1 transition breaks up into a set of four independent $J = \frac{1}{2} — J' = \frac{1}{2}$ transitions, each labelled by an invariant m_I (one of $-\frac{3}{2}, -\frac{1}{2}, +\frac{1}{2}, +\frac{3}{2}$). The complication of having *four* distinct classes of $J = \frac{1}{2} — J' = \frac{1}{2}$ 'atoms ', each interacting independently with the propagating beam, appears inconvenient; but is fundamentally no more difficult than the Doppler broadening problem where there is a continuous distribution of atom classes distinguished by their thermal velocity component parallel to the laser propagation direction. The point is that the physical behaviour is now determined purely from a $J = \frac{1}{2} — J' = \frac{1}{2}$ perspective.

5. Quantitative Test of a Laser–Atom Interaction Model

I would now like to apply the above ideas to discuss an experiment we have recently performed involving use of the nonlinear Voigt effect [14] in laser propagation.

The *nonlinear Voigt effect* relates to the birefringence of an atomic vapour in a magnetic field transverse to the laser propagation direction. When the laser intensity is sufficiently low, one has the simple (*linear*) Voigt effect [15] in which two modes of orthogonal polarisation, parallel and perpendicular to the transverse magnetic field, propagate with different velocities producing a polarised beam of varying ellipticity throughout the medium; *nonlinearity* enters via saturation, optical pumping, and more general mode–mode competition effects induced in the vapour. Calculations for laser propagation in a $J = \frac{1}{2} — J' = \frac{1}{2}$ atomic vapour, driven nonlinearly in the presence of a transverse magnetic field, have been carried out by McClelland and Ballagh [16,17].

There are two main reasons why the nonlinear Voigt effect is useful for providing a good experimental test of a laser interaction model: firstly, nonlinear birefringence intrinsically includes the presence of a magnetic field, needed for decoupling I and J and thus simplifying the D_1 line behaviour; and secondly, the effect is essentially an *interference* one in which the two modes, propagating with different velocities (and absorptions), produce an exit polarisation which is highly sensitive to the details of this behaviour.

The experiment is illustrated in Fig. 3. A continuous–wave ring dye laser, providing light linearly polarised at 45° to a 3.82 kG magnetic field, is tuned to the D_1 transition of sodium in a vapour cell.

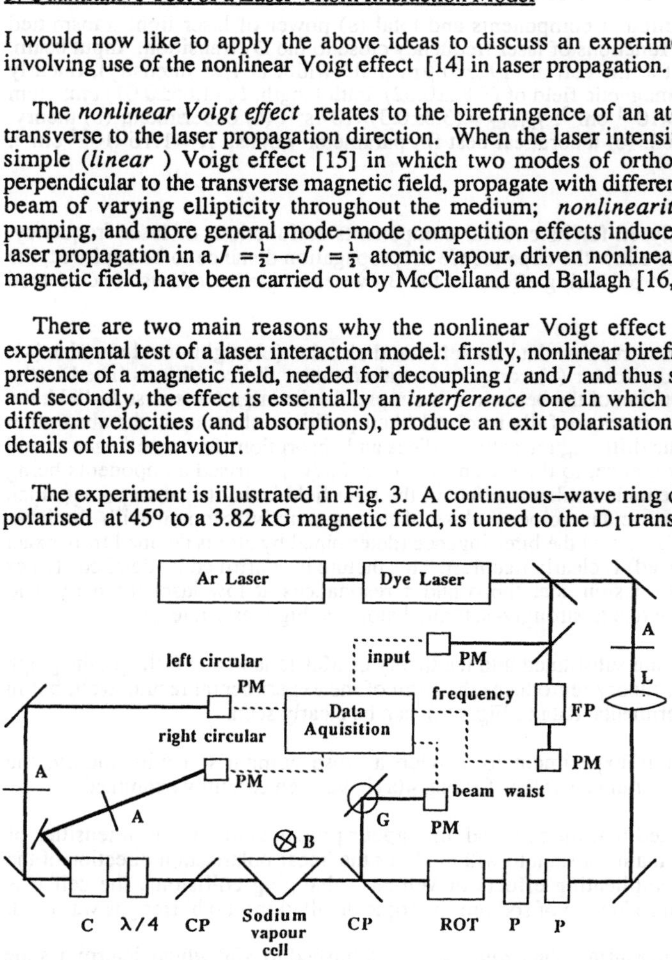

Figure 3. Experimental arrangement: A refers to attenuators; L to lenses; P to linear polarisers; CP to correction plates; C to a calcite prism; and PM to the photomultiplier detectors, used with integrating spheres. ROT is a polarisation rotator; G refers to a scanning galvanometer that was used in measuring the beam waist; and the Fabry Perôt FP provided an accurate calibration of the dye laser scan linearity

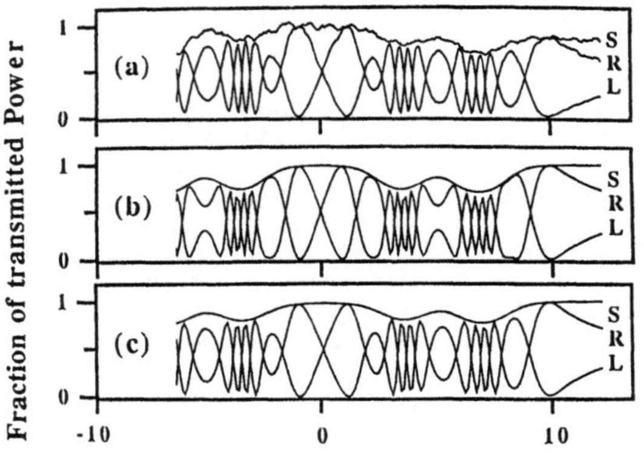

Figure 4. Left (L) and right (R) circular components and total (S) power of laser light transmitted through atomic sodium vapour as functions of laser frequency close to the D_1 transition. Input beam, (12±1) mW collimated to a waist at the cell of (192±5) μm (half–width at $1/e^2$ intensity), linearly polarized at 45° to the transverse magnetic field of (3.82±0.02). Path length, L, (1.65±0.01) cm; atom number density, $N = (1.33\pm0.02)\times10^{11}$ cm^{-3}; $\gamma_{lower} = (3.3\pm0.3)\times10^6$ s^{-1}: (a) experiment; (b) theory, with parameter values as measured; (c) theoretical best fit, parameter values: $N = 1.16\times10^{11}$ cm^{-3}; $\gamma_{lower} = 1.75\times10^6$ s^{-1}

The emergent polarisation is analysed for its circular components as functions of laser frequency, scanned over 20 GHz. [In order to allow the dependence on propagation distance to be explored, the cell is trapezoidally shaped, and the polarising effects of the Brewster–angled windows are corrected by compensating plates].

Figure 4(a) presents experimental results: data for left–hand circular (L), right–hand circular (R), and summed left– and right–hand circular (S) components taken under the conditions listed in the figure caption. The basic interpretation of the behaviour is simple. In the large transverse B field, the $J = \frac{1}{2}$ — $J' = \frac{1}{2}$ linear σ and π resonances are fully resolved (their positions show up in the absorption dips for the S transmission), and the differing refractive indices and absorptions for these *linear* σ and π components lead, through interference, to the transmitted *circularly* polarised components being strongly frequency dependent. The main nonlinear effect of the laser field is that the absorption (which arises predominantly from the particular velocity class of atoms exactly resonant with the laser frequency) is significantly bleached, so that the birefringence (determined by atoms detuned from exact resonance and therefore less saturated) is clearly manifest. This picture is confirmed by detailed studies [17] which show very little transmission near the σ and π resonances at low laser intensity, and 'washed out' birefringence structure due to strongly–saturated atoms at high laser intensity.

When the measured parameters are substituted into the theory of McClelland and Ballagh, the graph Fig. 4(b) results: qualitatively, the theory reproduces the form of the experimental results well; but in detail, disagreement from the experimental data of Fig. 4(a) can be clearly seen.

Because of the potential of this experiment to provide a close comparison with theory, the procedures for obtaining data and parameter values for Fig. 4(a) have been carefully examined.

For instance, it has been checked that (at zero sodium vapour pressure) the output intensities of circularly polarised components remain constant to within 5% as the linear polarisation direction of the input is rotated. To avoid the complicating effects of velocity–changing collisions, the cell was thoroughly outgassed before the introduction of sodium (by triple distillation); no buffer gas was used.

Particular care has been taken in measuring: the *beam waist* (established to 3%) which determines the effective lower–level relaxation rate as well as affecting the saturation level of the laser field; the *homogeneous line width* (using saturated absorption [18] — this measurement confirmed not only the validity of the use of the "standard" value $\gamma_{radiative}/2$ but also the absence of velocity–changing collisions, the latter conclusion following from the small size of "Doppler pedestals"); and the *lower–*

level (magnetic–dipole) relaxation rate [19] (using the method of transient decay of lower–level beats). The parameter which most sensitively influences the experimental results, and which therefore needs the most critical attention, is the atom–number density. Measurements of this were interleaved with data taken for the nonlinear birefringence experiment to minimise effects of drift. Short term density measurement variation was kept below 2% and an overall reproducibility to better than 5% achieved by the following procedure. Varying only the magnetic field, which was reduced to less than 1 gauss, and the laser intensity (to less than 10 nW), we recorded the total absorption as a function of the carefully–calibrated laser frequency; the atom–number density was then extracted by fitting to the zero–magnetic–field, Doppler–broadened, hyperfine–split, low–power absorption profile, combined with a direct measurement of beam path length in the cell.

The net effect of these precautions is that we are convinced there is a small, persistent departure from the model. In order to explore the detailed nature of this disagreement, we have investigated whether simple adjustment of one or more parameters will provide a good 'fit' of theory to experiment. The results of this survey are shown in Fig. 4(c). The fit is achieved by combining an adjustment of the atom number density to an 'effective' value 12% smaller, with an adjustment of the lower–level relaxation rate to a value some 50% smaller, but leaving all other parameters unchanged at their measured values. We find that it is not possible to satisfactorily reproduce the experimental curve with other adjustments. For example, changing the laser intensity to a substantially different value modifies different parts of the curve in different ways (in physical terms the relative degree of saturation between nearer–resonant and less–resonant regions is adversely affected). However, adjustments of effective atom number density and lower–level relaxation rate can to a degree be traded off against each other. The parameter values used for Fig. 4(c) are those from the point of minimum χ^2 in this fitting procedure.

We believe that two significant conclusions can be drawn. The first is that the measured value of lower–level relaxation rate is approximately a factor of two too large. In hindsight this is not too surprising: lower–level relaxation has no uniquely defined meaning under conditions of transit–time broadening, and the measurement applies to the time taken by atoms prepared within the laser beam at $t=0$ to *leave* the interaction region, travelling typically one beam *radius*, while the experiment involves the time taken by atoms to *transit* the interaction region, travelling typically one beam *diameter*.

The second conclusion is that the effective atom number density is significantly (10 to 15%) smaller than the value measured. Such a consequence has been predicted [21] for the saturation of a *two–state* atom under the combined influence of a primary saturating beam and the scattered fluorescent radiation.

Thus it would appear that the effect of radiatively trapped, and therefore multiply scattered, fluorescent radiation needs to be taken into account if a precise quantitative understanding is to be achieved of quantum and nonlinear optics experiments carried out in a vapour cell. However, what is remarkable is that even with total neglect of radiative trapping, good agreement has been found between experiment and theory in a two–mode situation, where the sodium D_1 line behaviour has been simplified to the point that it can be analytically modelled.

Acknowledgements

We are pleased to acknowledge and thank Dr Rob Ballagh for valuable discussions and assistance. We are also grateful to Dr Peter Hannaford, Mr Martin Lowe and Dr Russell McLean, from the Commonwealth Scientific and Industrial Research Organisation, Australia, for providing a direct measurement of the lower–level decay rate, and to Professor John Cooper for critical advice and comment. We are indebted to Messrs Robin Gledhill and Dale Watts for painstaking glasswork. Finally, we acknowledge financial support from the Research Committees of the New Zealand University Grants Committee and the University of Otago.

References

1. P.D.Lett, R.N.Watts, C.I.Westbrook , W.D.Phillips, P.L.Gould and H.J.Metcalf, Phys. Rev. Lett. **61**, 169 (1988)
2. A.Aspect, E.Arimondo, R.Kaiser, N.Vansteenkiste and C. Cohen–Tannoudji, Phys. Rev. Lett. **61**, 826 (1988)

3. For a discussion of these points see, e.g., G.K.Woodgate, *Elementary Atomic Structure (2nd Ed.)* (Clarendon Press, Oxford, 1980)
4. R.Schieder, H.Walther and L.Wöste, Opt. Commun. **5**, 337 (1972)
5. F.Schuda, C.R.Stroud and M.Hercher, J.Phys.B: At.Mol.Phys. **7**, L198 (1974)
6. H.Walther, in *Proceedings of the Second Laser Spectroscopy Conference, Megève, France, 1975* (Springer, Berlin,1975)
7. F.Y.Wu, R.E.Grove and S.Ezekiel, Phys. Rev. Lett. **35**, 1426 (1975)
8. J.A.Abate, Opt. Commun. **10**, 269 (1974)
9. W.Hartig, W.Rasmussen, R.Schieder, and H.Walther, Z. für Phys. A **278**, 205 (1976)
10. R.E.Grove, F.Y.Wu and S.Ezekiel, Phys. Rev. A **15**, 227 (1977)
11. H.J.Kimble, M.Dagenais, and L.Mandel, Phys. Rev. Lett. **39**, 691 (1977)
12. D.E.Grant and H.J.Kimble Optics Lett. **7**, 353 (1982)
13. F.Mitschke, R.Deserno, W.Lange and J.Mlynek, Phys. Rev. A **33**, 3219 (1986)
14. K.H.Drake, W.Lange and J.Mlynek, Opt. Commun. **66**, 315 (1988)
15. W.Voigt and H.M.Hansen, Phys. Zeit. **13**, 217 (1912)
16. D.E.McClelland and R.J. Ballagh, to be published
17. D.E.McClelland, *Ph.D. Thesis, University of Otago, Dunedin, New Zealand* (unpublished) (1987)
18. Measurements carried out with the collaboration of Dr P.Hannaford and co–workers at the Division of Material Science, Commonwealth Scientific and Industrial Research Organisation, Clayton, Victoria, Australia.
19. Measurements were carried out as in Ref.[20].
20. R.M.Lowe, D.S.Gough, R.J.McLean and P.Hannaford, Phys. Rev. A (RC) **36**, 5490 (1987)
21. A.W.McCord, R.J.Ballagh, and J.Cooper, Opt. Commun. **68**, 375 (1988)

Part IV

Laser Physics

Single Photon Transition Correlated Emission Schemes and Their Dependence on Pump Fluctuations

C. Benkert[1] *and M.O. Scully*[2]

[1]Max-Planck-Institut für Quantenoptik, D-8046 Garching, Fed. Rep. of Germany
[2]Center for Advanced Studies and Department of Physics and Astronomy, University of New Mexico, Albuquerque, NM 87131, USA

Over the last few years the generation of noise reduced and even squeezed light has been of particular interest in modern quantum optics. Among others, examples involve the degenerate parametric oscillator,[1] four-wave mixing,[2] atomic memory effects[3] and the correlated spontaneous emission laser (CEL).[4] We here discuss a new type of correlated-spontaneous-emission devices which exhibit noise quenching and squeezing. These systems constitute a very simple correlated spontaneous emission scheme in that they involve only single photon transitions, whereas all previous studies were based on correlated photon pairs. However, the amount of noise reduction in these single photon correlated emission schemes are found to depend crucially on the statistical properties of the pump mechanism. In the ordinary laser theory, as well as in many maser and laser related problems, one commonly neglects the effects of the statistical properties of the pump mechanism. But recent work on micromasers[5] and lasers[6] indicates that there is an important relationship between the amplitude fluctuations of the radiation field and the fluctuations of the pump. It was found[7] that in a micromaser a complete pump-noise suppression can lead to a reduction of photon number noise of up to 50% below the shot noise limit. In the laser case, in which the atoms decay to an inert gound state within a characteristic decay time Γ^{-1}, the maximum amount of photon number noise reduction is 25%. However, in both, the ordinary maser and laser case, the statistical properties of the pump only play a significant role under an important requirement: the atoms have to make an appreciable transition between the two lasing levels during the interaction with the field. This implies that for the micromaser the Rabi angle for the atoms has to become of the order π during their transit through the cavity. In the laser case we have to be far above threshold, i.e. involve high intensities, to see an influence of the pump statistics. In contrast, the single photon correlated emission schemes are already strongly influenced by a noise suppression of the pump in the low intensity regime. It is found that even for very small Rabi angles the amplitude fluctuations can be reduced through a combination of atomic coherence and pump-noise suppression. Before we start our discussion of this new type of masers and lasers by putting it into the perspective of the family of correlated spontaneous emission devices.

The first concept of CEL operation was proposed in 1985 and involved atoms with two different lasing transitions (see Fig. 1). In the quantum-beat laser[8a] (Fig. 1a) these two atomic transitions differ in frequency while in the Hanle laser[8b] (Fig. 1b) the difference is given by opposite polarizations. In both cases the photons from the transitions $a-c$ and $b-c$ are emitted in different modes of the radiation field. If there is no coherence between the upper atomic levels a and b, the two modes of the radiation field are completely uncorrelated. Thus the phase of both fields diffuse independently through spontaneous emission so that the relative phase between them gradually takes on all values between 0 and 2π. However, if we establish a coherence between the two upper atomic levels a and b the situation changes drastically. Such a coherence is generated through an injected microwave signal in the quantum beat laser and through an appropriately polarized pump source in the Hanle laser. The coherence between the excited levels a and b leads to a strong correlation of the spontaneous emission noise in the two modes of the radiation field. Whenever a spontaneous emission event occurs in one of the modes, the other mode experiences a corresponding event as depicted in Fig. 2. Therefore

Springer Proceedings in Physics, Vol. 41 **Quantum Optics V**
Editors: J.D. Harvey and D.F. Walls

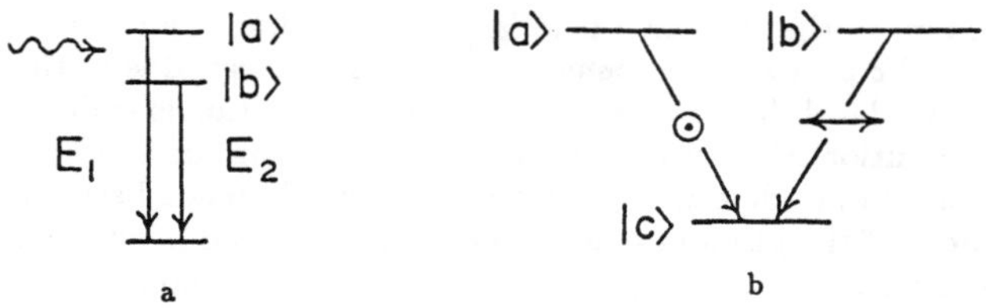

Fig. 1. Two mode CEL schemes: In the quantum beat laser (a) the coherence is induced by a microwave field while in the Hanle laser (b) the coherence is produced through a polarization of the pump source.

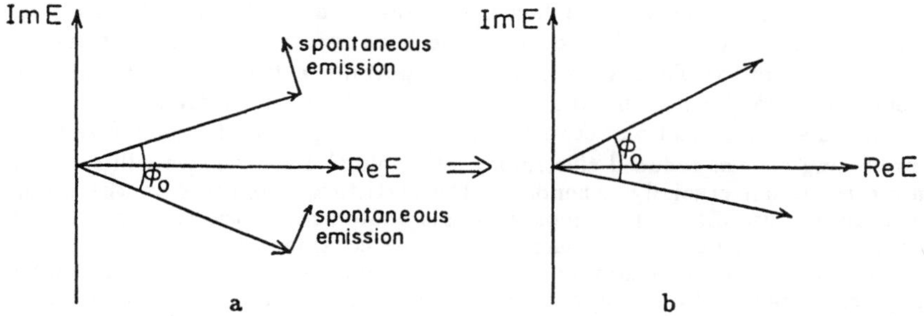

Fig. 2. The effect of a spontaneous emission event on the relative phase between the two modes. Note that the relative phase remains the same before (a) and after (b) the spontaneous emission event.

Fig. 3

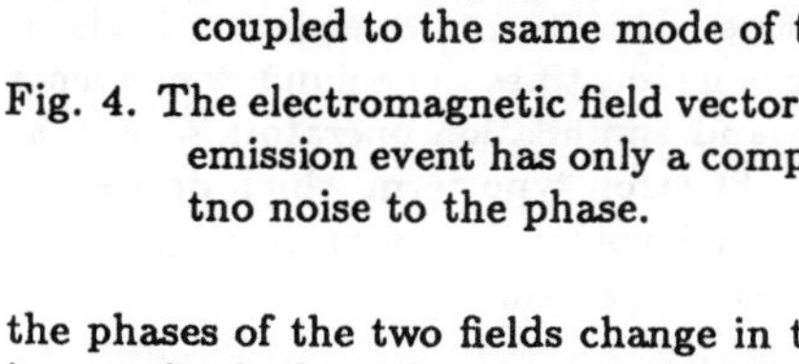

Fig. 4

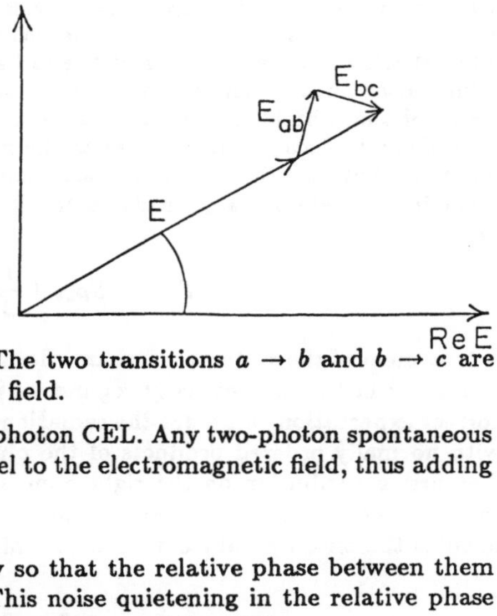

Fig. 3. A coherently prepared three level atom. The two transitions $a \rightarrow b$ and $b \rightarrow c$ are coupled to the same mode of the radiation field.

Fig. 4. The electromagnetic field vector for the two-photon CEL. Any two-photon spontaneous emission event has only a component parallel to the electromagnetic field, thus adding tno noise to the phase.

the phases of the two fields change in the same way so that the relative phase between them is completely free of spontaneous emission noise. This noise quietening in the relative phase between the two radiation modes is the most important feature of the CEL.

We next turn to the two-photon CEL[9] in which a three level atom in a cascade configuration is coupled to the radiation field (Fig. 3). In contrast to the ordinary CEL discussed above, the two photons from the lasing transitions $a-b$ and $b-c$ are emitted into the same mode of the radiation field. If the atoms are initially prepared in a superposition between the upper and lower atomic level, spontaneous emission events from the cascade transition $a-b$ and $b-c$ are again correlated. This is depicted in Fig. 4. If a spontaneous emission event occurs in the first transition, the noise contribution to the radiation field is totally random. However, a subsequent transition from level b to level c partially compensates this randomness, due to the atomic coherence between upper and lower level. The photon from the $b-c$ transition is emitted in such a way that the combined contribution of both transitions to the electromagnetic field always has the same phase. This phase is only determined by the atomic coherence between upper and lower level and is independent of the random phase which was produced in the first transition. We therefore see that in the case in which the phase of the cascade spontaneous emission event coincides with the phase of the radiation field, no noise is added to the phase of the electromagnetic field. Thus the atomic coherence between levels a and c can again lead to a complete quietening of spontaneous emission noise in one quadrature of the field.

While both types of correlated spontaneous emission lasers, discussed above, involved two photons for a noise reduction in the emitted fields, we now turn to the discussion of the polarization CEL,[10] in which only one photon at a time is necessary. The idea for such a single-photon CEL stems from the observation that the important feature of the two photon CEL was a coherence between the starting and finishing level of an atomic transitions. It is therefore plausible that a single transition between two, coherently prepared atomic levels has the potential of producing a noise-reduced electromagnetic field.[11] However, we will see that the amount of noise reduction crucially depends on the statistical properties of the pump mechanism. We will discuss two different concepts of single photon correlated emission. In the case of the polarization correlated emission maser (Fig. 5a) we consider two-level atoms which are injected into a microwave cavity according to some injection statistics. The atoms enter the cavity in a coherent superposition of upper and lower atomic level and start to interact with one mode of the radiation field. After a well-defined interaction time τ they leave the cavity again and have no further influence on the radiation field. We assume that during the whole interaction time the atoms do not loose their excitation or coherence through any decay process. In the case of the polarization correlated emission laser (Fig. 5b) we consider three-level atoms in which the upper two levels constitute the lasing transition. Again, the atoms are injected into the cavity in a coherent superposition of the two excited levels and start to interact with a resonant mode of the radiation field. Instead of removing the atoms from the cavity after an interaction time τ as in the maser case, the interaction is now limited by a decay of the atomic excitation to a lower-lying, inert ground state. We will show how such an additional decay at rate Γ affects the noise fluctuations of the field and leads to a physical difference between the maser and laser concept. In both cases a Langevin analysis up to linear order in the field variable leads to the following equation of motion for the electromagnetic field

$$\dot{\mathcal{E}} = -iS\rho_{ab} + \frac{1}{2}[\alpha(\rho_{aa} - \rho_{bb}) - \gamma]\mathcal{E} + F_{\mathcal{E}} \tag{1}$$

A treatment of the systems in linear order is sufficient because in this context we are specifically interested in the low intensity regime, as mentioned above. This c-number equation leads to correct expectation values for the radiation field, provided one identifies the c-number moments with normally ordered products of the photon creation and annihilation operators $a^\dagger$ and a. The first contribution on the right hand side of (1) is a klystron-type term which drives the radiation field through the injected atomic polarization. It involves the parameter ρ_{ab} which denotes the off-diagonal element of the initial atomic density operator, i.e.

$$\rho_{\text{atom}} = \begin{pmatrix} \rho_{aa} & \rho_{ab} \\ \rho_{ba} & \rho_{bb} \end{pmatrix} \tag{2}$$

The factor S determines the strength of the driving term and is given by $S_M = gR\tau$ and $S_L = gR/\Gamma$ in the maser and laser case respectively. Here g is the coupling constant between atoms and field and R is the mean injection rate of the atoms. The second term on the right hand side of (1) is the familiar linear gain and loss contribution to the radiation field. The parameter α is the linear gain coefficient, given by $\alpha_M = g^2R\tau^2$ and $\alpha_L = (2g^2R/\Gamma^2)$ for

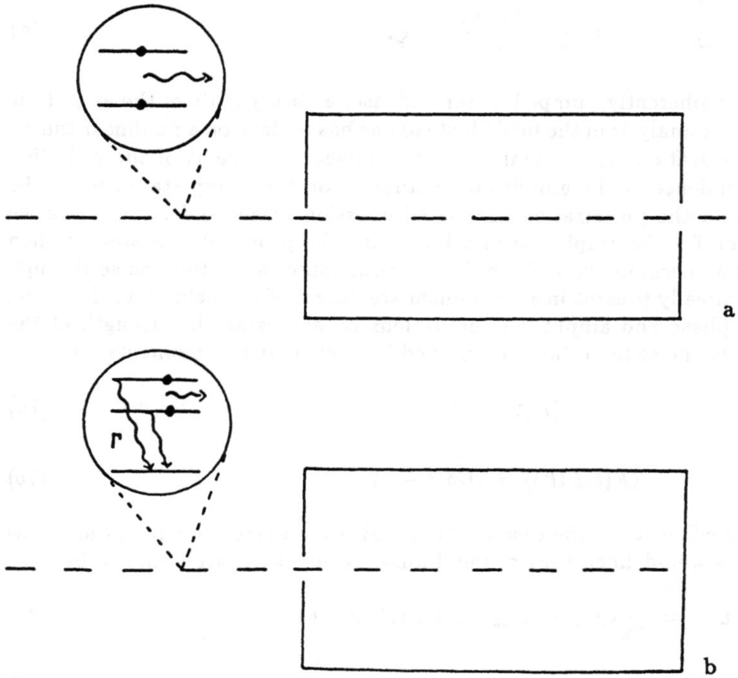

Fig. 5. (a) Polarization correlated emission maser: coherently prepared atoms pass through a microwave cavity within a time τ. (b) Polarization correlated emission laseer: coherently prepared atoms are injected into a cvity and decay to an inert ground state with a rate Γ^{-1}.

the maser and laser case, while γ is the photon decay rate due to cavity losses. Finally the last term in (1) is the Langevin noise force for the electromagnetic field which determines its fluctuations due to spontaneous emission.

For a discussion of phase and amplitude of the radiation field we change to a polar coordinate system and make the identification $\mathcal{E} = re^{i\varphi}$. Substituting this into (1) we obtain the following equations of motion

$$\dot{\varphi} = -\frac{S}{r}|\rho_{ab}|\sin\left[\varphi - \left(\theta - \frac{\pi}{2}\right)\right] + F_\varphi \tag{3}$$

$$\dot{r} = S|\rho_{ab}|\cos\left[\varphi - \left(\theta - \frac{\pi}{2}\right)\right] + \frac{\alpha(\rho_{aa} - \rho_{bb}) - \gamma}{2} r + F_r . \tag{4}$$

Here we defined phase and amplitude of the atomic coherence by $\rho_{ab} = |\rho_{ab}|e^{i\theta}$. Equation (3) is a typical locking equation for the phase and we find the steady state value to be

$$\varphi_0 = \theta - \frac{\pi}{2}. \tag{5}$$

Thus the phase of the radiation field is purely determined by the coherence between upper and lower atomic level. The factor of $\pi/2$ in (5) is due to the fact that the electromagnetic field $\mathcal{E}$ is driven by the imaginary part of the atomic polarization. Substituting the locking phase φ_0 into the equation for r and setting the time derivative equal to zero we find the steady state amplitude in linear order to be

$$r_0 = \frac{2S|\rho_{ab}|}{\gamma - \alpha(\rho_{aa} - \rho_{bb})}. \tag{6}$$

Note that in an ordinary, incoherently pumped maser and laser a steady state of the amplitude cannot be found from a linear analysis in the field. Instead one has to develop a nonlinear theory to obtain the necessary saturation terms. Only such a nonlinear treatment of the radiation field usually shows a dependence of the amplitude fluctuation on the pump statistics.[12] The situation is very different in the polarization correlated emission maser and laser. Here we find a steady state solution for the amplitude in a linear anlysis, provided α is smaller then the cavity decay rate γ. Furthermore we will see that a reduction of amplitude noise through pump noise suppresion is already present in such a linear treatment of the field. We, therefore, turn to the discussion of phase and amplitude fluctuations by analyzing the strength of the Langevin noise forces. These noise forces are determined by their first two moments, i.e.

$$\langle F(t) \rangle = 0 \tag{7a}$$

$$\langle F(t)F(t') \rangle = 2D\delta(t - t') \tag{7b}$$

where D is the diffusion coefficient. In the case of the polarization maser, in which the atoms interact with the field for a well-defined time τ, the diffusion coefficients are found to be

$$D_{\varphi\varphi} = \frac{\alpha}{4\bar{n}}\{\rho_{aa} - p|\rho_{ab}|^2[1 + \cos 2(\varphi - \theta)]\} \tag{8a}$$

$$D_{rr} = \frac{\alpha}{4}\{\rho_{aa} - p|\rho_{ab}|^2[1 - \cos 2(\varphi - \theta)]\} \tag{8b}$$

Here $\bar{n}$ denotes the average number of photons inside the cavity. The parameter p in (8a) and (8b) is characteristic for the particular pumping mechanism and clearly shows the effect of pump fluctuation on the diffusion coefficients. For a poissonian injection statistic, in which the probablility distribution for the time Δt between two consecutive atoms is given by $P(\Delta t) = \exp[-\Delta t/R]/R$, the parameter p has the value $p = 0$. In an other limiting case, in which the atoms are injected regularly, we find $p = 1$. The last case corresponds to a complete noise suppressed pump mechanism because there are no fluctuations in the number of injected atoms. From (8a) and (8b) we see that the atomic coherence between upper and lower level can only lead to a reduction of noise-fluctuations because the terms in the square brackets are always non-negative. Substituting the steady state value for the phase we obtain our final solution in the maser case

$$D_{\varphi\varphi} = \frac{\alpha}{4\bar{n}}\rho_{aa} \tag{9a}$$

$$D_{rr} = \frac{\alpha}{4}\{\rho_{aa} - 2p|\rho_{ab}|^2\}. \tag{9b}$$

Equation (9a) is the well-known Schawlow-Townes result for phase diffusion in an ordinary maser. We therefore see that the phase diffusion in steady state is neither affected by the atomic coherence nor by the pump fluctuation. However, the phase is locked so that the fluctuations are smaller than in a free running laser. The really interesting features of the polarization maser lie in the diffusion coefficient of the amplitude. We find a reduction of the spontaneous emission noise which depends on the amount of atomic coherence and the statistical properties of the pump mechanism. In the case of a poissonian injection statistic, in which $p = 0$, we recover the result of an ordinary, incoherently pumped maser. However, if the pump-noise is suppressed, i.e. $p = 1$, we find a significant reduction of the amplitude fluctuation of the field. For example in the case that upper and lower atomic levels are equally populated, i.e. $\rho_{aa} - \rho_{bb}$, the diffusion coefficient is zero, indicating a complete noise quietening. If ρ_{bb} is larger than ρ_{aa} the diffusion coefficient even becomes negative which corresponds to a squeezing of the fluctuations beyond the vacuum noise level. Thus such a maser-device can produce squeezed states of the radiation field.

We now turn to the analysis of the laser case in which the atoms can decay to a lower lying ground state. From a naive standpoint one could expect that all the maser results would also be valid for the laser case especially as the laser is often denoted as an "optical maser." Indeed, as mentioned before, the drift contribution for the phase and amplitude of the radiation field are the same and are given by (3) and (4). The noise functions, however, are different and we find for the diffusion coefficients in the laser case

$$D_{\varphi\varphi} = \frac{\alpha}{4\bar{n}}\{\rho_{aa} - \frac{p}{2}|\rho_{ab}|^2[1 + \cos 2(\varphi - \theta)]\} \tag{10a}$$

$$D_{rr} = \frac{\alpha}{4}\{\rho_{aa} - \frac{p}{2}|\rho_{ab}|^2[1 - \cos 2(\varphi - \theta)]\} \tag{10b}$$

When comparing these equations with the corresponding ones in the maser case ((8a) and (8b)) we see that the coherence factors differ by a factor of 2. This difference can be best understood by writing the diffusion coefficients in a more persuasive way. For example, the diffusion coefficient for the amplitude can be written as

$$D_{rr} = \frac{\alpha}{8}\{\rho_{aa} + \rho_{aa} - p|\rho_{ab}|^2[1 - \cos 2(\varphi - \theta)]\} \tag{11}$$

the last two terms in the curly brackets are essentially the same as is the maser case. In the laser case we have an additional decay of the atoms which introduces another noise source into the system. This additional noise source is independent of the atomic coherence and is given by the first term in the curly brackets in (11). Therefore the reason for the difference between the maser and laser results is not a decrease in the coherence contribution but an increase of the coherence-independent noise term. We next substitute the steady state phase into (10) and obtain our final result for the diffusion coefficients in the laser case

$$D_{\varphi\varphi} = \frac{\alpha}{4\bar{n}}\rho_{aa} \tag{12a}$$

$$D_{rr} = \frac{\alpha}{4}(\rho_{aa} - p|\rho_{ab}|^2). \tag{12b}$$

Again we find no effect of the atomic coherence or the pump fluctuation on the steady state diffusion coefficient for the phase. In contrast,the amplitude diffusion, can be reduced by the atomic coherence, depending on the statistical properties of the pump. However, the reduction is smaller than in the maser case as discussed before. We do not find squeezing in a linear order analysis in the laser case. This is a significant, physical difference between the laser and maser.

The results, presented in this paper, are based on a Langevin analysis. Further insight into the problem of the polarization maser and laser and the role of pump fluctuation can be gained through a different approach, involving the density matrix for the radiation field. A detailed discussion[7] of the statistics of the incoming atoms leads to the following master equation for the density matrix of the field.

$$\dot{\rho}(t) = \frac{R}{p}\ln[1 + p(M - 1)]\rho(t) \tag{13}$$

Here R is again the mean injection rate. The operator M specifies the evolution of the field due to the interaction with a single atom and includes the effect of initial atomic coherence. This generalized master equation also emphasizes the important influence of pump fluctuations on the field. One can now use (13) in order to discuss the fluctuations in amplitude and phase of the electromagnetic field. Such an analysis produces the same results as the Langevin approach, presented here, and will be given elsewhere.

So far we have only discussed the resonant case in which the atomic transition frequency coincides with the frequency of the cavity. We will now discuss the effects that a broad frequency distribution over the atomic transition frequencies has on the polarization maser and laser results. As we saw before, a poissonian injection statistic leads back to the results of an ordinary, incoherently pumped maser of laser. We therefore restrict the following analysis to the interesting case of a regular injection, i.e. $p = 1$. From a naive standpoint one could expect that inhomogeneous broadening of the atomic line would destroy the noise reducing features of the atomic coherence. However, a careful analysis yields the surprising result that inhomogeneous broadening essentially redistributes the noise reducing terms between phase and amplitude. For a broad frequency distribution we find the steady state diffusion coefficients in the maser case to be

$$D_{\varphi\varphi} = \frac{\tilde{\alpha}}{4\bar{n}}\{\rho_{aa} - |\rho_{ab}|^2\} \tag{14a}$$

$$D_{rr} = \frac{\tilde{\alpha}}{4}\{\rho_{aa} - |\rho_{ab}|^2\}. \tag{14b}$$

Here $\tilde{\alpha}$ is the linear gain coefficient for large inhomogeneous broadening. Comparing the results in (14) with the ones for the resonant case in (9) we see that the noise reduction has been shifted from the amplitude to the phase. This can be most easily understood through the general expressions for the diffusion coefficients in (8). The noise reducing terms proportional to $|\rho_{ab}|^2$ consist of a phase-independent and a phase-dependent contribution. In the phase locked case these terms conspire such that they cancel each other in the phase diffusion but add up in the amplitude diffusion. This leads to the results in (9). The effect of inhomogeneous broadening is essentially an average over all possible phases. Therefore, in the case of a broad atomic frequency distribution the phase-dependent terms vanish, leaving an equal coherence contribution for phase and amplitude diffusion.

The laser result is modified in an analogous way through inhomogeneous broadening. In the limit of a broad atomic line we find the steady state diffusion coefficients for the laser to be

$$D_{\varphi\varphi} = \frac{\tilde{\alpha}}{4\bar{n}}\left(\rho_{aa} - \frac{1}{2}|\rho_{ab}|^2\right) \tag{15a}$$

$$D_{rr} = \frac{\tilde{\alpha}}{4}\left(\rho_{aa} - \frac{1}{2}|\rho_{ab}|^2\right). \tag{15b}$$

We have summarized our findings for the polarization correlated emission maser and laser in Table 1. In an analysis which is linear in the radiation field, we have found that significant noise reduction due to the initial atomic coherence between the two lasing levels is possible. The amount of noise reduction crucially depends on the statistical properties of the pump mechanism. For a poissonian injection statistic of the atoms the diffusion coefficients for phase and amplitude were found to be indentical to the ones of ordinary, incoherently pumped masers and lasers. However, if the pump noise is suppressed the atomic coherence leads to a reduction of amplitude fluctuations of the field. While in the laser case we can achieve at maximum a perfect noise quietening, the maser case even exhibits squeezing of spontaneous emission noise in the amplitude. In that sense the laser and maser behave very differently. Finally we found that inhomogeneous broadening resulted in a redistribution of the noise-reducing coherence terms. The noise reduction was partially shifted from the amplitude to the phase diffusion.

Table 1. Diffusion coefficients for phase and amplitude of the radiation field for maser and laser case.

Diffusion Coefficients		Maser	Laser
General	$D_{\varphi\varphi}$	$\frac{\alpha}{4\bar{n}}\{\rho_{aa} - p\lvert\rho_{ab}\rvert^2[1+\cos 2(\varphi-\theta)]\}$	$\frac{\alpha}{4\bar{n}}\{\rho_{aa} - \frac{1}{2}p\lvert\rho_{ab}\rvert^2[1+\cos 2(\varphi-\theta)]\}$
	D_{rr}	$\frac{\alpha}{4}\{\rho_{aa} - p\lvert\rho_{ab}\rvert^2[1-\cos 2(\varphi-\theta)]\}$	$\frac{\alpha}{4}\{\rho_{aa} - \frac{1}{2}p\lvert\rho_{ab}\rvert^2[1-\cos 2(\varphi-\theta)]\}$
Locked	$D_{\varphi\varphi}$	$\frac{\alpha}{4\bar{n}}\rho_{aa}$	$\frac{\alpha}{4\bar{n}}\rho_{aa}$
	D_{rr}	$\frac{\alpha}{4}(\rho_{aa} - 2p\lvert\rho_{ab}\rvert^2$	$\frac{\alpha}{4}(\rho_{aa} - p\lvert\rho_{ab}\rvert^2)$
Broadened	$D_{\varphi\varphi}$	$\frac{\alpha}{4\bar{n}}(\rho_{aa} - p\lvert\rho_{ab}\rvert^2)$	$\frac{\alpha}{4\bar{n}}(\rho_{aa} - \frac{1}{2}p\lvert\rho_{ab}\rvert^2)$
	D_{rr}	$\frac{\alpha}{4}(\rho_{aa} - p\lvert\rho_{ab}\rvert^2)$	$\frac{\alpha}{4}(\rho_{aa} - \frac{1}{2}p\lvert\rho_{ab}\rvert^2)$

Acknowledgment

We would like to thank J. Bergou, L. Davidovich, N. Lu, and M. Orszag for helpful discussions. This work was supported in part by the Office of Naval Reseach.

References

1. B. Yurke, Phys. Rev. **A29**, 408 (1984); M. J. Collett and C. W. Gardiner, Phys. Rev. **A30**, 1386 (1984), ibid. **A31**, 3761 (1985); C. W. Gardiner and C. M. Savage, Opt. Commun. **50**, 173 (1984); M. J. Collet and D. F. Walls, Phys. Rev. **A32**, 2887 (1985); and L. Wu, M. Xiao, and H. J. Kimble, J. Opt. Soc. Am. **B4**, 1465 (1987).
2. H. P. Yuen and J. H. Shapiro, Opt. Lett. **4**, 334 (1979); R. E. Slusher, L. W. Hollberg, B. Yurke, J. C. Mertz and J. F. Valley, Phys. Rev. **A31**, 3512 (1985); M. D. Reid and D. F. Walls, Phys. Rev. **A31**, 1622 (1985); D. A. Holm, M. Sargent III, and B. A. Capron, Opt. Lett. **11**, 443 (1986); D. A. Holm and M. Sargent III, Phys. Rev. **A35**, 2510 (1987); and R. E. Slusher, B. Yurke, P. Grangier, A. La Porta, D. F. Walls and M. Reid, J. Opt. Soc. Am. **B4**, 1453 (1987).
3. M. O. Scully, G. Süssmann, and C. Benkert, Phys. Rev. Lett. **60**, 1014 (1988).
4. M. O. Scully, Phys. Rev. Lett. **55**, 2802 (1985); and W. Schleich and M. O. Scully, Phys. Rev. **A35**, 1261 (1988).
5. P. Filipowicz, J. Javanainen, and P. Meystre, Opt. Commun.. **58**, 327 (1986); P. Filipowicz, J. Javanainen, and P. Meystre, Phys. Rev. **A34**, 3077 (1986); J. Krause, M. O. Scully, and H. Walther, Phys. Rev. **A34** 2032 (1986); L. A. Lugiato, M. O. Scully and H. Walther, Phys. Rev. **A36**, 740 (1987); and L. Davidovich, J. M. Raimond, M. Brune, and S. Haroche, Phys. Rev. **A36**, 3771 (1987).
6. Y. Yamamoto, S. Machida, and O. Nilsson, Phys. Rev. **A34**, 4025 (1986); Y. Yamamoto, and S. Machida, Phys. Rev. **A35**, 5114 (1987); and M. A. Marte, H. Ritsch, and D. F. Walls, Phys. Rev. Lett. **61**, 1093 (1988).
7. J. Bergou, L. Davidovich, M. Orszag, C. Benkert, M. Hillery, and M. O. Scully, to be published.
8. (a) M. O. Scully and M. S. Zubairy, Phys. Rev. **A35**, 752 (1987); J. Bergou, M. Orszag, and M. O. Scully, Phys. Rev. **A38**, 754 (1988); J. Bergou and M. Orszag, Phys. Rev. **A38**, 763 (1988).
8. (b) J. Bergou, M. Orszag, and M. O. Sxully, Phys. Rev. **A38**, 768 (1988).
9. M. O. Scully, K. Wódkiewicz, M. S. Zubariy, J. Bergou, N. Lu, and J. Meyer ter Vehn, Phys. Rev. Lett. **60**, 1832 (1988).
10. C. Benkert, M. O. Scully, J. Bergou and N. Lu, submitted to Phys. Rev. Lett.
11. This idea was also presented in a discussion of the Jaynes-Cummings model in which the atoms start from a coherent superposition of states, see, for example, K. Wódkiewicz, P. L. Knight, S. J. Buckle, and S. M. Barrett, Phys. Rev. **A35**, 2567 (1987).
12. In an incoherently pumped laser the statistical properties of the pump do not affect terms of linear order in the field. Insted, only the saturation terms are affected. See, for example, Ref. 7 or Y. Yamamoto, S. Machida and O. Nilson, Phys. Rev. **A34**, 4025 (1986).

Stochastic Position Noise, or Why a Laser Beam Cannot Go Straight

M.D. Levenson, W.H. Richardson, and S.H. Perlmutter

Research Division, IBM Corporation, San Jose, CA 95120, USA

The centroid of a single transverse mode laser beam fluctuates in position because of spontaneous emission of the laser medium into higher transverse cavity modes. Such effects can confuse a variety of measurements in quantum optics, but also point the way to new phenomena worthy of study.

I. Introduction

I am sometimes asked where new effects come from. If you are an experimentalist, new things sometimes just jump out at you. Often, one is not amused, but if you keep your wits, unexpected new effects can be valuable and interesting, even if they keep you from getting the results you expected! Of course, it does help to talk with people - especially theorists - to prepare one's mind for the unexpected. In this project, Dr. Aram Mooradian made key suggestions early on.

Stochastic position noise of single mode laser beams appeared in experiments intended to show that semiconductor diode lasers can produce squeezed light. We had set up a standard dual differential homodyne detector and found that the noise on the difference current was well above the noise on the sum. Normally, that is a sure signature of squeezing as the noise on the difference current results only from vacuum noise input through the detector beam splitter [1]. However, an independent measurement of the vacuum noise level showed that both the sum and difference current signals were above the vacuum! What was going on?

The photodiodes we had used in that experiment had very small active areas, and it turned out that the laser beam was off-center. Motion of the laser beam might produce excess noise in the difference current, but how could the laser beam move at multi-megahertz frequencies? Answering that question led us to study stochastic position noise in well-characterized lasers.

II. Transverse Modes and Spontaneous Emission

Lasers typically have cavities formed by mirrors around the gain medium, and these cavities have rich spectra of excitation modes. In the case of most lasers, these modes are well described by the Li-Kogelnik TEM_{mnq} spatial wave functions [2]. The frequency of the TEM_{mnq} mode is given by $\nu_{mnq} = q\nu_L+(m+n)\nu_T$, where the longitudinal mode spacing is $\nu_L = c/2L$ and the transverse mode spacing depends on the focussing properties of the mirrors and laser medium. The laser spots produced by different modes are diagrammed schematically in figure 1.

Modern lasers are designed to oscillate in the lowest order transverse mode, the TEM_{00q}. The cavities of single mode lasers also prevent oscillation in modes other than one with $q=q_0$. Spontaneous emission by the laser medium into that oscillating mode

Springer Proceedings in Physics, Vol. 41 Quantum Optics V
Editors: J.D. Harvey and D.F. Walls

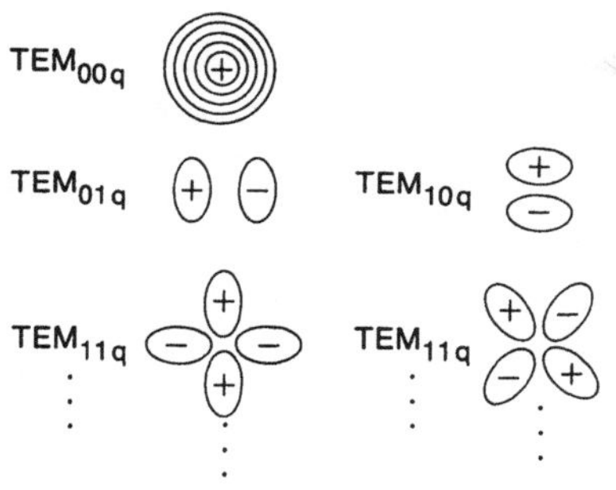

Fig. 1. The symmetries of the various modes of an axially symmetric laser cavity. The plus and minus signs indicate relative phases.

causes the familiar phase diffusion and Schawlow Townes linewidth [3]. What about spontaneous emission into other cavity modes? In particular, what about the higher transverse modes, where spontaneous emission can add momentum perpendicular to the axis of the laser beam? Might that cause the laser spot to move on the detector surface?

Spontaneous emission into TEM$_{ooq}$ modes where $q \neq q_o$ beats with the strongly oscillating TEM$_{ooq_o}$ mode on the surface of a uniform photodetector, producing current noise at multiples of the longitudinal mode frequency ν_L. The wavefunctions of the higher order transverse modes are orthogonal to that of the TEM$_{ooq}$. The heterodyne efficiency thus vanishes - but only for a uniform detector.

A split detector system such as that in Fig. 2 can detect emission in the TEM$_{01q}$ modes. In such a system, the current produced by one side of the detector is electronically subtracted from the current of the other side. When the TEM$_{oo}$ mode is centered on such a system, the noise peaks corresponding to emission into the TEM$_{ooq}$ modes vanish. There still is noise due to vacuum fluctuations, but those fluctuations correspond to odd symmetry modes such as the TEM$_{01}$ [4]. Figure 2 illustrates how the products of the wave functions which appear in the theory of heterodyne detection are altered by the use of a split differential detector. The even symmetry terms are made to cancel out by the odd symmetry of the detector response. The mutual orthogonality of the TEM$_{mnq}$ modes insures that only the TEM$_{01}$ overlaps properly with the TEM$_{oo}$ to produce a substantial heterodyne signal.

Detectors with other symmetries can detect the beats between the TEM$_{oo}$ local oscillator and spontaneous emission in other transverse modes. In particular, a quadrant detector with diagonally opposite segments summed and then differenced is sensitive to the TEM$_{11q}$ modes. All these detectors produce the same vacuum noise level,

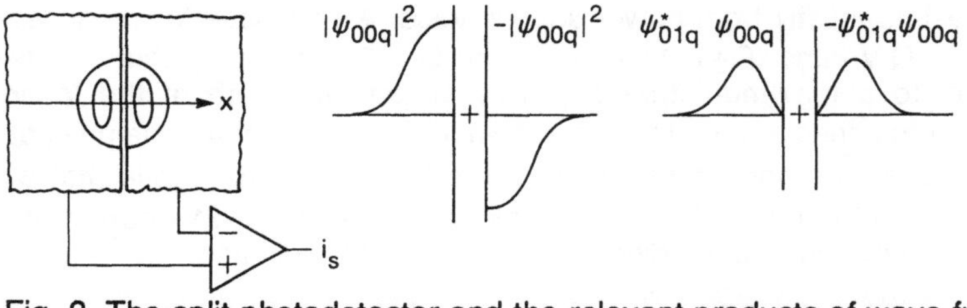

Fig. 2. The split photodetector and the relevant products of wave functions for heterodyne detection of TEM$_{01}$ emission. The split detector output current i_s has no contribution from even-symmetry products of wave functions since the contributions from the left and right sides cancel out. Odd-symmetry overlaps are differenced by the detector to produce a net contribution to the current.

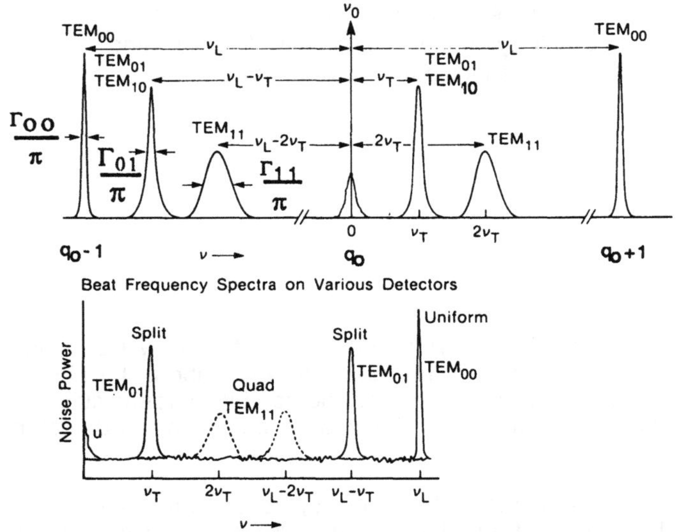

Fig. 3. The emission and heterodyne beat spectrum of a typical single mode laser. The TEM_{oo} mode at ν_o (the TEM_{ooq_o} mode) oscillates strongly and acts as a local oscillator for the detection of the other modes. The shift frequencies for the higher order transverse modes are indicated. The linewidths of the emissions vary because of differences in the diffraction loss. The beat frequency spectra differ for the various possible detectors as shown below.

independent of the summing or differencing and proportional only to the total D.C. photocurrent. This illustrates the equality of the vacuum fluctuations of all normalized modes.

Figure 3 shows the expected optical frequency spectrum of spontaneous emission into various cavity modes. The mode at ν_o oscillates and acts as the local oscillator for the detection of the very much weaker fields in the other modes. Diffraction loss varies with the transverse mode indices m and n. The TEM_{oo} modes have the least loss and the narrowest linewidths.

Figure 3 also shows the r.f. spectrum of current fluctuations due to the heterodyne detection of the stochastic amplitudes of modes below threshold. A uniform detector produces signals at multiples of ν_L due to the TEM_{ooq} modes as well as a wing centered at zero frequency that results mostly from power supply noise. A split detector produces heterodyne peaks at $\nu_T + Q\,\nu_L$ and $(Q+1)\,\nu_L - \nu_T$, where $Q \geq 0$ is an index that counts axial mode spacings up to a maximum that depends on the linewidth of the laser transition. Only the part of the spectrum up to ν_L is shown in Fig. 3. Note that the peak at ν_T corresponds to a transverse mode on one side of the local oscillator, while that at $\nu_L - \nu_T$ represents a mode on the other side. The peaks at $2\nu_T$ and $\nu_L - 2\nu_T$ due to the TEM_{11} mode, detected by a quadrant photodiode system, are also indicated.

<u>III. Stochastic Position Noise Theory</u>

The spectral features due to the beats between the TEM_{01} and TEM_{oo} modes can be considered as arising from displacement of the laser beam from its average position [5].

The wave function for a TEM_{oo} beam displaced slightly from the axis can be expanded as a Taylor series in the displacement.

$$\psi_{oo}(x+\delta, y) = \psi_{oo}(x,y) - \frac{\delta}{w}\,\psi_{01}(x,y) + \ldots \qquad (1)$$

where $\psi_{oo}(x,y) = \sqrt{\frac{2}{\pi}}\,\frac{1}{w}\exp\left(-\frac{x^2+y^2}{w^2}\right)$ is the TEM_{oo} wave function in a given plane, and $\psi_{01} = 2\sqrt{\frac{2}{\pi}}\,\frac{x}{w^2}\exp\left(-\frac{x^2+y^2}{w^2}\right)$ is the TEM_{01} wave function in the same plane.

The instantaneous displacement can be written in terms of the split detector difference current:

$$\delta(t) = w\,\frac{i_s(t)}{i_u(t)} \approx w\,\frac{i_s(t)}{i_o} \qquad (2)$$

where the current $i_s(t)$ is defined in figure 2, $i_u(t)$ is the uniform detector response, and i_o is the total detector photocurrent. Now, both the displacement and split detector current are zero average stochastic quantities.

$$< \delta(t) > = < i_s(t) > = 0 \qquad (3)$$

$$< |\delta(t)|^2 > \neq 0\,, \quad < |i_s(t)|^2 > \neq 0 \qquad (4)$$

The spectrum of the displacement can be defined and expressed in terms of the spectrum of the difference current noise:

$$S_\delta(\omega) = \int <\delta(t)\delta^*(0)> e^{-i\omega t}\,dt = \frac{w^2}{i_o}\int <i_s(t) i_s^*(o)>\, e^{-i\omega t}\,dt$$
$$= \frac{w^2}{i_o}\, i_s^2(\omega) \qquad (5)$$

Thus the observation of the spectral components predicted in Fig. 3 on the output of a split detector must be interpreted as random motion of the laser beam: stochastic displacement. Now, there will always be some noise on the output of a split detector, if only the vacuum noise. The presence of these vacuum fluctuations implies a quantum limit on the precision with which the center of the laser beam can be determined.

Since both the spontaneous emission and the vacuum fluctuations propagate at the speed of light along the output beam, the laser beam is never instantaneously exactly straight! Figure 4, however, illustrates a major subtlety: the fluctuations causing displacement in the far field are independent of those which alter the beam location in the near field. Spontaneous emission and vacuum fluctuations have two quadratures; one beats with the amplitude in the near field while the other alters the phase front. In the far field, the roles are reversed. Phase shifts in the near field become beam displacements in the far field and vice versa [2].

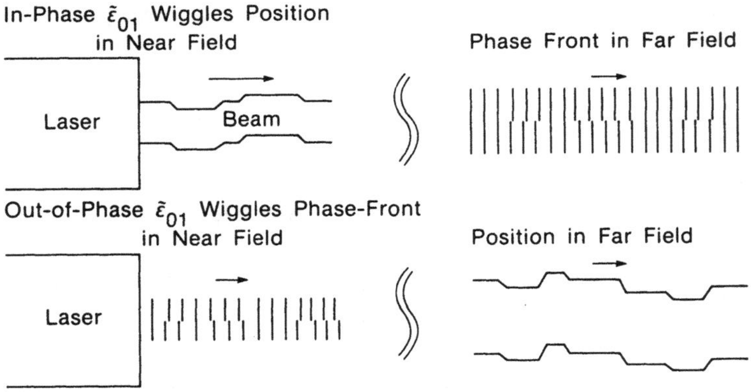

Fig. 4. The two phase quadratures of the TEM_{01} spontaneous emission have different effects on the TEM_{00} beam. In the near field one quadrature beats with the field amplitude, translating the beam, while the other affects the phase front, causing the beam direction to vary. In the far field, diffraction reverses these roles.

IV. Experimental Features of Stochastic Position Noise

Figure 5 portrays the actual experiment, which employed a hybrid junction to take the sum and difference of the currents from two detector segments [4]. Figure 6, shows the spectrum analyzer output for data taken using our well-characterized Kr^+ laser oscillating at 647 nm [6]. When the sum of the currents from the two segments of the detector is resolved on the spectrum analyzer, a peak appears at ν_L = 83 MHz along with a wing near zero frequency. When the difference current is resolved, two peaks appear one at ν_T = 25 MHz and another at $\nu_L - \nu_T$ = 58 MHz. Below the peaks is a constant noise floor due to vacuum fluctuations.

The vacuum noise floor corresponds to a mean square displacement of the beam of

$$<|\delta_V|^2> = w^2 \frac{\hbar\Omega}{P_o} \frac{\Delta\omega}{2\pi} \tag{6}$$

where P_o is the TEM_{00} laser power and w is the $1/e^2$ beam radius. For a 30 mW laser having a 1 mm diameter beam, the mean square displacement is 2.5 $(nm)^2$ per Hz of bandwidth due to vacuum fluctuations! The presence of spontaneous emission into TEM_{01} modes enhances the displacement; in general:

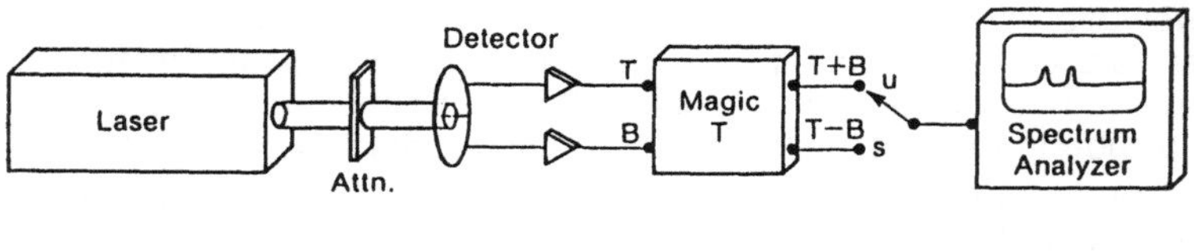

Fig. 5. Schematic of the experiment. The laser beam passed through a calibrated attenuator and fell on a split photodiode. The two detector outputs were amplified and either summed to produce i_u or differenced to yield i_s in a hybrid junction magic T. The electrical currents were then resolved in frequency with an H.P. 8568B spectrum analyzer.

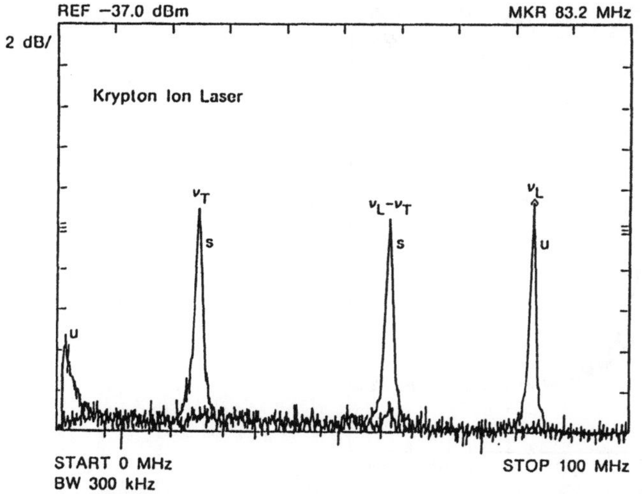

Fig. 6. Spectrum analyzer traces for the uniform (u) and split (s) detectors. The uniform detector signal shows a peak at ν_L at 83.2 MHz, while the split detector shows the expected two peaks. Both give the same vacuum noise level at the bottom of the trace. The vertical scale is logrithmic, at 2 dB per unit.

$$S_\delta(\omega) = w^2 \frac{\hbar\Omega}{P_o}\left[1 + F_{01}(\omega)\right] \tag{7}$$

where the spontaneous emission terms reponsible for the peaks in Fig. 6 are represented by $F_{01}(\omega)$, which will be defined in (19). We find that the maximum value of $F_{01}(\omega)$ is between 1 and 1000. Similar formulae apply to to modes of other symmetry.

Figure 7 compares spectra obtained with a split and a quad detector. The shift between the TEM_{01} and TEM_{11} peaks is $\nu_T = 25.45$ MHz. The TEM_{11} peaks are

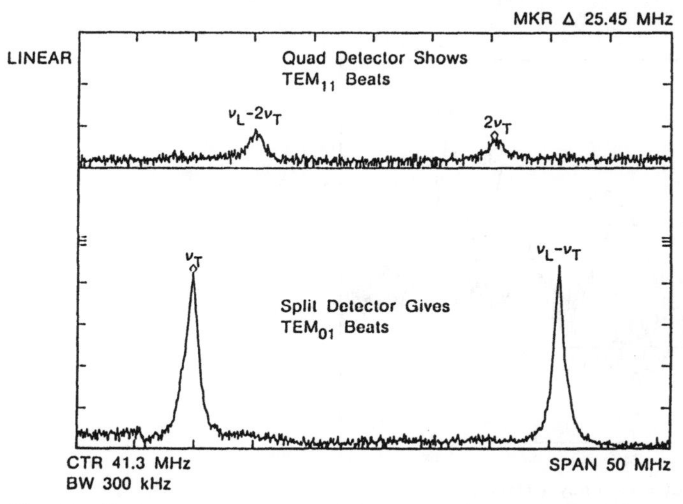

Fig. 7. Spectrum analyzer traces showing signals from split and quad detectors. The shift between the ν_T peak and the $2\nu_T$ peak is 25.45 MHz. The vertical scale is linear, with vacuum noise at bottom.

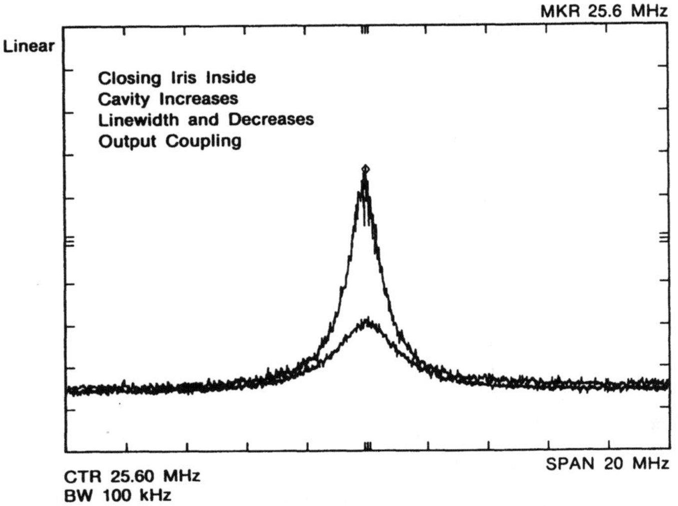

Fig. 8. The effect of closing the intracavity iris. The laser power P_0 did not change, and hence the vacuum noise level is constant. However, the increased loss for the TEM_{01} mode at 25.6 MHz clearly broadened and weakened the heterodyne peak on this linear plot.

weaker and broader because less of the spontaneous emission couples out through the laser mirror and more is lost to diffraction. Figure 8 shows the effect of increasing the diffraction loss for the TEM_{01} by closing the intra-cavity iris. The effect is to weaken and broaden the peak, but not to eliminate it.

The noise features due to excess laser wiggle can be eliminated by spatially filtering the output with a single mode optical fiber. When the fiber is properly centered on the beam, there is no component at ν_T. When the fiber is misaligned, the noise peak

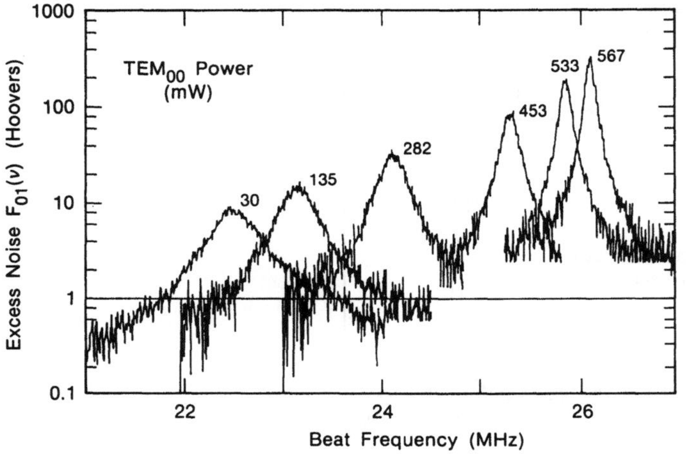

Fig. 9. Stochastic position excess noise traces normalized to the vacuum level and plotted for various values of the laser power P_0. The vertical scale is in units of the vacuum noise, defined as 1 Hoover.

reappears, but only for a uniform detector. The fiber truly does not transmit excess noise in higher transverse modes.

As the discharge current is increased, the TEM_{00} output also increases, but the stochastic position noise peaks increase in amplitude, narrow and shift in frequency. Figure 9 shows some traces illustrating these effects. The increase in amplitude and decrease in linewidth are well explained by the theory that follows. The shift remains a bit of a mystery.

V. Spontaneous Emission Dynamics

The physics of the spontaneous emission terms $F_{mn}(\omega)$ is best discussed in a extension of the Risken van der Pol oscillator theory of the laser. The strongly oscillating TEM_{ooq_o} obeys the usual van der Pol oscillator equation:

$$\beta = \left(G - \kappa_o\right)\beta - G\frac{|\beta|^2}{n_o}\beta + \frac{\mu}{\gamma_\perp}\Gamma_v{}^{oo} \tag{8}$$

where β is the amplitude in the cavity in units where $|\beta|^2$ is photon number, κ_o is the loss for this mode,

$$G = \frac{\mu^2}{\gamma_\perp}\left(N_2{}^{oo} - N_1{}^{oo}\right) \tag{9}$$

is the unsaturated gain in frequency units and the saturation parameter is:

$$n_o = \frac{\gamma_{||}\,\gamma_\perp}{4\mu^2} \tag{10}$$

The atomic medium is assumed homogeneously broadened, with dipole coupling rate μ, and longitudinal and transverse relaxation rates $\gamma_{||}$ and $\gamma_\perp$, respectively. The population inversion is $N_2{}^{oo} - N_1{}^{oo}$ for this mode, and the noise input to the cavity has the property:

$$<\Gamma_v{}^{oo}(t)\,\Gamma_{v^*}{}^{oo}(t')> \;=\; N_2{}^{oo}\,\gamma_\perp\delta\,(t - t') \tag{11}$$

Physically the noise term at the far right of (8) corresponds to spontaneous emission from the upper laser level. It has been treated in the laser literature as vacuum fluctuations amplified by the gain medium with a correction factor to account for the fact that only the upper level population and not the inversion is effective in emission [3]. The relevant populations are those within the modal volume of the individual modes, which may vary. In terms of the unsaturated gain factor G, the noise source terms in (8), (14) and (20) have the correlation:

$$<\frac{\mu}{\gamma_\perp}\Gamma_v{}^{mn}(t)\,\frac{\mu^*}{\gamma_\perp}\Gamma_{v^*}{}^{mn}(t')> \;=\; G\,\frac{N_2{}^{mn}}{N_2{}^{mn} - N_1{}^{mn}}\,\delta(t - t') \tag{12}$$

The average amplitude of the TEM_{ooq_o} mode is then

$$|\beta_o|^2 = n_o\,(1 - \kappa_o/G) \tag{13}$$

The higher order transverse modes obey equations of the form:

$$\alpha = (G - \kappa_o - \eta)\,\alpha - (1-\phi)\,G\frac{|\beta_o|^2}{n_o}\,\alpha + \frac{\mu}{\gamma_\perp}\,\Gamma_v{}^{mn} \qquad (14)$$

The second term on the right results from cross saturation by the TEM_{ooq_o} mode [8]. The cross saturation is much less than for sub-threshold TEM_{oo} modes as there are substantial numbers of atoms that can interact with larger diameter TEM_{mnq} modes but not with the TEM_{ooq_o}. The loss may also be substantially larger because of diffraction from the edges of the laser medium. Hence the excess loss term η may be as large or larger than the TEM_{ooq_o} loss κ_o. The spectral densities of these modes are formally:

$$<|\alpha(\Omega)|^2> = \frac{G/(1 - N_1{}^{mn}/N_2{}^{mn})}{(\eta - g)^2 + (\Omega - 2\pi\nu_{mnq})^2} \qquad (15)$$

where $g = \phi G|\beta_o|^2/n_o$ is the excess gain of the TEM_{mnq} mode above that of the TEM_{ooq_o} mode, the gain of which is clamped at threshold by the saturation of the laser medium. This theory predicts that the excess gain is linear in the circulating power of the TEM_{ooq_o} field. The fields outside the laser cavity depend on the output coupling $\kappa_{out} = \kappa_o$ as well as the fields inside. Applying the usual formulae, we find the TEM_{ooq_o} output power to be [9]:

$$P_o = \hbar\Omega\kappa_{out}\,|\beta_o|^2 = \hbar\Omega\kappa_{out}\,n_o\,(1 - \kappa_o/G) \qquad (16)$$

The output spectral fluxes of the sub-threshold TEM_{mnq} modes are

$$F_{mnq}(\Delta\Omega) = \frac{\kappa_{out}G/\left(1 - N_1{}^{mn}/N_2{}^{mn}\right)}{\eta^2\left(1 - g/\eta\right)^2\left(1 + \Delta\Omega^2/\Gamma_{mn}{}^2\right)} \qquad (17)$$

where the linewidth is:

$$\Gamma_{mn} = \eta(1 - g/\eta) \qquad (18)$$

and the enhancement factor from (7) is:

$$F_{01}(\omega) = \sum_{Q=0}^{Q_{max}} \left\{ F_{01q_o+Q}\left(\omega - 2\pi\left[\nu_T + Q\nu_L\right]\right) + F_{01q_o-Q}\left(\omega - 2\pi\left[(Q+1)\nu_L - \nu_T\right]\right)\right\} \qquad (19)$$

where the r.f. noise frequency is $\omega = |\Omega - 2\pi\nu_o|$ and an appropriate sum has been taken over the longitudinal index q. The dummy variable Q counts from the strongly oscillating q_o mode to the edge of the gain band of the medium, hence $0 \leq Q \leq Q_{max} = \nu_g/2\nu_L$.

The loss coefficients for the other TEM_{oo} modes is higher than for the TEM_{ooq_o}, and they do not oscillate. Their amplitudes obey equations of the form:

$$\zeta = (G - \kappa)\zeta - G\frac{|\beta_o|^2}{n_o}(1 - \varepsilon)\zeta + \frac{\mu}{\gamma_\perp}\ \Gamma_V^{oo} \tag{20}$$

where the term proportional to $|\beta_o|^2$ results from cross saturation by the strongly oscillating TEM_{ooq_o} mode. In a standing wave laser cavity, the coefficient of this cross-saturation is typically only slightly less than that in equation (8) because relatively few atoms lie at nodes of the TEM_{ooq_o} field. The spectral density of these sub-threshold TEM_{ooq} modes is:

$$<|\zeta(\Omega)|^2> = \frac{G/(1 - N_1^{oo}/N_2^{oo})}{(\kappa - \kappa_o - g_{oo})^2 + (\Omega - 2\pi\nu_{ooq})^2} \tag{21}$$

The population factors in the numerators correspond to TEM_{oo} mode profiles. The excess gain is $g_{oo} = \varepsilon G|\beta_o|^2/n_o$ which is small. Note that in the representation where the average amplitude of the TEM_{ooq_o} mode has zero phase, the two orthogonal quadrature amplitudes of the sub threshold modes are equal in this approximation.

VI. Heterodyne Detection of Higher Order Modes

The photodetector noise currents depend on the overlap intregals for the various modes with the TEM_{ooq_o} which acts as local oscillator, the local oscillator power P_o and the photon content. The heterodyne efficiency for the uniform photodetector (H_U) is unity for the TEM_{oo} modes and zero for the hiqher order modes. For the TEM_{01} mode and the split detector, the heterodyne efficiency is $H_S=2/\pi$. This quantity is zero for even symmetry modes.

The spectrum analyzer measures a vacuum noise r.f. power spectral density of

$$i_V^2(\omega) = ei_o\frac{\Delta\omega}{2\pi} \tag{22}$$

for both uniform and split detector configurations. The D.C. detector current is related to the TEM_{ooq_o} power by $i_o = eP_o/\hbar\Omega$. The r.f. noise power is then:

$$i_j^2(\omega)/i_V^2(\omega) = 1+H_j \sum_{mnQ}^{Q_{max}}{}' \Big\{F_{mnq_o+Q}\Big(\omega - 2\pi\big[m+n\big]\nu_T+Q\nu_L\Big) + F_{mnq_o-Q}\Big(\omega - 2\pi\big[(Q+1)\nu_L-(m+n)\nu_T\big]\Big)\Big\} \tag{23}$$

where the prime on Σ' indicates that only certain terms in the sum over m and n need be considered for a particular choice of detector symmetry (i.e. H_U, H_S or H_Q etc.). The detuning of the r.f. noise frequency ω from the heterodyne beat frequency has been shown explicitly. The longitudinal mode index is $Q = q - q_o$ where $0 \le Q \le Q_{max} < \nu_g/2\nu_L$. Note that this is a generalization of (5) and (7).

Similar formulae apply to all modes and detector systems. Equations (17-23) make very specific predictions about the peak heights and linewidths observed in stochastic position noise spectra traces as in Fig. 8. The linewidth is predicted to decrease linearly with TEM_{ooq_o} output power:

$$\Gamma_{mn} = \eta\left(1 - \frac{\phi G P_o}{\eta n_o \hbar\Omega \kappa_{out}}\right) \tag{24}$$

and the product of peak height and linewidth squared is predicted to increase slightly since $G = \kappa_o + g$ between TEM_{ooq_o} threshold and TEM_{mnq} threshold.

In contrast, the linewidth of the noise due to sub-threshold TEM_{oo} modes should vary little with TEM_{ooq_o} power. In practice, we find that the loss of these modes varies as the laser structure heats and cools, moving cavity resonances in and out of the etalon pass-band, and making reproducible data difficult to obtain.

While vacuum fluctuations of the TEM_{01} and TEM_{10} modes can cause the laser beam to fluctuate in position at all frequencies, the effects of spontaneous emission are limited by the bandwidth of the laser medium. Integrating the emission terms $F_{01q}(\omega)$ over the gain band gives a formula for the excess mean square uncertainty in the position of the centroid of the TEM_{oo} laser beam due to TEM_{01} emission:

$$<|\delta|^2> = \frac{w^2 \kappa_{out} \nu_g \hbar\Omega G}{\eta \nu_L P_o (1 - g/\eta)(1 - N_1^{01}/N_2^{01})} \tag{25}$$

where ν_g is the gain bandwidth of the laser medium and and vacuum fluctuations have been omitted. This equation originates in the same sort of spontaneous emission physics that gives rise to the similar, but better known, Schawlow-Townes linewidth formula [3].

VII. Comparison with Experiment

Figure 10 plots the linewidth and peak heights measured from a series of traces as in Fig. 9 as a function of the TEM_{ooq_o} power P_o. Just above TEM_{ooq_o} threshold, the stochastic position noise linewidth is η/π = 520kHz. The output coupling for the TEM_{oo} mode was nominally 1.2%, which yields a predicted TEM_{ooq_o} cold cavity linewidth of κ_o/π = 540 kHz. The measurements of position noise peak height and width near TEM_{ooq_o} threshold imply that $\kappa_o \approx \eta \approx \kappa_{out}$. Between TEM_{ooq_o} and TEM_{01q} threshold: $\kappa_o < G = \kappa_o + g < \kappa_o + \eta$.

The linewidth plotted in Fig. 10 evidently falls linearly with TEM_{ooq_o} output power, reaching zero near the observed TEM_{01q} threshold at 680mW. At TEM_{01q} threshold, the stochastic position noise suddenly increases by several orders of magnitude. Below that threshold, the increase in peak noise power follows (17) quite nicely.

The absolute level of the noise depends on the ratio of lower to upper laser level population N_1^{01}/N_2^{01} (i.e. within the TEM_{01} mode volume) at TEM_{ooq_o} threshold. This quantity can be extracted from the data in figure 10 and appears plotted as a function of P_o in Fig.11. While the value of this ratio might appear somewhat high, it is indeed constant as predicted by the theory here.

The theory thus far appears to explain the features of the laser noise spectrum. In addition to the linewidth and peak-height, it explains the dependence on internal diffraction loss η illustrated by Fig. 8. The theory does not, however, account for the shift of the transverse mode frequency ν_T from the value ν_T = 19 MHz predicted by the mirror radii [2]. This shift, which varies with TEM_{ooq_o} power, more likely results from a lens action due to non-uniform current in the Kr^+ arc than from cross saturation of the medium

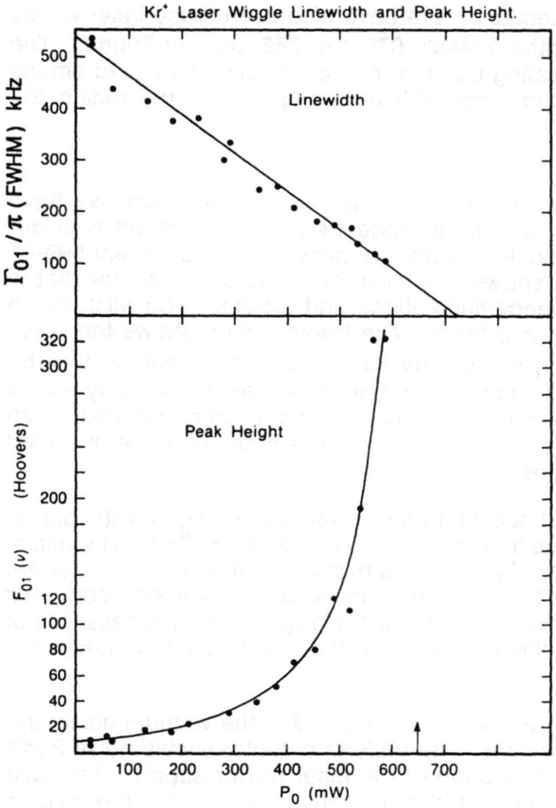

Fig. 10. Linewidth and peak height for stochastic position noise traces similar to those in figure 9 as a function of laser power.

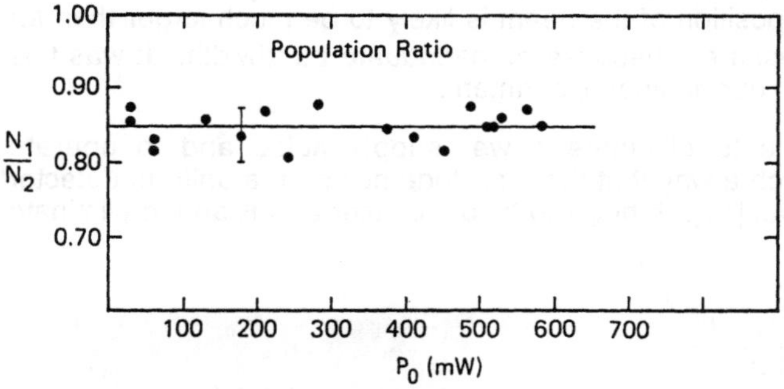

Fig. 11. The population ratio N_1^{01}/N_2^{01} for the TEM_{01} mode as calculated from the data in Figs. 9 and 10 using (15), (17), and (23).

by the TEM_{00q_0} power. If saturated dispersion were dominant, the ν_T and $\nu_L - \nu_T$ peaks would shift in the same direction rather than the observed opposite directions.

For our Kr^+ laser at 647 nm and with a 0.5 mm beam waist, the RMS uncertainty in the position of the centroid of the 30 mW TEM_{00} beam is $<|\delta|^2>^{1/2} = 28$ nm, when only the spontaneous emission is considered (as in (25)) and $<|\delta|^2>^{1/2} = 92$ nm when the vacuum fluctuations within the 3GHz gain bandwidth are included. As small as these effects may be, they are quite easily detectable. We have also seen these laser wiggle phenomena at other Kr^+ laser wavelengths, notably 676 nm, 756 nm and 799nm. The linewidths and peak heights varied, indicating that the ratio of populations varied among the lines and that the diffraction loss also changes with wavelength and mirror radius [2].

VIII. Other Lasers

We have found stochastic position noise in the output of every laser which we have studied and which has a low enough transverse mode frequency to detect with our electronics. Figure 12 shows spectrum analyzer traces obtained with a Coherent 599-21 single mode dye laser. The top figure shows the noise on a uniform detector that is dominated by the well known low frequency fluctuations and a peak at 83 MHz which results from axial mode beating in the pump laser. The lower picture shows the noise spectrum of a split detector. There clearly is a stochastic position noise peak at $\nu_T = 69$ MHz. This peak decreases in frequency as the pump power increases, evidently as the result of thermal lens action in the dye jet. At lower pump power, the peak also broadens. In fairness to Coherent, I must point out that careful alignment of the dye laser can greatly reduce all the noise it produces.

Semiconductor diode lasers have flat facets forming nominally degenerate planar cavities [2]. The axial mode spacing is quite high - tens to hundreds of GHz. The actual transverse mode spacing depends critically on the structure of the device. Modern lasers of this type are expected to oscillate in a single transverse (waveguide) mode, but the cavity may support higher order modes below threshold [10]. The high excess loss of these modes may make stochastic position noise detectable with our low bandwidth electronics.

Figure 13 shows characteristic laser noise spectra normalized to the vacuum noise and obtained with a Liconix 40 diode laser system. The uniform detector shows noise 25dB above the vacuum, due mostly to poor regulation of the laser power supply. The split detector shows none of the the much-suppressed power supply noise, but the level is still 10dB above the vacuum over the entire bandwidth of the detector system. The mean square uncertainty of the center position of the beam is likely to be much larger than for the Kr^+ laser, but we cannot measure it because of inadequate bandwidth. It was this noise that mimicked squeezing in our original experiment.

With great care, it is possible to eliminate power supply noise and to operate semiconductor diode lasers in such a way that the amplitude noise on a uniform detector is less than the vacuum noise level [11]. It helps to hold the laser at 4 K and to eliminate

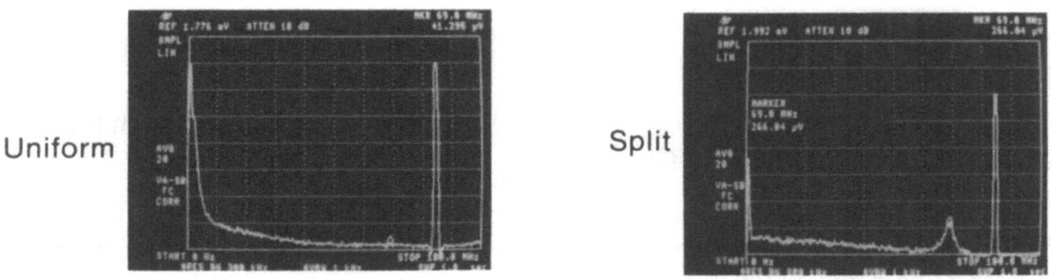

Fig.12. Uniform and split detector noise spectra obtained using a CR-599-21 single mode dye laser. The vertical scale is linear in $i^2(\omega)$ and horizontal scale goes from 1 to 100 MHz. Vacuum noise is at the bottom of each trace, the strong peak near 83MHz results from pump modes beating. The split detector clearly shows less low frequency noise, but also a significant position noise peak at 69MHz.

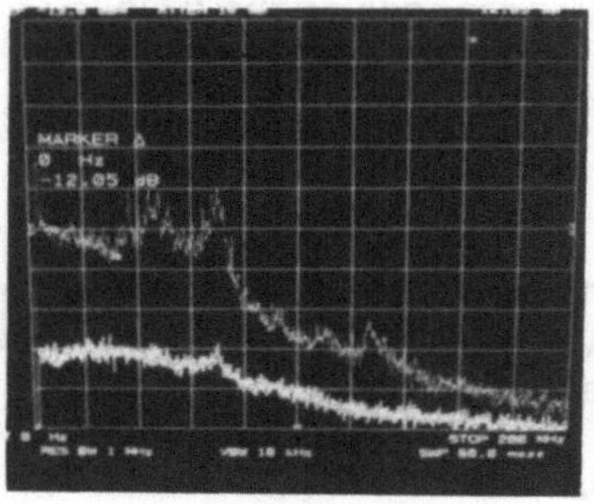

Fig. 13. Uniform and split detector noise traces for a typical semiconductor diode laser system. The vacuum level is at the bottom, and the vertical scale is 5 dB per box. The horozontal span is 0 to 200 MHz even though the detector-amplifier system loses sensitivity above 100 MHz. The uniform detector trace shows a noise level 25 dB above the vacuum while the split detector gives a noise level 10 dB above the vacuum.

all reflections back into the laser cavity [12]. However, even under these ideal conditions, the noise on the split detector can rise above the vacuum level!

IX. Squeezing of Higher Transverse Modes

We have now shown that the TEM_{01} modes of a conventional laser act below threshold as conventional laser theory might predict. The Hamiltonian we have discussed, however, is insensitive to the relative phase of the TEM_{ooq_o} and TEM_{mnq}. That does not mean that the actual laser interaction is insensitive to phase. In fact, four wave mixing effects are well-known in saturated absorption spectroscopy and should affect the stochastic position noise in real lasers [13].

But what about the parametric mixing devices used to produce squeezed states, especially those employing optical cavities? Might those devices be squeezing the higher order transverse modes? Might that effect be detectable?

In order to squeeze the TEM_{01q} modes, it is necessary that the parametric process phase match and that the cavity structure not impose excess diffraction loss on the higher order modes. The phase-conjugate four wave mixing geometry used by SLUSHER [14] fulfills those criteria as does the micro-cavity device used by KIMBLE [15]. The IBM fiber-optic experiments support no higher-order modes, and phase matching considerations complicate the case of the degenerate parametric oscillator below threshold [6]. Still, there are quite a number of experiments where the stochastic position noise might be squeezed.

Detecting the squeezing of the TEM_{01} modes might even be easier than detecting squeezing of TEM_{oo} modes. If one considers, for example, a cavity with a strong pump beam in the TEM_{oo} mode and a four wave mixing medium, the TEM_{oo} output can act as a local oscillator for detection of the squeezing of the TEM_{01}. One need only direct that light onto a split detector as described here. The heterodyne detection efficiency is almost guaranteed to have its maximum value of $2/\pi = 64\%$, which is better than can usually be obtained by overlapping separate beams.

It is necessary to vary the relative phases of the TEM_{oo} local oscillator with respect to the TEM_{01} quantum field in order to verify squeezing. This phase shift can be conveniently achieved by moving the detector through a beam waist. There is a net phase shift between the TEM_{00} and TEM_{mn} that depends on the distance from the beam waist:

$$\Phi = -(m+n)\tan^{-1}\left(\frac{z\lambda}{\pi w_0^2}\right) \tag{26}$$

where w_0 is the TEM_{00} beam waist radius. Thus axially translating the detector probes different quadratures of the TEM_{01} field. At some distance from the beam waist, the stochastic position noise should fall below the vacuum level.

The New Zealand groups have outlined a number of effects of squeezed light on atoms [16]. To observe these effects, it is necessary to surround the atoms with squeezed vacuum, which is very difficult experimentally if only a single mode can be squeezed. But what if many transverse modes of a cavity can be squeezed? Might a substantial effect on the atoms remain? Might new effects appear? Is anyone willing to explore this idea?

X. Postscript: The original experiment worked

While all of this was going on, one of the authors (WHR) continued trying to suppress the amplitude noise of a semiconductor diode laser. Eventually, this effort lead to success, and will be published separately [12].

Several innovations were necessary: First, large area, high bandwidth, high quantum efficiency photodiodes had to be indentified and built into suitable pre-amplifiers. Our diode lasers had serious 1/f noise which masked squeezing below a few hundred MHz. The laser had to be operated at liquid helium temperature in order to raise its quantum efficiency and suppress noise due to light scattering. It was essential to eliminate all feedback into the laser diode, even that from the collimating lens. And finally, we had to find a reliable way to find the vacuum noise level within a percent or so. This was achieved with the infra-red lines of our krypton laser.

Once all of this had been accomplished - and made to work simultaneously - the amplitude noise of the diode laser was measured to be 10 ±2% below the vacuum when the current was 10 to 20 times threshold. Since the quantum efficiency of the whole experiment was only 20%, this value is fully half of the maximum possible reduction. Why only half? We do not know yet, but it appears that the noise level increases far above threshold due to effects that are not yet included in the theory.

Thus, as always in science, there is more interesting and challenging work to be done.

The authors acknowledge seminal conversations with Dr. Aram Mooradian, Dr. R.M. Shelby and Prof. D.F. Walls. This research was partly supported by the U.S. Office of Naval Research under contract N00014-82-C-0694.

References:

1. B.L. Schumaker, *Opt. Letters*, **9**, 189, (1984).
2. A. E. Siegman, *Lasers*, ..(Science Books, Mill Valley, California, 1986) Chapters 14-17.
3. *Ibid*, Chapter 11.
4. M.D. Levenson, W.H. Richardson, S.H. Perlmutter, Opt. Letters, (submitted).

5. D.H. Auston, *IEEE J. Quantum Electron.*, **QE-4**, 420, (1968); P.W. Smith, *Appl. Phys. Letters,* **13**, 235 (1968).
6. G.J. Milburn, M.D. Levenson, R.M. Shelby, S.H. Perlmutter, R.G. Devoe, and D.F. Walls, *J. Opt. Soc. Am. B*, **4**, 1476 (1987).
7. H. Risken, *The Fokker-Planck Equation*, (Springer Verlag, Berlin, 1984).
8. M.D. Levenson and S.S. Kano, *Introduction to Nonlinear Laser Spectroscopy, Revised Edition*, .(Academic Press, Boston, 1988) Chapter 3.
9. M.J. Collett and C.W. Gardner, *Phys. Rev. A* **30**, 1386 (1984) and C.W. Gardiner and M. J. Collett, *Phys. Rev. A,* **31**, 3761 (1985).
10. R. Lang, *IEEE J. Quantum Electron.* **QE-15**, 718 (1979); W. Streifer, R. D. Burnham and D. R. Scifres, *Ibid,* 136.
11. S. Machida, Y. Yamamoto, and Y. Itaya, *Phys. Rev. Lett.*, **58**, 1000 (1987).
12. W.H. Richardson and R.M. Shelby, *Phys. Rev. A.* (To be submitted).
13. B.J. Feldman and M.S. Feld, *Phys. Rev. A* **1**, 1375 (1970); **5**, 899 (1972). also G. Camy, C.J. Borde, and M. Ducloy, *Opt. Commun.* **41**, 325 (1982).
14. R.E. Slusher, B. Yurke, P. Grangier, A. LaPorta, D.F. Walls and M.Reid, *J. Opt. Soc. Am. B.* **4**, 1450 (1987).
15. L.A. Orozco, M.G. Raizen, Min Xiao, R.J. Brecha, and H.J. Kimble, *J. Opt. Soc. Am. B,* **4**, 1490 (1987).
16. C.W. Gardiner, *Phys. Rev. Lett.* **56**, 1917 (1986), also H.J. Carmichael, A.S. Lane, and D.F. Walls, *Phys Rev. Lett.* **58**, 2539 (1987), *J. Modern Optics* **34**, 821 (1987) and private communications.

The Generation of Mode-Locked Pi Pulses

J.D. Harvey[1], *J. Dudley*[1], *R. Leonhardt*[1], *S. Carter*[2], *and P.D. Drummond*[2]
[1]Physics Department, University of Auckland, Auckland, New Zealand
[2]Physics Department, University of Queensland, St. Lucia 4076, Australia

1. Introduction

This paper has its origin in a rather surprising observation made in our laboratory about two years ago. It was discovered that a cavity dumped mode-locked argon laser was capable of producing pulses much shorter than those from a conventional mode-locked argon laser system [1]. This was surprising for two reasons. Firstly, it is known that a cavity dumped mode-locked dye laser system generally produces pulses which are broader than those from a non cavity dumped mode-locked system, (thereby partially defeating the original motivation for cavity dumping the system - to obtain increased peak power). The second, and more arresting reason, concerns the fact that conventional mode-locked gas lasers, which have been investigated and used extensively now for over 25 years [2,3], generally produce pulses whose duration is determined by the bandwidth of the inhomogeneously broadened transition involved (previously reported pulse durations range from 70-120ps) [4]. In our case however, pulses as short as one third of the apparent limit set by the atomic linewidth were observed. In the period following this observation we have investigated this phenomenon both theoretically and experimentally, and the results of these investigations are summarised below.

2. Experimental Studies

The pulse duration of a mode-locked argon ion laser can be measured by crosscorrelation of the argon pump pulse with the pulses from a mode-locked dye laser synchronously pumped by the same argon laser using sum frequency mixing in KDP in an optical correlator. This technique yields essentially the argon laser pulse shape since the duration of the dye laser pulses is very short compared to those of the pump laser. A series of measurements performed in this way on a conventional mode-locked argon ion laser (Coherent CR12) yielded results consistent with those obtained by others. The pulse lengths obtained varied with the gain and detuning of the laser, and were generally in the range 80-100ps, while the minimum pulse width of 70ps was obtained at maximum plasma current.

The original apparatus used to observe the short pulses was a short frame argon ion laser (spectra Physics 165) mounted in an optical cavity defined by an acousto-optic mode-locker (Coherent model 467) and a cavity dumper (Coherent model 7200). The pulse durations were measured by optical autocorrelation. In the initial experiments, a two photon fluorescent solution was used as the nonlinear medium [1] in the autocorrelator which operated in a colinear configuration. More recently it has been possible to replace this solution with a beta barium borate crystal and perform noncolinear autocorrelation at 514nm wavelength. Apart from the advantage of collecting background free autocorrelation functions, the greater sensitivity of this arrangement permits faster alignment and data collection. The spectrum of the mode-locked pulse train was observed using a 30GHz free spectral range scanning confocal Fabry Perot interferometer (Coherent model 240), while the pulses were also monitored on both sampling and real time oscilloscopes. In the original experiments we obtained pulse

Springer Proceedings in Physics, Vol. 41 Quantum Optics V
Editors: J.D. Harvey and D.F. Walls

widths of about 35ps which were largely independent of the cavity dumping ratio. In fact, short pulses and broad spectra were also obtained when the cavity dumper was used in "CW" mode, in which configuration it acts as a variable (but relatively low) loss output coupler. Further experiments were performed on the small frame linear laser to determine the small signal gain of the laser. For these experiments a glass plate was installed in the cavity and the cavity losses adjusted by rotating the plate away from Brewster's angle until the laser was at threshold.

In the initial experiments, the two photon fluorescence autocorrelation technique was not sufficiently sensitive to measure the autocorrelation function of the pulses in the CW mode, but pulsewidth measurements in this mode were possible at maximum plasma current, using the crosscorrelation technique described above. The observed crosscorrelation width of 37ps was consistent with the results obtained by autocorrelation of pulses with similar spectra at higher cavity dumping ratios, and with the results obtained later by noncolinear autocorrelation using beta barium borate. A typical autocorrelation function of the cavity dumped pulse train and the corresponding spectrum are shown in Fig. 1. For all of the results obtained, for both cavity dumped and non cavity dumped operation, no significant variation of the time bandwidth product $\Delta\nu\ \Delta t$ was detected. Although the widths of the spectra (which were generally double peaked at high gain and single peaked at low gain) were difficult to determine precisely, the value obtained in our work was generally about 0.5, a value consistent with good mode-locked pulses (a value of 0.44 is expected for Gaussian pulse shapes and spectra).

We have now been able to obtain short pulses with a conventional linear mode-locked cavity in which the cavity dumper was replaced by a low transmission output coupler (3.8%). In this context it is important to note that the mode-locking of argon lasers is generally achieved with a relatively lossy (20-30%) output coupler, and indeed, experiments with such an output coupler have generated only long (~ 100ps) pulses in our system. We have also performed experiments with the large frame argon laser fitted with a low loss output coupler, but although we do observe frequency broadening with increased cavity Q values, one is generally unable to produce good mode-locked output pulses in such a system. Results obtained for the pulsewidth of the mode-locked laser are shown in Fig. 2 where the pulsewidth is plotted as a function of the mean intracavity power.

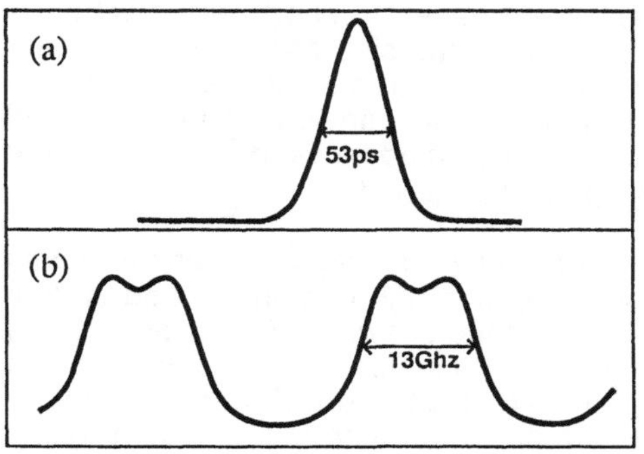

Figure 1 *(a) Autocorrelation function of the pulse train from the cavity dumped argon ion laser (dumping ratio 1:8, output power 160mW). Full width half maximum of this function is 53ps implying a pulse width of about 37ps.*

(b) Corresponding spectrum of pulse train measured on a Fabry-Perot interferometer with free spectral range of 30GHz.

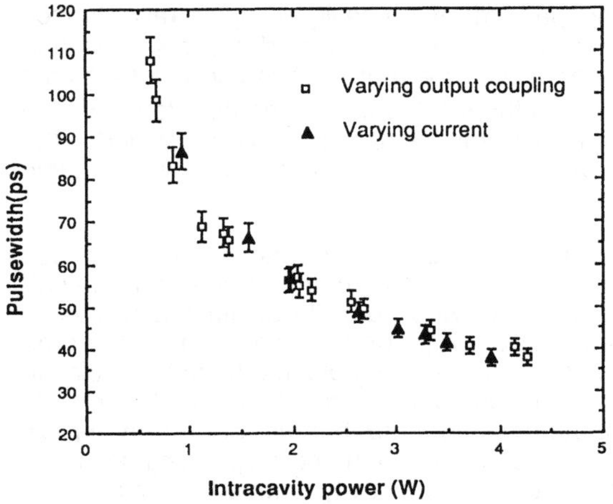

Figure 2 *Pulse width of the mode-locked argon laser as a function of intracavity power. The intracavity power was varied both by altering the plasma current, (holding the output coupling constant) and by varying the output coupling at constant plasma current.*

3. Theoretical Studies

We have considered three possible contributing factors to the increased bandwidth of the high Q mode-locked system. These include the acousto-optic effect of the cavity dumper, self phase modulation in the glass block of the cavity dumper, and power broadening of the atomic transition in the laser medium. It is now possible to eliminate the first two possible explanations since we have also obtained short pulses in a simple linear cavity. In addition, the first two effects were unable to explain the magnitude of the spectral broadening which was observed. The effect of the acousto-optic interaction is to re-inject a frequency shifted pulse into the cavity, but the frequency is always decreased, leading to an anticipated asymmetry in the spectrum. The effect of self phase modulation is to generate a chirp on the pulse, but numerical estimates indicate that the expected phase shift is an order of magnitude too small to explain the observed broadening with a focussed beam waist in the silica block. In addition, to generate short pulses with self phase modulation, it is necessary to have a means for dispersive compensation of the chirped pulses, and we have not been able to identify a plausible mechanism for this compensation. It is possible that these effects need to be considered in a detailed quantitative study of the performance of the cavity dumped mode-locked system, but the effects are clearly not the dominant cause of the spectral broadening.

In contrast, a simple estimate of the magnitude of the Rabi frequency for the atoms exposed to the peak intracavity powers of the pulses in our high Q mode-locked system gives a result in excess of the inhomogeneously broadened linewidth. Power broadening of the atomic transition is therefore expected to be important in this laser. In any system in which the interaction with the gain medium is primarily via coherent Rabi cycling rather than the usual adiabatic rate equation coupling, the pulse duration is not limited by the atomic linewidth. Pulses considerably shorter than the inverse of the linewidth have been generated by superfluorescent emission from a pencil shaped inverted medium in the far infrared [5]. We propose therefore that the high Q mode-locked argon laser system harnesses the pulse shortening mechanism operative in swept gain superfluorescence to generate pulses which are significantly shorter than those obtained in conventionally mode-locked systems.

The atomic system within the laser was modelled using the standard Maxwell-Bloch equations [6,7] for a two level gain medium that are used in superfluorescence calculations:

$$\frac{\partial \Omega}{\partial x} = g \int \rho(\Delta) P(\Delta) d\Delta$$

$$\frac{\partial P(\Delta)}{\partial t} = -(\gamma_2' + i\Delta) P(\Delta) + \Omega D(\Delta)$$

$$\frac{\partial D(\Delta)}{\partial t} = -\gamma_p \big(D(\Delta) - 1\big) - \mathrm{Re}\big(\Omega^* P(\Delta)\big)$$

Here t,x, are co-moving time and space coordinates, Ω is the local Rabi frequency, g is the gain constant, $\rho(\Delta)$ the polarization and $D(\Delta)$ the inversion. The constants γ_2' , γ_p are the transverse relaxation rate and pumping rate respectively, while Δ is the inhomogeneous detuning. These equations ignore the radial variation in the mode, in a first approximation.

In initial calculations, the system was approximated as a fully homogeneously broadened medium. Under these conditions, the required constants in the equations are all determined by the observed linewidth (~6GHz) and the lifetime of the excited state (~5ns) together with the small signal gain per pass. These initial calculations generated a satisfying qualitative agreement with experiment, but in order to model the argon laser atomic system realistically, it is necessary to include the effects of Doppler broadening.

The inhomogeneously broadened system was modelled numerically by subdividing the total number of atoms into N distinct classes. This yields an equivalent number of coupled equations to solve and increases the computation time proportionally. In these calculations the Zeeman contribution is assumed lumped into the inhomogenous linewidth. The numerical simulation was allowed to proceed as follows. Upon entering the gain medium the pulse (initiated as a small seed pulse of variable width) interacts with the atomic variables which are initially saturated (inversion) or relaxed (polarization). The equations for the polarization and inversion are numerically integrated and the values obtained during this integration are used to update the field's time dependence at the next spatial point in the propagation through the medium.

The mode-locker is modelled as a sinusoidally varying loss which generates a maximum loss of 40% per double pass. Dispersion within the mode-locker was also investigated and found to be so small as to have no observable effect on the simulations. After numerical integration of the set of equations to simulate one pass through the cavity, the resulting pulse was re-injected into the gain medium, and the simulation was continued until the calculation converged and the pulse stabilized. In order to obtain convergence, it was necessary to adjust the frequency of the sinusoidally varying loss to compensate for the tendency of the pulse to "walk" out of the time window of the calculation due to gain shaping. This is analogous to the experimental procedure of adjusting the cavity length to optimise the pulses.

Once the numerical simulation has stabilized, the bandwidth of the resultant pulses and also the pulse area (the time integral over Ω) can be calculated. This latter quantity is of particular interest in view of the prediction of π pulses as the long distance limit of superfluorescent pulse generation in the case of a swept gain medium [6]. The results of a typical run are shown in Fig.3, where the evolution of the pulse and its spectrum are plotted over 1000 round trips through the cavity. In this simulation, the inhomogeneous medium was subdivided into 8 different doppler classes, but very similar results are obtained with more classes.

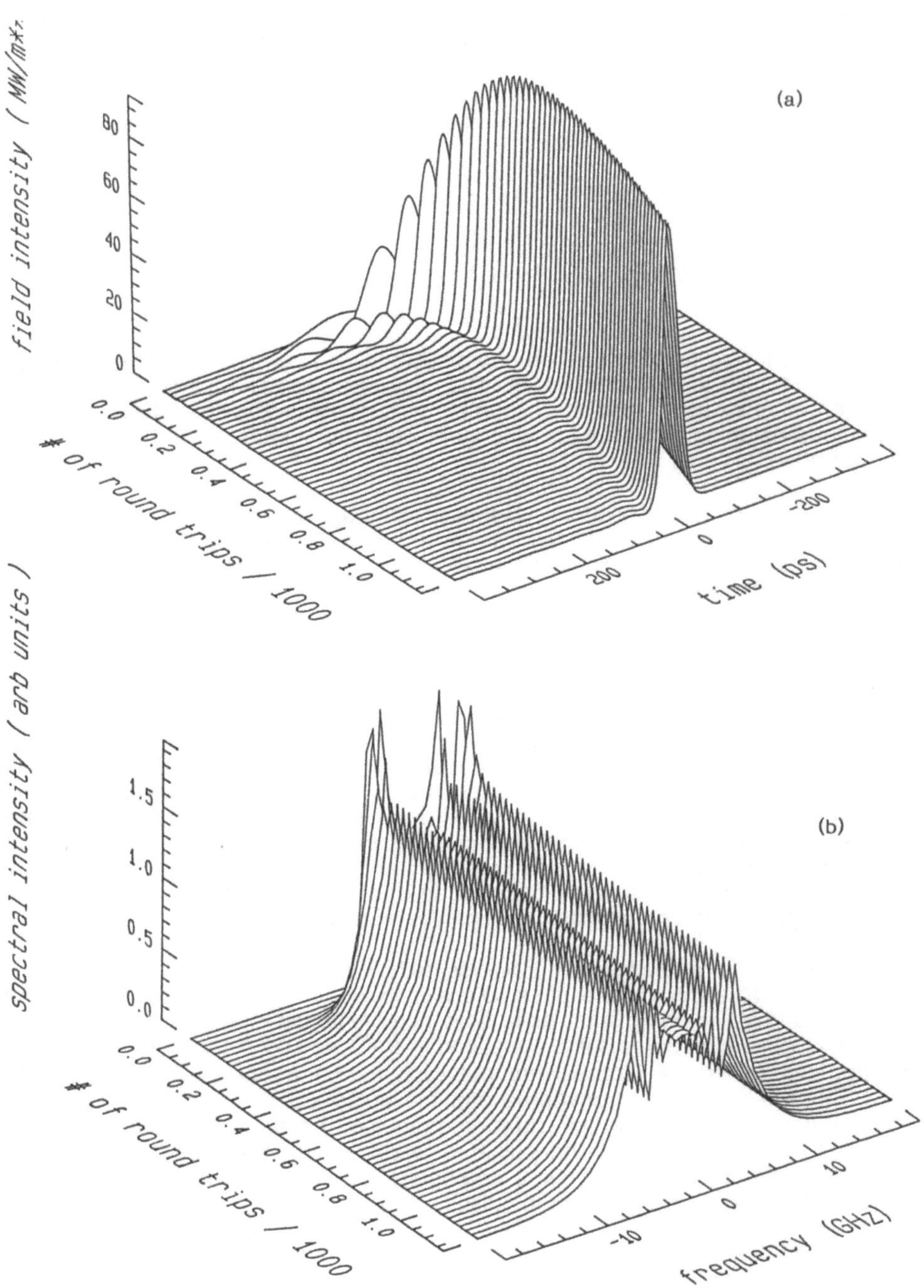

Figure 3 *Time development of (a) pulse and (b) spectrum of the mode-locked argon laser for a numerical simulation over 1000 round trips starting with a small broad seed pulse. Output coupling was 0.1, resultant stabilised characteristics of pulse train: Pulse width 35ps, pulse area 2.51 (0.8π), spectral width 13GHz.*

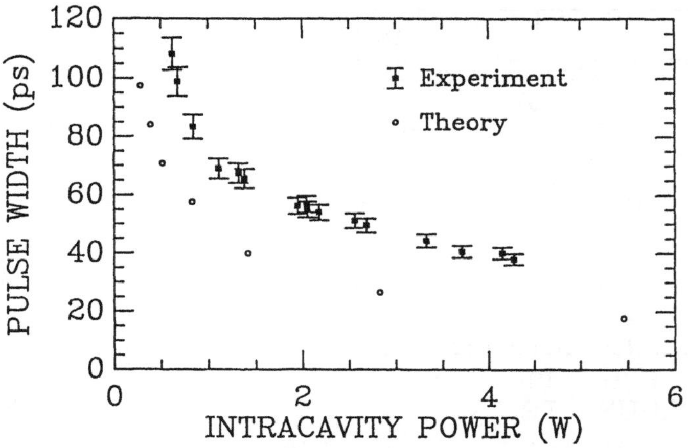

Figure 4 *Comparison of experimentally measured pulse widths and those generated by the numerical simulations.*

4. Comparison of Theoretical and Experimental Results

A comparison between the experimentally observed pulse widths and the results of the numerical simulations is shown in Fig. 4. The most significant features of this graph are summarised below:

(i) The pulse widths observed in the experiments depend primarily on the intracavity power, as can be seen from the experiments performed at constant plasma current, and those performed at constant output coupling, which yield results falling on the same curve (Fig. 2).

(ii) The simulations provide a satisfactory explanation of the qualitative features of the variation of pulse width with intracavity power.

It should be emphasised that the calculations leading to the theoretical predictions in Fig. 4 have no adjustable parameters, but an exact fit to the experimental data should not be expected in view of the approximations made in the simulations. The discrepancy between theory and experiment is reasonable however, given that the transverse variation of the field has been neglected. Semiclassical calculations [8] and fully quantized calculations [9] including transverse fields and diffraction in superfluorescence have generated longer pulse lengths than plane wave calculations, and this result is consistent with our results.

Our equations are the same as those used to describe superfluorescent pulse generation and propagation in the semiclassical model. We interpret the resulting pulses from the high Q mode-locked system as those generated from a swept gain superfluorescent system. The action of the modelocker is simply to suppress counterpropagating or independently triggered pulses in the infinite length gain medium seen by the propagating pulse. In this picture then, in order to see superfluorescent operation, we simply require that the pulse width be significantly shorter than the relaxation times characteristic of the medium. Since the shortest such relaxation time (T_2, the transverse relaxation time) is about 60ps in the argon laser, the transition from conventional mode-locking characterised by the inhomogeneously broadened linewidth of the transition, to superfluorescent generation of short coherent pulses with increased bandwidth can readily be achieved. The large Rabi frequencies present in our high Q mode-locked laser cavity broaden the gain band, and drive the pulse towards a stable shape and pulse area close to π.

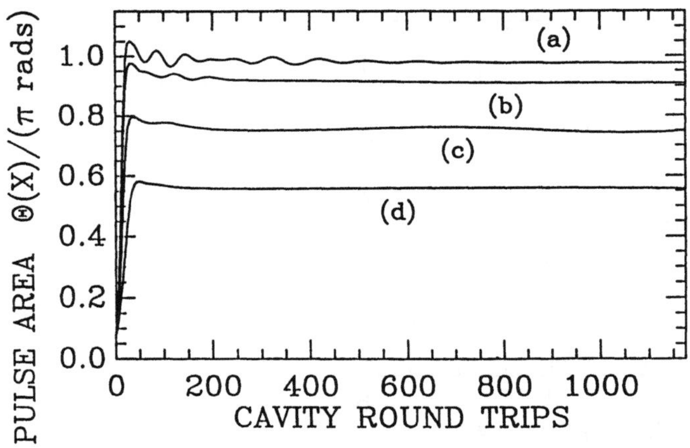

Figure 5 *Evolution of the pulse area for different cavity losses. The pulse area rapidly approaches π for a high Q cavity.*

The evolution of the pulse area is shown in Fig. 5. It is clear from this figure that the pulse rapidly approaches a pi pulse in the high Q cavity, and this is the result predicted in the experimentally inaccessible long path length limit of superfluorescent pulse propagation. In fact, the idea that the pulses circulating in a mode-locked laser might approach the intensity necessary for pi pulse propagation was first proposed long ago [10]. Nearly all current treatments of mode-locked lasers, however, ignore the possibility of a strong coherent interaction between the propagating field and the gain medium, and utilise a simple rate equation analysis. The origin of superfluorescence is the fast decay of excited atoms by cooperative emission, in a pencil shaped volume. It is generally difficult to achieve superfluorescence in the visible region, due to the usual requirements of high gain and an overall length less than the Arecchi-Courtens cooperation length [11]. In our work, these difficulties are overcome through the use of a high-Q cavity to effectively extend the total gain length. In addition, a mode-locker is used to modulate the cavity reflectance, thus producing an effective swept-gain [6] via loss modulation. This technique permits a superfluorescent-pulse output to occur even with a relatively short cooperation length.

Our method appears to provide a universal technique for reducing the pulse length of a mode-locked laser below the apparent limit set by the continuous wave gain bandwidth. In addition, the coherent interaction of the field and the gain medium drives the pulse towards the stable pi pulse, characterised by increased stability of operation of the laser. Finally it should be noted that apart from the intrinsic scientific interest of these superfluorescent mode-locked pulses, the reduction in pulse width is accompanied by a corresponding increase in peak intensity. This effect is particularly useful in the cavity dumped system where kilowatt peak power pulses have been obtained from a small frame laser [1]. These pulses can be expected to find several applications.

References

1. J.D. Harvey, M.J. Proctor and C.A. Steed: Appl.Phys.Lett. **52**, 688 (1988).

2. M. Di Domenico Jun.: J.Appl.Phys. **35** 2870 (1964).

3. S.E. Harris: Appl.Optics **5** 1639 (1966).

4. V. Sundstrom and T. Gillbro: Appl.Phys. **35** 2870 (1964).

5. A.T. Rosenberger, H.K. Chung and T.A. DeTemple: IEEE J.Q.Electron. **QE-20**, 523 (1984).

6. F.T. Arecchi and R. Bonifacio: IEEE J.Q.Electron. **QE-1**, 169 (1965).

7. D. Polder, M.F.H. Schnurmans and Q.F.H. Vrehen: Phys.Rev. **A19** 1192 (1979); F. Haake, H. King, G. Schroeder, J. Haus and R. Glauber: Phys.Rev. **A20** 2047 (1979).

8. F.P. Mattar, H.M. Gibbs, S.L. McCall and M.S. Feld, Phys.Rev.Lett. **46**, 1123 (1981).

9. P.D. Drummond and J.H. Eberly, Phys.Rev. **A25**, 3446 (1982).

10. A.G. Fox and P.W. Smith: Phys.Rev.Lett. **26**, 826, (1967).

11. F.T. Arecchi and E. Courtens: Phys.Rev. **A2**, 1730 (1970).

Spontaneous and Induced Solitons in Stimulated Raman Scattering

C.M. Bowden[1] *and J.C. Englund*[2]

[1]Research Directorate, AMSMI-RD-RE-OP Research, Development, and Engineering Center US Army Missile Command, Redstone Arsenal, AL 35898-5248, USA
[2]Department of Physics, Southern Methodist University, Dallas, TX 75275, USA

I. INTRODUCTION

Soliton solutions to the classical equations of stimulated Raman scattering (SRS) have been known for more than a decade [1]. However, it has been only recently that solitons were observed in SRS experiments [2-4], where DRÜHL, WENZEL and CARLSTEN (DWC) [2], used CO_2 pumped parahydrogen with a Stokes "seed" and observed occasional solitary pulses as nearly complete repletions in the depletion zone of the pump pulse. The frequency of occurrence of the solitary pulses appeared to be stochastic, suggesting the origin as noise-generated.

Solitary pulses were produced deterministically in subsequent experiments by DWC [3-4] by inducing a π-phase change in the Stokes "seed" prior to its entry into the Raman gain cell. The experimental results were simulated numerically using the classical SRS equations [1] with the initial and boundary conditions of the experiments [3-4]. Previous theoretical treatments [1] dealt mainly with existence and characterization of solitons in SRS, but were quite indefinite concerning initial and boundary conditions which cause them. The experimental and theoretical analysis of DWC demonstrated sufficient conditions for the generation of Raman solitons deterministically. Later, ACKERHALT and MILONNI [5] discussed a variety of induced Raman soliton effects, including second Stokes, anti-Stokes and the use of two-pump Raman scattering to study solitons in four-wave mixing.

This left open the question of noise-generated Raman solitons observed in the first experimental observation [2]. They speculated that stochastic phase fluctuations in the Stokes seed may have been the source of these solitons. Subsequently, ENGLUND and BOWDEN [6] showed that Raman solitons can be generated stochastically from quantum noise, and depend entirely upon amplification of quantum phase fluctuations in a pump field-induced, spontaneously generated Stokes field (no "seed"). The process in the linear regime of quantum initiation and coherent pulse buildup is identical to that in superfluorescence ([7] and swept-gain superfluorescence [8], and gives rise to phase waves, first discussed by HOPF [8]. The fundamental relationship among amplified quantum fluctuations in superfluorescence, swept-gain superfluorescence and spontaneous Raman soliton generation have been discussed by BOWDEN and ENGLUND [9-11]. The distinguishing feature of SRS lies in how the phase waves affect the nonlinear regime to induce solitons. The stochastic properties of spontaneous soliton generation in terms of system parameters and phase wave statistics has been further discussed by ENGLUND and BOWDEN [12-13]. Spontaneous and induced Raman solitons, and possible realizations in Raman spectroscopy have also been discussed [14]. Recently, spontaneously generated Raman solitons have been observed in hydrogen by CARLSTEN and MACPHERSON [15], and stochastic properties studied experimentally [16-17].

Here, we present the fundamental model and underlying physics for discussing spontaneous as well as induced Raman soliton generation and present some new results which characterize the stochastic properties. The next section will be used to present and discuss the model. Section III will contain the discussion of the basis for our numerical simulation based upon

Springer Proceedings in Physics, Vol. 41 **Quantum Optics V**
Editors: J.D. Harvey and D.F. Walls

quantum-classical correspondence. Some new results of the calculations will be presented and discussed in the final section, together with a summary and conclusion.

II. MODEL

Since our motivation is to discuss effects of amplified quantum stochastic processes in SRS, we formulate the problem on a completely quantum mechanical foundation. The simplest model for the medium is a collection of spacially distributed, three-level systems, each having identical lambda energy-level configurations, shown in Fig. 1. The detuning, Δ, is used to adiabatically remove the upper level and to isolate the two-photon Raman process (of course, any number of levels can be adiabatically eliminated in this manner). The assumption is made that the pump field, A_L, and the Stokes field, A_S, are separate and distinguished, and that any higher order Raman processes such as higher Stokes or anti-Stokes are suppressed.

The calculational procedure is identical to that already reported in the literature [18] - [20]. After adiabatic elimination of the upper level by taking Δ very large, Fig. 1, the normally ordered equations of motion and propagation in the plane wave approximation and retarded time, become [6], [9]

$$\frac{\partial A_L^{(-)}}{\partial z} = -k_{LS}\, A_S^{(-)} R^{+} - k_{LS}\, f^{(-)} R^{+} \tag{1a}$$

$$\frac{\partial A_S}{\partial z} = k_{LS}^{*}\, A_L^{(-)} R \tag{1b}$$

$$\frac{\partial R}{\partial \tau} = -k_{LS}\, A_S^{(-)} W A_L^{(+)} - \gamma R + G - k_{LS}\, f^{(-)} W A_L^{(+)} \tag{1c}$$

$$\frac{\partial W}{\partial \tau} = 2k_{LS}\, A_L^{(-)} R^{+} A_S^{(+)} + 2k_{LS}\, A_L^{(-)} R^{+} f^{(+)} + \text{h.c.} \;. \tag{1d}$$

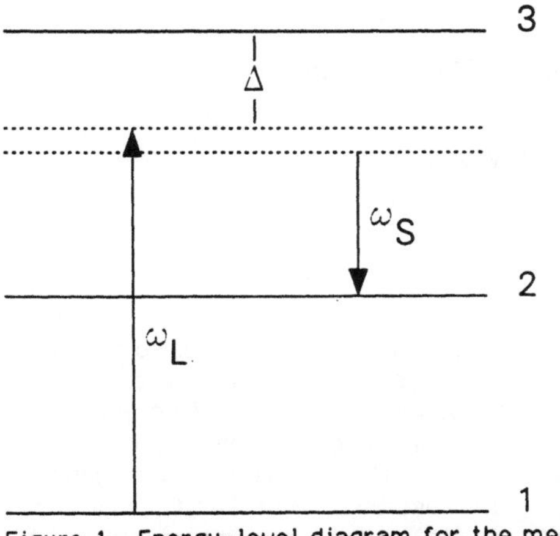

Figure 1. Energy-level diagram for the medium. The pump field and Stokes field frequencies are ω_L and ω_S, respectively. The detuning, Δ, from the upper level is taken as very large so that the two-photon Raman transition remains. Levels 1 and 2 are not radiatively connected.

Here, $R \equiv R_{12}$ is the polarization operator and $W \equiv R_{22} - R_{11}$, is the population difference operator. All operators are functions of the propagation variable z and retarded time τ, except the free Stokes field operators which under these conditions is a function of τ only, explicitly $f^{(+)} = f^{(+)}(\tau)$. The constant k_{LS} is the usual coupling parameter and we have ignored Stark shift contributions which do not cause qualitative effects and make only minor quantitative contributions. In (1c), we have added the nonHamiltonian decay term, γR, and corresponding fluctuating source term G, to take into account collisional dephasing. The even moments of the source terms, G and f, completely factorize and the source contributions are delta-correlated. For the initial state of the system, the ground state of the atoms and the vacuum for the Stokes field [6], [9], we have,

$$\langle f^{(+)}(\tau)\ f^{(-)}(\tau')\rangle = \frac{1}{N\tau_R}\ \delta(\tau-\tau') \ ,$$

$$\langle f^{(-)}(\tau)\ f^{(+)}(\tau')\rangle = \langle f^{(-)}(\tau)\ f^{(-)}(\tau')\rangle = \langle f^{(+)}(\tau)\rangle = 0 \ , \qquad (2)$$

$$\langle G(z,\tau)\rangle_{\mathcal{R}} = \langle G(z,\tau)\ G(z'\tau')\rangle_{\mathcal{R}} = 0 \ ,$$

$$\langle G^{(+)}(z,\tau)\ G(z',\tau')\rangle_{\mathcal{R}} = \gamma\left[N + W(z,\tau)\right]\delta(z-z')\,\delta(\tau-\tau')$$

$$\langle G(z,\tau)\ G^{(+)}(z',\tau')\rangle_{\mathcal{R}} = \gamma\left[N - W(z,\tau)\right]\delta(z-z')\,\delta(\tau-\tau') \qquad (3)$$

where N is the number of atoms composing the sample and $\mathcal{R}$ indicates that the corresponding average is over the reservoir variables. The time τ_R is given by $\tau_R = \frac{u\gamma_0^{-1}}{N}$ where γ_0^{-1} is the natural lifetime of the $3 \rightarrow 2$ transition and u is a structure factor [21]. Equations (1) constitute a set of operator Langevin equations with multiplicative noise source terms. Quantum initiation is generated by these source terms if expectation values of moments are taken in antinormally ordered form. Discussion of the ordering of the operators in expectation values of their products will be addressed later on in the context of the quantum-classical correspondence.

The initial and boundary conditions for the variables of (1) are taken as

$$\langle A_L^{(+)}(0,\tau)\rangle = \mathcal{F}(\tau) \ , \ \langle R(z,0)\rangle = 0 \ , \ \langle W(z,0)\rangle = -N$$

$$\langle A_S(0,\tau)\rangle = 0 \ , \ \langle A_S(z,0)\rangle = 0 \ . \qquad (4)$$

These conditions indicate that the atomic system is in the ground state at $\tau = 0$, no Stokes "seed" is introduced at the input and that the input pump field is in a coherent state. Thus, the Stokes field builds up by quantum initiation. It should be noted that for antinormal ordering, (3), together with (4), indicates no contributions from the collisional source terms precisely at $\tau = 0$. Thus, in the antinormal ordering convention, the free field source can be interpreted as generating polarization stochastically by the last term in (1c) in the presence of the pump field. This in turn drives the Stokes field via (1b) and the process couples with the pump field via (1a) and the inversion via (1d). These equations, together with the initial and boundary conditions, convey all of the dynamical information about the system.

A. Classical Nonlinear Regime: Induced Solitons

In the nonlinear regime of Stokes signal buildup and amplification, the stochastic effects are dynamically unimportant. This is indicated by taking expectation values of (1), in which case the Langevin source terms vanish under the same conditions which led to (2) and (3). Thus, factorization of products of the operators in the nonlinear equations of motion is justified, and the result is,

$$\frac{\partial a_L}{\partial z} = -K_{LS}\, a_S\, r^* \quad , \quad \frac{\partial a_S}{\partial z} = K_{LS}^*\, a_L\, r \qquad (5a,b)$$

$$\frac{\partial r}{\partial \tau} = -K_{LS}\, a_S\, W\, a_L^* - \gamma r \quad , \quad \frac{\partial W}{\partial \tau} = 2\, K_{LS}\, a_L\, r^*\, a_S^* + \text{h.c.} \quad . \qquad (5c,d)$$

where

$$a_{\substack{S\\L}} = \left\langle A^{(-)}_{\substack{S\\L}} \right\rangle \quad , \quad r = \langle R \rangle \quad , \quad W = \langle W \rangle \; .$$

With the exception of the last equation for the ground state depletion, this set of equations is identical to the classical SRS equations which have been widely used [2-5] and studied extensively for their mathematical properties [1]. The set of equations (5a) - (5d), have known soliton solutions [1]. Initial and boundary conditions that result in soliton solutions have been studied numerically and experimentally [2-5].

Neglecting ground state depletion and assuming that the collisional dephasing, γ, is sufficiently large that the polarization, r, can be eliminated adiabatically, (5a) - (5c) combine to give

$$\frac{\partial a_L}{\partial \tilde{z}} = -\left|a_S\right|^2 a_L \qquad \text{Pump Loss} \qquad (6a)$$

$$\frac{\partial a_S}{\partial \tilde{z}} = \left|a_L\right|^2 a_S \qquad \text{Stokes Gain} \qquad (6b)$$

which is the normal condition for a Raman amplifier (here, the coordinate z has been normalized for convenience, $\tilde{z} = (z|K_{LS}|^2)/\gamma$). If, however, again neglecting ground state depletion, we make the transformation $P \equiv a_L^* a_S$, $Q = |a_L|^2 - |a_S|^2$, (5a) - (5c) take the form,

$$\frac{\partial P}{\partial \tilde{z}} = \tilde{r} Q \quad , \quad \frac{\partial Q}{\partial \tilde{z}} = -2\, \tilde{r}^* P - 2 \tilde{r} P^* \quad , \quad \frac{\partial \tilde{r}}{\partial \tilde{\tau}} = P - \tilde{r} \; , \qquad (7,a,b,c)$$

where $\tilde{\tau} = \gamma\tau$ and $\tilde{r} = r/(NK_{LS}/\gamma)$.

Equations (7) are identical in form to the optical Bloch equations at atomic resonance, coupled to the linearized Maxwell equation in the slowly varying envelope and phase approximation (SVEA) with linear loss, but with the roles of the space-like variable, $\tilde{z}$, and time-like variable, $\tilde{\tau}$ interchanged. These equations can be transformed to a single second-order partial differential equation in the form of a Sine-Gordon equation, which is well known to have soliton solutions in the asymptotic regime of pulse propagation [22]. The transformation and the resulting equations have been used in connection with the Bäcklund transformation for the wave-wave scattering problem to obtain one- and two-soliton solutions [23].

To discuss conditions which cause solitons, it is useful to write the variables in (7) in amplitude and phase,

$$P = |P|\, e^{i\phi_p} \quad , \quad \tilde{r} = |\tilde{r}|\, e^{i\phi_r} . \tag{8}$$

In terms of (8), (7b) becomes

$$\frac{\partial Q}{\partial \tilde{z}} = -4\, |\tilde{r}|\, |P| \cos\phi \tag{9}$$

where $\phi = \phi_p - \phi_r$ is the relative phase. For conditions which led to (6), $\phi = 0$. However, if $\phi \neq 0$ due to stochastic effects or deterministically induced [3-4], then we have, from (9)

$0 \leq \phi \leq \pi/2$	Pump loss-Stokes gain	(9a)
$\pi/2 < \phi \leq \pi$	Stokes loss-Pump gain .	(9b)

So, if condition (9b) exits in the pump depletion region in the nonlinear regime of amplification, and within the dephasing time of the medium, the pump experiences a coherent repletion quite similar to the coherent process between medium and field in self-induced transparencies [24]. A phase change of $\phi = \pi$ in the Stokes signal within a temporal regime γ^{-1} can be viewed as a close approximation to a hyperbolic tangent temporal dependence of the Stokes amplitude. Such a functional dependence of the Stokes signal is a soliton solution corresponding to (5a) - (5c) and to (7), and the associated pump soliton envelope would then be a hyperbolic-secant [14]. Induced phase flips of this nature in a Stokes signal "seed" have been studied both numerically, using (5a) - (5c), and experimentally [2-5] and show a one-to-one correspondence between soliton generation in the pump depletion region and Stokes field phase changes $\phi > \pi/2$ [14].

B. Quantum Initiation: Spontaneously-Induced Solitons

Since induced changes in the Stokes amplitude play a crucial role in producing solitons deterministically in SRS, one would suppose that amplified processes stemming from stochastic phase noise may cause solitons to occur "spontaneously", with a corresponding stochastic distribution of occurrence and other properties.

Amplified phase noise from quantum initiation was first treated by HOPF [7] in the context of superfluorescence (SF) and by HOPF and OVERMAN [8] in the context of swept-gain superfluorescence (SGSF) [22]. It was shown by them that effects of phase noise are manifest as phase waves in the linear regime of pulse buildup and in the nonlinear region of pulse amplification. We have shown, for the first time, the role of phase noise and phase waves in spontaneous generation of Raman solitons, SGRS [6], and the connection of phase wave statistics in SRS with SF and SGSF [9]. Phase wave and SGRS statistics were further investigated by us [10-14] theoretically and by CARLSTEN and MACPHERSON experimentally [15, [17].

For the purpose of discussing dynamics of amplified quantum statistics, it is convenient to regard the dynamical evolution of the system in terms of three essential regimes:

a. The linear regime of stochastic quantum initiation
b. The amplified linear regime of coherent pulse buildup
c. The amplified nonlinear regime of pulse evolution.

We are interested primarily in the nonlinear regime, where solitons occur, and associated amplified, macroscopic stochastic properties which stem from quantum initiation.

In the linearized regimes a. and b. above, analytical results are easily obtained from (1) - (3), together with initial and boundary conditions, (4), (the problem in the linearized regime is identical to that in the linearized regimes of SF and SGSF). To obtain results in the nonlinear regime of temporal and spacial evolution requires numerical analysis. We have developed a simulation algorithm based upon (1) - (3) and quantum-classical correspondence, which is used to treat the entire evolution of the system. This will be discussed in the next section. Here, we analyze the linear regime from which the quantum statistics are derived analytically. These results can be used to validate the simulation results in the linear regime.

The operator equations of motion and propagation, (1), are linear in the regime of negligible pump depletion, $\partial A_L^{(-)}/\partial z = 0$, and ground state depletion, $\partial W/\partial \tau = 0$. Thus, for $W(z,\tau) \approx -N$, $A_L^{(-)}(z,\tau) = A_L(\tau)$ where $A_L(\tau)$ is a c-number (coherent state), (1) becomes

$$\frac{\partial A_S^{(-)}}{\partial z} \approx K_{LS}^{*} A_L R \tag{10a}$$

$$\frac{\partial R}{\partial \tau} \approx N K_{LS} A_L A_S^{(-)} - \gamma R + N K_{LS} A_L f^{(-)} + G \,. \tag{10b}$$

These equations are manifestly linear and can therefore be solved without further approximation. The solution to these equations is obtained in a straightforward manner using the Laplace transform technique.

It is noted that for initial and boundary conditions, (4), that expectation values of normally ordered products of the operators, have nonzero contributions for $A_L(\tau) \neq 0$ entirely from the terms in the initial polarization operators, $R(z,0)$, i.e., the radiation reaction quantum initiation terms. On the other hand, expectation values of antinormally ordered products of the operators, have nonzero contributions entirely from the terms in the free field operators, $f^{(-)}(\tau)$, i.e., the vacuum field quantum initiation terms. The particular choice of ordering will have no effect on the results in the amplified nonlinear regime where quantum fluctuations no longer contribute to the dynamical evolution of the system.

We use the antinormal ordering convention to calculate expectation values of operator products. This is done in order to maintain consistency with the quantum-classical correspondence intrinsic to the algorithm we obtain from the quantum mechanical model to numerically calculate the dynamical behavior of the system from initial and boundary conditions through the nonlinear regime of pulse amplification. This is discussed in the next section. In fact, when one is interested in effects of amplified quantum fluctuations in the nonlinear region, the choice of antinormal ordering is quite reasonable, since the medium amplifies quantum fluctuations which originate in the linear regime, and since the particular ordering cannot affect results in the nonlinear regime.

Since all the stochastic behavior of the system originates in the linear regime, it is important to analyze the system in terms of relevant expectation values of operator products in this region. For the purpose of discussing spontaneously generated solitons and phase wave statistics, of particular importance is the correlation function,

$$C(z,\tau) = \frac{\left| \langle A_S^{(+)}(z,\tau)\, R(z,\tau) \rangle \right|}{\left[\langle A_S^{+}(z,\tau)\, A_S^{(-)}(z,\tau) \rangle \;\; \langle R^{+}(z,\tau)\, R(z,\tau) \rangle \right]^{1/2}} \tag{11}$$

where from the solution of (10a) and (10b),

$$\langle A_S^{(+)}(z,\tau)\, A_S^{(-)}(z,\tau) \rangle = N^2 K_{LS}^4 \left| A_L(\tau) \right|^2 \int_0^{\tau} d\tau' \, e^{-2\gamma(\tau-\tau')}$$

$$\times \left| A_L(\tau') \right|^2 \left[\frac{z}{F(\tau) - F(\tau')} \right] I_1^2 \left[2\sqrt{(F(\tau) - F(\tau'))z} \right] , \tag{12}$$

$$\langle R^{+}(z,\tau)\, R(z,\tau) \rangle = N^2 \left| K_{LS} \right|^2 \int_0^{\tau} d\tau' \, e^{-2\gamma(\tau-\tau')} \left| A_L(\tau') \right|^2$$

$$\times I_0^2 \left[2\sqrt{(F(\tau) - F(\tau'))z} \right] \tag{13}$$

and

$$\langle A_S^{(+)}(z,\tau)\, R(z,\tau) \rangle = N K_{LS} A_L(\tau) \int_0^{\tau} d\tau' \, e^{-2\gamma(\tau-\tau')} \left| A_L(\tau') \right|^2$$

$$\times \sqrt{\frac{z}{F(\tau) - F(\tau')}} \; I_1 \left[2\sqrt{(F(\tau) - F(\tau'))z} \right] I_0 \left[2\sqrt{(F(\tau)-F(\tau'))z} \right] . \tag{14}$$

Here, $F(\tau) = N \left| K_{LS} \right|^2 \int_0^{\tau} d\tau' \left| A_L(\tau) \right|^2$ is the integrated pump intensity function. It is noted that, from (3), the collisional source term in (1c) and (10b) does not contribute to the moments, (12) - (14), due to the conditions, (4).

All of the statistical information regarding spontaneous soliton generation and phase wave statistics is controlled by the spacial and temporal dependence of the function C, defined by (11). For example, from the discussion in the previous section, a soliton is expected to form if a sudden phase change, $\phi > \pi/2$, occurs in the spontaneously generated Stokes field in the linear region of pulse buildup (generated from stochastic phase fluctuations in the linear regime of quantum initiation). When the Stokes field with the phase change becomes amplified into the nonlinear regime, a soliton is expected to occur as a repletion in the pump depletion region, just as in the induced deterministic situation discussed in the last section. Since quantum fluctuations are unimportant in the nonlinear regime, the frequency of solitons is expected to be intimately associated with the frequency of generation of phase waves in the linear regime. Here, we define a phase wave as a propagating quantum fluctuation initiated peak in the relative phase ϕ of the field amplitude which originates in the linear regime.

Since the statistics of the quantum initiation process is bivariate Gaussian with zero mean, it is straightforward to calculate the probability that the phase ϕ is greater than a certain value ϕ_0 [25], [7-8], [6], and [9],

$$P\left(|\phi| > \phi_0, z,\tau\right) = 1 - \frac{\phi_0}{\pi} - \frac{C(z,\tau)\sin\phi_0}{\sqrt{1 - C^2(z,\tau)\cos^2\phi_0}} \times \left[1 - \frac{\arccos\left(C(z,\tau)\cos\phi_0\right)}{\pi}\right] \quad . \tag{15}$$

Note that for C = 1, which occurs when the correlation function in (11) factorizes, P = 0 from (15) (except at $\phi_0 = 0$, since P is normalized to unity). It is reasonable and useful to label as the classical regime the region where C = 1, and $0 < C < 1$ denotes quantum stochastic behavior. Another example is the joint probability density of a Stokes intensity, I_S, $\langle I_S \rangle = \langle A_S^+(z,\tau)\, A_S^{(-)}(z,\tau)\rangle$ and relative phase ϕ,

$$P\left(I_S, \phi ; z,\tau\right) = \frac{e^{-x^2}}{2\pi \langle I_S\rangle}\left[1 - \pi\chi\lambda e^{\chi^2\lambda^2}\,\mathrm{erf}\,(\chi\lambda)\right] \tag{16}$$

where

$$\lambda = -C\cos\phi \quad ; \quad \chi = \sqrt{\frac{I_S}{\left[\langle I_S\rangle\left(1 - C^2\right)\right]}} \quad . \tag{17}$$

This function is plotted in Fig. 2 for two values of the quantity C, C = 0 and C = 0.95. Figure 2a shows the isotropic distribution for C = 0 and corresponds to the stochastic regime of quantum initiation where $A_S^{(-)}$ and R are uncorrelated, whereas Fig. 2b demonstrates the

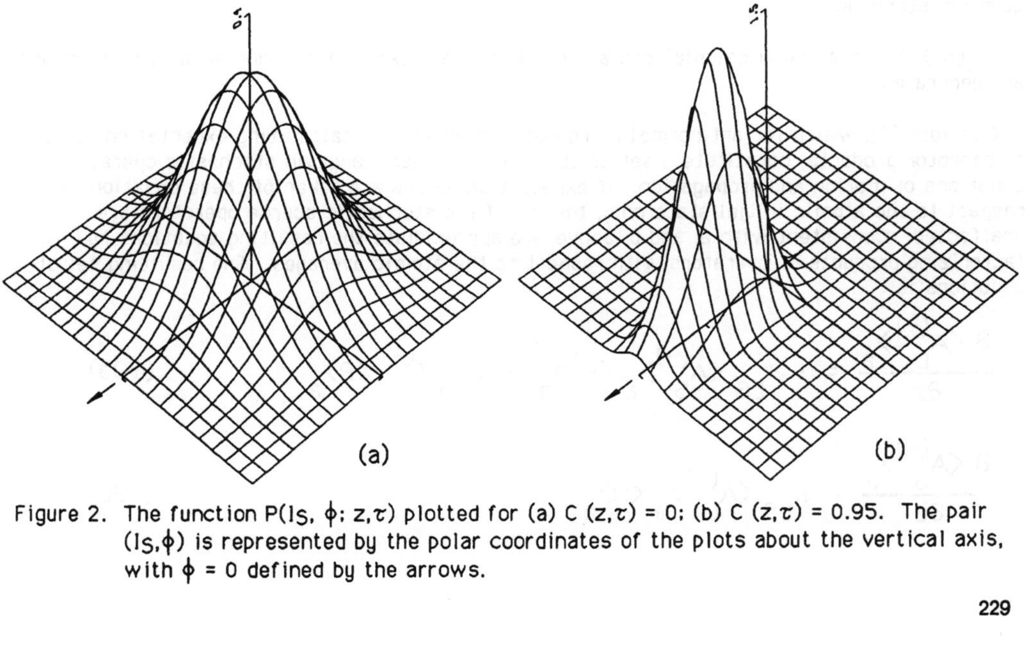

(a) (b)

Figure 2. The function $P(I_S, \phi; z,\tau)$ plotted for (a) $C(z,\tau) = 0$; (b) $C(z,\tau) = 0.95$. The pair (I_S,ϕ) is represented by the polar coordinates of the plots about the vertical axis, with $\phi = 0$ defined by the arrows.

tendency to phase locking as $A_S^{(-)}$ and R becomes highly correlated, corresponding to the linear regime of coherent pulse buildup. The latter can be readily seen by taking the limit of (16),

$$\lim_{C\to 1} P(I_S , \phi ; z,\tau) = \langle I_S \rangle^{-1} \exp\left[-\frac{I_S}{\langle I_S \rangle}\right] \delta(\phi) , \qquad (18)$$

and illustrates the phase locking to zero relative phase, which is the normal classical result, as discussed in the last section.

III. Calculational Scheme: Quantum-Classical Correspondence

All dynamical and propagational information and stochastic behavior concerning the system in the linearized regimes of quantum initiation and coherent pulse buildup is obtained analytically from the solution of (10a) and (10b). We are interested in the evolution of the system in the amplified nonlinear regime where solitons occur and the quantum statistical behavior, which has origin in the linear regime, is amplified and manifest macroscopically. Since the nonlinear regime defies analytical solution, the procedure must be calculational.

For the calculational scheme, we require the following:

a. Must reproduce analytical results in linearized regime of quantum initiation and pulse buildup.

b. Must reproduce analytical and calculational results in classical nonlinear regime.

With the conditions a and b in mind, our procedure consists of the following:

(i) A self-consistent set of partial differential equations is obtained from the model consistent with a. and b. above which intrinsically manifests the quantum-classical correspondence.

(ii) The resulting set of coupled nonlinear equations is numerically integrated in conjunction with the appropriate initial and boundary conditions and according to the intrinsic quantum statistics.

(iii) The numerical calculations are treated as an "experiment", and ensembles of results are generated.

Consider (1), which for antinormally ordered convention for calculating expectation values of operator products, constitute a set of coupled, nonlinear, Langevin stochastic operator equations of motion and propagation. If expectation values are taken of these equations with respect to the atomic initial state only, the free field stochastic source operators are unaffected. Consistent with a. and b. above, we approximate the resulting equations by factorizing products of operators with respect to the atomic averages. The resulting set of equations is,

$$\frac{\partial \langle A_L^{(-)} \rangle_a}{\partial z} = - K_{LS} \langle A_S^{(-)} \rangle_a \langle R^+ \rangle_a - K_{LS} f^{(-)} \langle R \rangle_a \qquad (19a)$$

$$\frac{\partial \langle A_S^{(-)} \rangle_a}{\partial z} = K_{LS}^* \langle A_L^{(-)} \rangle_a \langle R \rangle_a \qquad (19b)$$

$$\frac{\partial \langle R\rangle_a}{\partial \tau} = - K_{LS} \langle A_S^{(-)}\rangle_a \langle W\rangle_a \langle A_L^{(+)}\rangle_a - \gamma \langle R\rangle_a$$

$$+ G - K_{LS} f^{(-)} \langle w\rangle_a \langle A_L^{(+)}\rangle_a \qquad (19c)$$

$$\frac{\partial \langle w\rangle_a}{\partial \tau} = 2 K_{LS} \langle A_S^{(-)}\rangle_a \langle R^{+}\rangle_a \langle A_L^{(+)}\rangle_a$$

$$+ 2 K_{LS} f^{(-)} \langle R\rangle_a \langle A_L^{(+)}\rangle_a + \text{h. c.} \quad . \qquad (19d)$$

Here, $\langle\ \rangle_a$ indicates average with respect to the atomic initial state only. It is emphasized that (19) are still operator equations due to the presence of the stochastic operators $f^{(-)}$ and G, and therefore, the ordering of products in the equations should be maintained. The only operator character to this set of equations are the operator stochastic source terms.

In the nonlinear regime of pulse amplification, the effects of the source terms in (19) are negligible and the corresponding source terms can be ignored. Equations (19) then reduce precisely to the set of classical equations, (5), and carry all the information concerning the system in deterministic classical evolution.

On the other hand, (19) reduce precisely to the quantum mechanical equations, (10), in the linearized regime of quantum initiation and pulse buildup, provided we agree to calculate expectation values of operator products in antinormal ordered form. Equations (19) carry all the information concerning the system in the linear regimes of stochastic quantum initiation and pulse buildup.

Thus, (19) completely and precisely describe the system in both linear and nonlinear regimes and satisfy the conditions of a. and b. above. At this point, we can use (19) together with the stochastic properties expressed by (2) and (3) and proceed with numerical calculations of ensemble averages of products of the system variables according to (i) - (iii) above. Due to the stochastic properties expressed by (2) and (3) and the fact that all even order moments of the stochastic operators completely factorize, the stochastic properties of the sources are bivariate Gaussian with zero mean. We can therefore, with validity, replace the operator source terms in (19) by c-number, classical stochastic variables with classical stochastic properties dictated by (2) and (3). Then (19) become classical Langevin equations with multiplicative noise source terms. Thus,

$$f^{+} \rightarrow |f|\, e^{i\theta} \quad , \qquad P\left(|f|^2\right) = \frac{1}{\pi\sigma} \exp\left(- \frac{|f|^2}{\sigma}\right) \qquad (20a)$$

$$0 \leq \theta \leq 2\pi \qquad (20b)$$

where $\sigma = \left(\frac{N\tau_R L}{c}\right)^{-1}$ and L is the length of the medium. The fluctuating amplitude of f corresponds to a Gaussian distribution, P, with zero mean and the phase is defined on a domain from zero to 2π. A similar requirement pertains to the collisional source term G, however, we find that for the initial and boundary conditions, (4), and condition, (3), the collisional sources play no role in the stochastic properties of the system. Thus, at each temporal step in the numerical calculation, the amplitude and phase of the temporally stochastic source $f(\tau)$

is evaluated by random number generator according to (20). We use the initial and boundary conditions, (4), and run ensembles of numerical calculations to obtain stochastic averages and results of amplified quantum effects.

IV. RESULTS AND DISCUSSION

Some results from a 1,000 member ensemble are depicted in Figs. 3-5. The system parameters used are close to those corresponding to the experiments in CO_2 pumped hydrogen [2] - [4]. The pump pulse shape introduced was taken from digitized data from the experiments. The values of the parameters used were FWHM = 1.3, $A_{L\ max} = 7.2\times10^7$, $|K_{LS}| = 5.8\times10^{-17}$, $N = 8.5\times10^{19}$, $\gamma = 32$. For the specific normalization of parameters, see reference [9].

Figure 3 shows the location and amplitude of soliton events, (crosses), in this ensemble relative to the pump pulse depletion zone at the end of the propagation length, (solid curve). The dotted curve is the outline of the incident pump pulse. The horizontal dotted line is the amplitude cutoff, below which soliton events were not counted.

Figure 4 is a histogram which depicts the frequency of fractional heights of the soliton events from the pulse maximum. There is evident a bias to small amplitudes which corresponds to the tendency for soliton decay for phase changes $\phi \gtrsim \pi/2$. There is also a bias to a fractional height of approximately 0.6 which corresponds to a frequency of soliton occurrence with nearly complete repletion near the inflection point at the leading edge of the pump pulse. We have little explanation for this phenomenon at present, except the fact that solitons near that point have only recently entered the depletion zone and have had little time to decay.

Figure 5 shows the frequency distribution in soliton delay times from pulse turn-on. This shows that most of the solitons occur near the beginning of the depletion zone. This may also be a consequence of soliton decay.

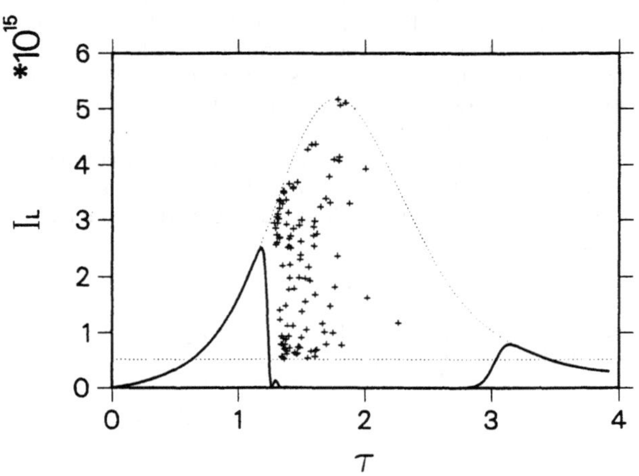

Figure 3. Soliton peak locations, crosses, relative to the pump depletion region at z = L for an ensemble of 1,000. The solid line is the pump intensity profile at z = L; the dotted curve is the incident pump intensity profile which was taken from experiments (see text). The horizontal dotted line indicates the cutoff soliton peak height intensity below which solitons were not counted. For parameter values used, see text.

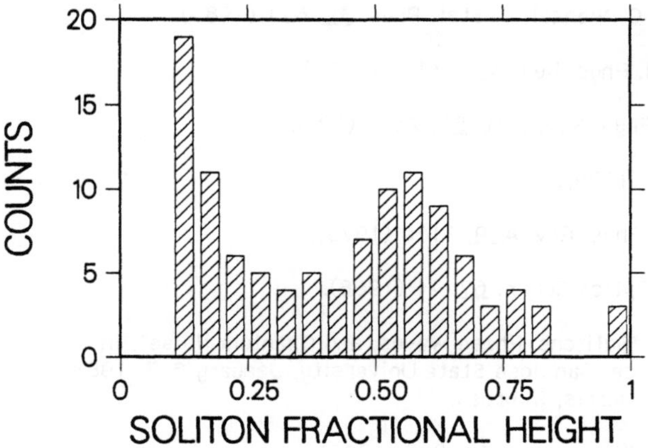

Figure 4. Soliton fractional height distribution corresponding to the ensemble of Fig. 3.

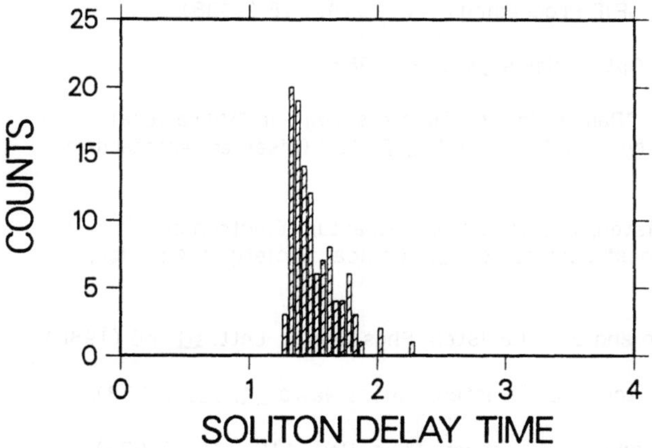

Figure 5. Soliton delay time distribution corresponding to the ensemble of Fig. 3.

The soliton count in the numerical ensemble was 11% to 12%. Some occurrences of multiple (double) solitons were noted. Preliminary comparisons have all tended to support the validity of the simple, one-dimensional model for the aforementioned experiments.

Further statistical analysis of spontaneously generated Raman solitons is in progress and analysis in conjunction with current experiments is in process [17], [26].

REFERENCES

1. F. Y. F. Chu and A. C. Scott, Phys. Rev. A12, 2060 (1975); N. Tan-No, F. Shirakata, K. Kaloto and H. Inaba, Phys. Rev. A12, 159 (1975); T. M. Makhviladze, M. E. Sanychev and L. A. Shelepin, Zn. Eksp. Teor. Fiz. 69, 499 (1975) [Sov. Phys. JETP 42, 255 (1976)]; D. J. Kaup, Physica 6D, 143 (1983); H. Steudel, Physica 6D, 155 (1983); R. Meinel, Opt. Comm. 49, 224 (1984); T. M. Makhviladze and M. E. Sanychev, Zn. Eksp. Teor. Fiz. 71, 896 (1976) [Sov. Phys. JETP 44, 471 (1976)]; A. Hasegawa, Appl. Opt. 23, 3302 (1984); Y. Gürsey, Phys. Rev. A36, 4792 (1987).

2. K. J. Drühl, R. G. Wenzel and J. L. Carlsten, Phys. Rev. Lett. 51, 1171 (1983).

3. K. J. Drühl, J. L. Carlsten and R. G. Wenzel, J. Stat. Phys. 39, 615 (1985).

4. K. J. Drühl, J. L. Carlsten and R. G. Wenzel, J. Stat. Phys. 39, 621 (1985).

5. J. R. Ackerhalt and P. W. Milonni, Phys. Rev. A33, 3195 (1986).

6. J. C. Englund and C. M. Bowden, Phys. Rev. Lett. 57, 2661 (1986).

7. F. A. Hopf, Phys. Rev. A20, 2064 (1979).

8. F. A. Hopf and E. A. Overman, II, Phys. Rev. A19, 1180 (1979).

9. C. M. Bowden and J. C. Englund, Optics Comm. 67, 71 (1988).

10. C. M. Bowden and J. C. Englund, "Solitons, Phase Waves, and Quantum Noise", in "Proceedings of SOCOS Conference, San Jose State University, January 6-7, 1988, edited by D. Heffernan and H. C. Morris, in press.

11. C. M. Bowden and J. C. Englund, Optics News 14, 222 (1988).

12. J. C. Englund and C. M. Bowden, SPIE Proceedings, VOL. 874, 218 (1988).

13. J. C. Englund and C. M. Bowden, Optics News 14, 224 (1988).

14. C. M. Bowden and J. C. Englund, "Raman Soliton Spectroscopy", in "Vibrational Spectra and Structure," edited by J. R. Durig, Vol. 17B, 107 (Elsevier, Amsterdam, 1989).

15. D. C. MacPherson and J. L. Carlsten, in International Quantum Electronics Conference 1987, Technical Digest Series, Vol. 21 (Optical Society of America, Washington, DC, 1987).

16. D. C. MacPherson, R. C. Swanson and J. L. Carlsten, Phys. Rev. Lett. 61, 66 (1988).

17. D. C. MacPherson, R. C. Swanson and J. L. Carlsten, Optics News 14, 222 (1988).

18. D. Polder, M. F. H. Schuurmans and Q. H. F. Vrehen, Phys. Rev. A19, 1192 (1979).

19. C. M. Bowden and C. C. Sung, Phys. Rev. A18, 1558 (1978).

20. C. M. Bowden and C. C. Sung, Phys. Rev. Lett. 50, 156 (1983).

21. N. E. Rehler and J. H. Eberly, Phys. Rev. A3, 1735 (1971).

22. R. Bonifacio, F. A. Hopf, P. Meystre and M. O. Scully, Phys. Rev. A12, 2568 (1975).

23. F. Y. F. Chu, Phys. Rev. 12A, 2065 (1975).

24. G. S. McCall and C. S. Hahn, Phys. Rev. 183, 457 (1969).

25. K. Miller, "Complex Stochastic Processes (Addison-Wesley, Reading, MA, (1974), Ch. 3.

26. J. L. Carlsten, D. C. MacPherson and R. C. Swanson, private communication.

Many-Body Theory of the Electron–Hole Plasma in Bulk and Quantum-Well Semiconductor Lasers

S.W. Koch [1,2], *H. Haug* [3], *C. Ell* [3], *M. Lindberg* [1], *and M. Sargent III* [1]

[1]Optical Sciences Center, University of Arizona, Tucson, AZ 85721, USA
[2]Department of Physics, University of Arizona, Tucson, AZ 85721, USA
[3]Institut für Theoretische Physik, Universität Frankfurt, D-6000 Frankfurt, Fed. Rep. of Germany

Generalized Bloch equations for semiconductors are discussed that include the most important many-body Coulomb effects. The equations are used to describe the high-density electron-hole plasma in semiconductor lasers. Using a Padè approximation for the optical susceptibility, the theory is evaluated for bulk and quantum-well GaAs. Gain/absorption and refractive index spectra are computed for various plasma densities and temperatures.

1. Introduction

The emission of light from an excited intrinsic semiconductor requires the decay of electrons from the conduction band into the valence band. Since the electrons obey Fermi-Dirac statistics, this decay process can occur only if electron vacancies (holes) in the valence band are available allowing energy and momentum conservation. Hence the light-semiconductor interaction involves pairs of conduction-band electrons and valence-band holes, i.e., pairs of quantum mechanical quasi-particles with opposite charges. These electrons and holes are strongly correlated through the Pauli exclusion principle and interact through the Coulomb potential. The Coulomb interaction is attractive for oppositely charged particles (interband interaction) and repulsive for equal charges (intraband interaction) and it is strong enough to cause modifications (renormalizations) of the semiconductor material properties. Both the linear optical properties and the nonlinear response of semiconductors are significantly influenced by the interactions in the system of electron-hole excitations. Consequences of these interactions are (i) the Coulomb enhancement of gain and absorption leading, e.g., to drastic deviations of the semiconductor absorption spectra from the square root-law expected for noninteracting Fermions, (ii) the electron-hole plasma screening of the Coulomb potential, and (iii) the reduction of the semiconductor bandgap with increasing carrier density [1,2].

In general the theoretical description of the electron-hole system in semiconductors leads to equations for the electron and hole populations and for the interband polarization, which are coupled to correlations of higher order [3-8]. However the theory can be simplified considerably if one makes the so-called quasi-equilibrium or adiabatic approximation. This approximation is well justified for single-mode and typical multimode semiconductor lasers, since the intraband carrier relaxation time due to the carrier-carrier scattering is less than 100 fs [9]. Within this short period the electrons and holes establish local quasi-equilibrium Fermi distributions within their bands and the higher-order operator products can be factorized into products involving population and polarization terms only. These factorized equations can be written in the form of generalized Bloch equations for semiconductors [4,5]. They are discussed in the next section of this paper.

Using the quasi-equilibrium approximation, we present in Sec. 3 an approximate solution of the equation for the interband polarization [10]. The final expression for the

Springer Proceedings in Physics, Vol. 41 **Quantum Optics V**
Editors: J.D. Harvey and D.F. Walls

optical susceptibility depends on the electron and hole distributions and on the screened Coulomb potential. In Sec. 4 we summarize the theory of plasma screening for two- and three-dimensional systems, which are idealizations of quantum-well and bulk semiconductors, respectively. In Sec. 5 we evaluate the expression for the optical susceptibility and compute the spectra of gain/absorption and refractive index. We present numerical evaluations for the examples of bulk and quantum-well GaAs at various plasma densities and temperatures [11].

2. Generalized Bloch Equations for Semiconductors

In this section we briefly summarize the microscopic theory for a semiconductor that is dipole coupled to a classical radiation field. We especially discuss the generalized Bloch equations for semiconductors as derived in Ref. [4].

In the two-band approximation, the Hamiltonian of the electron-hole system is

$$\mathscr{H} = \sum_{k} \left[\epsilon^{eff}_{e,k}\, a^{\dagger}_{k} a_{k} + \epsilon^{eff}_{h,k}\, b^{\dagger}_{-k} b_{-k} \right]$$

$$+ \frac{1}{2} \sum_{kk'q\neq 0} V_q \left[a^{\dagger}_{k+q} a^{\dagger}_{k'-q} a_{k'} a_{k} + b^{\dagger}_{k+q} b^{\dagger}_{k'-q} b_{k'} b_{k} - 2 a^{\dagger}_{k+q} b^{\dagger}_{k'-q} b_{k'} a_{k} \right]$$

$$+ \sum_{k} \left\{ \mathscr{V}_{cv}\, a^{\dagger}_{k} b^{\dagger}_{-k} + \mathscr{V}_{cv}^{*}\, b_{-k} a_{k} \right\}. \tag{1}$$

Here the interaction energy

$$\mathscr{V}_{cv} = - d_k \mathscr{E}(t) \tag{2}$$

describes the dipole coupling to the electric field $\mathscr{E}(t)$, d_k is the optical dipole matrix element, a_k is the electron annihilation operator for the wave vector k and b_{-k} is the hole annihilation operator for the wave vector -k. The energies $\epsilon^{eff}_{e,k}$ and $\epsilon^{eff}_{h,k}$ are the effective single-particle energies of the electron and hole

$$\epsilon^{eff}_{e,k} = \frac{\hbar^2 k^2}{2m_e} + \frac{E^0_g}{2} \equiv \epsilon_{e,k} \tag{3}$$

$$\epsilon^{eff}_{h,k} = \frac{\hbar^2 k^2}{2m_h} + \frac{E^0_g}{2} + \sum_{q\neq 0} V_q \equiv \epsilon_{h,k} + \sum_{q\neq 0} V_q \, , \tag{4}$$

where m_e and m_h are the electron and hole effective masses and E^0_g is the zero-field band-gap energy. We assume that the Coulomb interaction conserves the number of electrons in each band and we describe the dipole coupling to the laser field $\mathscr{E}(t)$ in the rotating wave approximation. It is straightforward to generalize this treatment to include the quantum statistics of the laser light, by quantization of the field.

We want to emphasize the relation of the Hamiltonian (1) to the Hamiltonian used in studying atomic two-level systems [12]. If we neglect the Coulomb interaction by setting V_q equal to zero, all k states become independent and the problem can be reduced to that of an inhomogeneously broadened two-level system without damping. Damping processes can be modeled phenomenologically [12]. Corresponding to any individual k vector, we have the set of four eigenstates

$$|0\rangle \,,\quad a_k^{\dagger}|0\rangle \,,\quad b_{-k}^{\dagger}|0\rangle \,,\quad a_k^{\dagger}b_{-k}^{\dagger}|0\rangle \,. \tag{5}$$

Notice however that two states $a_k^{\dagger}|0\rangle$, $b_{-k}^{\dagger}|0\rangle$ are optically inactive, i.e.,

$$a_k^{\dagger}b_{-k}^{\dagger} \;\; a_k^{\dagger}|0\rangle = 0 \,, \tag{6}$$

since the electron and hole operators obey Fermi-Dirac statistics. Absorption of a photon always creates an electron-hole pair.

To the extent that the optically inactive states can be neglected, we can identify the remaining two states as the ground state and excited states

$$|0\rangle \equiv |-\rangle \,,\quad a_k^{\dagger}b_{-k}^{\dagger}|0\rangle \equiv |+\rangle, \tag{7}$$

respectively. Similarly, the operators can be translated to the familiar two-level notation by studying their effect on the states. We obtain the correspondence

$$\epsilon_{e,k}^{eff}\, a_k^{\dagger}a_k + \epsilon_{h,k}^{eff}\, b_{-k}^{\dagger}b_{-k} \;\equiv\; \left[\epsilon_{e,k}^{eff} + \epsilon_{h,k}^{eff}\right] \sigma_+\sigma_- \,,$$

$$a_k^{\dagger}b_{-k}^{\dagger} \equiv \sigma_+$$

$$b_{-k}a_k \equiv \sigma_- \,. \tag{8}$$

The Hamiltonian for a single k state can then be written in the form

$$\mathscr{H}(k) = \left[\epsilon_{e,k}^{eff} + \epsilon_{h,k}^{eff}\right] \sigma_+\sigma_- + (\mathscr{V}_{cv}\sigma_+ + \mathscr{V}_{cv}^{\,*}\,\sigma_-) \tag{9}$$

familiar from quantum electronics, where we take the lower energy level to have zero energy. However the effect of the Coulomb interaction is to mix the different k states ultimately leading to thermal Fermi-Dirac distributions of the electrons and holes. In principle similar effects occur in two-level gaseous media due to velocity changing collisions.

Now we go back to our original Hamiltonian (1). To describe laser action in semiconductors, we derive the coupled equations of motion for the expectation values $n_{e,k} \equiv \langle a_k^{\dagger}a_k\rangle$ (probability of having electrons with the wave vector k), $n_{h,k} \equiv \langle b_{-k}^{\dagger}b_{-k}\rangle$ (the probability for holes), and $\rho_{cv,k} \equiv \langle a_k^{\dagger}b_{-k}^{\dagger}\rangle$ (the polarization of the momentum state k). The total carrier density is given by

$$N = \frac{1}{V}\sum_k n_{e,k} = \frac{1}{V}\sum_k n_{h,k} \,. \tag{10}$$

where V is the volume of the semiconductor. Since the Σ_k is proportional to V, the volume cancels out leaving a number density.

Using straightforward operator algebra, we obtain equations of motion for the two-operator expectation values in terms of themselves and of expectation values of four-operator products. Reference [4] closes this set of equations by quantum mechanical projection operator techniques. Here we are only interested in the limit that the carrier-carrier scattering time and the dipole dephasing time are much less than all other relaxation times. In this *quasi-equilibrium* or *hydrodynamic* regime the electrons and holes are within their bands in a quasi-thermal equilibrium, and their distribution functions are given by the Fermi distribution

$$n_{i,k} = \frac{1}{e^{\beta(\epsilon_{i,k} - \mu_i)} + 1} , \tag{11}$$

where μ_i is the quasi-chemical potential obtained from Eq. (10) for a given total carrier density N.

As derived in Refs. [2-8] the equation for the interband polarization can be written as

$$\frac{\partial \rho_{cv,k}}{\partial t} = -(i\omega_t + \gamma)\rho_{cv,k} + i\left[\mathcal{V}_{cv}(t) - \sum_{q\neq 0} V_{s,q}\rho_{cv,k+q}\right](n_{e,k} + n_{h,k} - 1) . \tag{12}$$

The transition energies are renormalized through the Debye shift and the screened exchange contribution, yielding the effective transition energy

$$\hbar\omega_t = \epsilon_{e,k} + \epsilon_{h,k} + \delta\epsilon_g \equiv \hbar\omega_k + \delta\epsilon_g , \tag{13}$$

with $\delta\epsilon_g$ given by

$$\delta\epsilon_g = \sum_{q\neq 0} \left[(V_{s,q} - V_q) - V_{s,q} (n_{e,q} + n_{h,q})\right] , \tag{14}$$

where we have assumed a rigid, i.e., k-independent band shift.

In Eq. (12) we recognize a variety of many-body contributions that have been discussed extensively in the context of high excitation semiconductor pump-probe spectroscopy [1-8], such as plasma screening (in V_s), bandgap renormalization (in $\delta\epsilon_g$), as well as phase-space filling ($1 - n_h - n_e$). As limiting cases, Eq. (14) reproduces well-known results from quantum optics and semiconductor physics. Neglecting all Coulomb-interaction terms yields the Bloch equation for the off-diagonal element of the density matrix

$$\frac{\partial \rho_{cv,k}}{\partial t} = -(i\omega_k + \gamma)\rho_{cv,k} + i\mathcal{V}_{cv}(t)(n_{e,k} + n_{h,k} - 1) . \tag{15}$$

used in atomic spectroscopy of inhomogeneously broadened systems. In semiconductors, the inhomogeneous broadening is an intrinsic consequence of the dispersion of the single particle energies, $\epsilon_{e,k}$ and $\epsilon_{h,k}$.

On the other hand, for one electron, one hole, and a vanishing optical field, we obtain

$$\frac{\partial \rho_{cv,k}}{\partial t} = -i\omega_k \rho_{cv,k} - i\sum_{q\neq 0} V_q \rho_{cv,k+q} . \tag{16}$$

Using the effective mass approximation, $\hbar\omega_k = \hbar^2k^2/2m$, where m is the reduced electron-hole mass, $m = m_e m_h/(m_e+m_h)$, and Fourier transforming to real space shows that Eq. (16) has exactly the form of the Schrödinger equation for the relative motion of electron and ion in a hydrogen atom. In the present context, this equation is known as *Wannier equation* describing the relative motion of an electron-hole pair. The bound-state solutions of the Wannier equation are called excitons and the ionization continuum describes the unbound, but Coulomb-correlated electron-hole pairs. This Coulomb correlation is responsible for the so-called *Coulomb enhancement* of the interband absorption causing the deviations from the $\sqrt{\hbar\omega}$ - absorption expected for non-interacting electrons and holes.

3. Padè Approximation

To solve the polarization equation (12) in the quasi-equilibrium regime, we split-off the rapid time variation with the central frequency ω of the light field

$$\mathcal{E}(t) = E(\omega)\, e^{-i\omega t} \tag{17}$$

and make the adiabatic approximation (often call the rate-equation approximation),

$$|\partial\rho_{cv}/dt| << \gamma|\rho_{cv}| .$$

Next we introduce the susceptibility function $\chi_k(\omega)$

$$\rho_{cv,k}(\omega) = \chi_k(\omega)\, E(\omega) , \tag{18}$$

and the optical susceptibility as

$$\chi(\omega) = \frac{1}{V}\sum_k d_k\, \chi_k(\omega) . \tag{19}$$

The absorption coefficient and the refractive index are given by the imaginary and real parts of χ, according to the semiclassical theory. Inserting (18) into (12) yields the following integral equation

$$\chi_k(\omega) = \chi_k^0(\omega)\left[1+\frac{1}{d_k}\,\frac{1}{V}\sum_{k} V_{s,k-k'}\, \chi_{k'}(\omega)\right] , \tag{20}$$

where

$$\chi_k^0(\omega) = \frac{d_k}{\hbar}\,\frac{n_{e,k} + n_{h,k} - 1}{\omega - \omega_t + i\gamma} . \tag{21}$$

Notice that the deviations from χ^0, the result for non-interacting electrons and holes, are induced by the interband Coulomb interaction. It is a well-known feature that a straightforward perturbative treatment is not effective for most Coulomb problems and one has to apply non-perturbative solution schemes. As discussed in Ref. [10], it is a good approximation for an inverted system to solve Eq. (20) by the first non-trivial Padè approximation. The result is

$$\chi(\omega) = \frac{1}{V} \sum_k \chi_k^0(\omega)\, Q_k(\omega), \tag{22}$$

where

$$Q_k(\omega) = \frac{1}{1-q_1(k,\omega)} \tag{23}$$

and

$$q_1(k,\omega) = \frac{1}{d_k} \frac{1}{V} \sum_{k'} V_{s,k-k'}\, \chi_0(k',\omega)\ . \tag{24}$$

The factor $Q_k(\omega)$ is the Coulomb enhancement factor mentioned at the end of Sec. 2. It describes the enhancement of the optical transitions caused by the interband attraction of electrons and holes. Schematically, the Coulomb enhancement factor and the inversion are plotted in Fig. 1. As discussed in detail in Ref. [10], the enhancement factor $Q_k(\omega)$ has a peak around the quasi-chemical potential $\mu = \mu_e + \mu_h$ and approaches unity for small and large energies. For low electron-hole densities, where no inversion occurs and bound states begin to exist, the Coulomb enhancement factor diverges indicating the occurrence of the bound state. In this limit, the Padè approximation is not satisfactory and one has to use more sophisticated solutions of Eq. (20) [1-8]. However for the purpose of discussing the electron-hole plasma in semiconductor lasers, the present approximation scheme is very well justified.

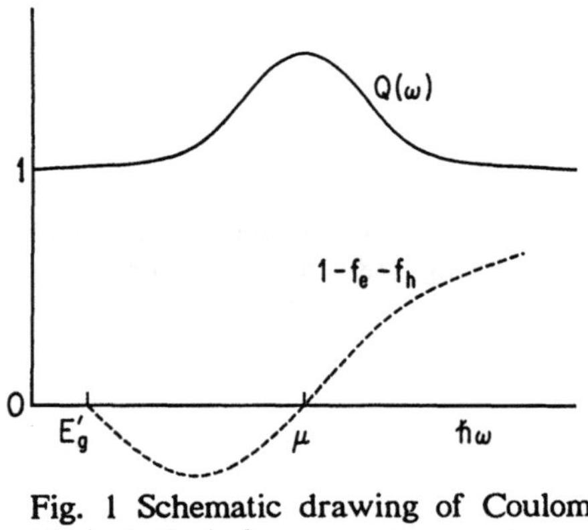

Fig. 1 Schematic drawing of Coulomb enhancement factor $Q(\omega)$ (full line) and inversion (dashed line) for room temperature conditions. We have also indicated the locations of the renormalized bandgap E'_g and the quasi-chemical potential μ, respectively.

4. Plasma Screening

To evaluate the optical susceptibility of Eq. (22), we need an expression for the screened Coulomb potential. In Ref. [10] we review the theory of screening in an electron-hole plasma. Following this discussion, we only summarize the main results here. Generally the screened Coulomb potential $V_{s,q}$ can be written in terms of the unscreened potential V_q as

$$V_s(q,\omega) = \frac{V_q}{\epsilon(q,\omega)}, \tag{25}$$

where $\epsilon(q,\omega)$ is the longitudinal dielectric function. As is discussed in many textbooks [13], the density-dependent dielectric function of a carrier plasma in the random phase approximation (RPA) is given by the Lindhard formula

$$\epsilon(q,\omega) = 1 - V_q \frac{1}{V} \sum_{i=e,h;k} \frac{n_{i,k} - n_{i,k+q}}{\hbar\omega + i\delta - \epsilon_{i,k+q} + \epsilon_{i,k}}, \tag{26}$$

where ϵ_0 is a background dielectric constant and the summation over the spin states is included in the k-summation. Since in general Eq. (26) can only be evaluated numerically, one obtains a considerable simplification if one replaces the multiple poles in (26) by a single effective plasmon pole. If one further disregards the dynamical aspect of screening contained in the frequency dependence of (26), one obtains the much simpler static plasmon pole approximation

$$\frac{1}{\epsilon(q)} = \left[1 - \frac{\omega_{pl}^2}{\omega_q^2}\right], \tag{27}$$

where ω_q is the dispersion of the effective plasmon mode.

In a three-dimensional (3d) bulk semiconductor one has

$$V_q = \frac{4\pi e^2}{\epsilon_0 q^2},$$

$$\omega_{pl}^2 = \frac{4\pi e^2 N}{\epsilon_0 m} = 16\pi\, N a_0^3 \left[\frac{E_R}{\hbar}\right]^2,$$

$$\omega_q^2 = \omega_{pl}^2 \left[1 + \frac{q^2}{\kappa^2}\right] + \frac{C}{4}\left[\frac{\hbar q^2}{2m}\right]^2, \tag{28}$$

where C is a numerical constant, and the plasma frequency ω_{pl} is expressed in terms of the exciton Bohr radius a_0 and the exciton Rydberg $E_R = e^2/2\epsilon_0 a_0 = \hbar^2/2ma_0^2$. The inverse screening length κ is given as

$$\kappa^2 = \frac{4\pi e^2}{\epsilon_0} \sum_i \frac{\partial N}{\partial \mu_i}. \tag{29}$$

In semiconductor quantum wells in which the width is smaller than the exciton Bohr radius, the translational motion perpendicular to the well is quantized in analog to a particle in a box. If one can limit oneself to the lowest subband, the system can be considered quasi-two-dimensional (2d). In statistical systems the spatial dimensionality strongly influences the system properties. In the two-dimensional case the physical parameters of the plasmon pole approximation are given by

$$V_q = \frac{2\pi e^2}{\epsilon_0 q},$$

$$\omega_{pl}^2 = \frac{2\pi e^2 N q}{\epsilon_0 m} = 8\pi \; N a_0^2 \; q a_0 \left[\frac{E_R}{\hbar}\right]^2,$$

$$\omega_q^2 = \omega_{pl}^2\left[1 + \frac{q}{\kappa}\right] + \frac{C}{4}\left[\frac{\hbar q^2}{2m}\right]^2, \tag{30}$$

and

$$\kappa = \frac{2\pi e^2}{\epsilon_0} \sum_i \frac{\partial N}{\partial \mu_i}, \tag{31}$$

where N in Eqs. (30) and (31) is the 2d plasma density.

5. Numerical Results for Bulk and Quantum-Well GaAs

From the real and imaginary part of the susceptibility χ, Eq. (22), we obtain the spectra of absorption (gain) and index of refraction [11]. Fig. 2 shows the computed results for bulk GaAs at $T = 300K$ and various plasma densities corresponding to various injection currents in a real semiconductor laser. In the top part of Fig. 2 we plot the absorption (gain) and in the bottom part we show the corresponding refractive index. A pronounced expansion of the gain region occurs with increasing plasma densities. As usual the cross-over from gain to absorption takes place at the total quasi-chemical potential, which shifts to higher energies for higher plasma densities due to bandfilling. At the same time the renormalization of the transition energies causes the bandgap to decrease with increasing plasma density. However since the quasi-chemical potential is defined with respect to the actual gap, the bandgap renormalization changes only the

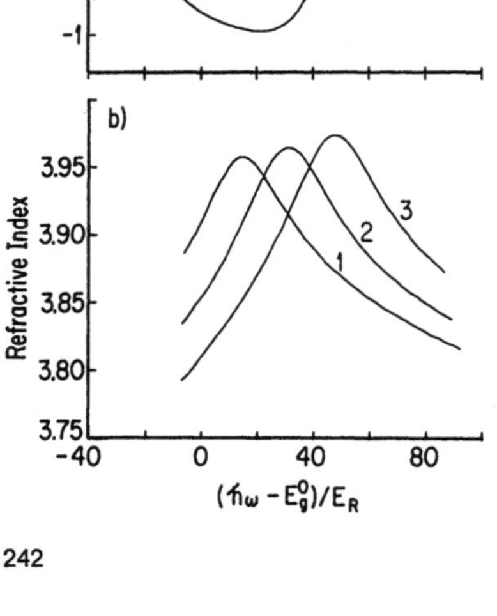

Fig. 2 Gain/absorption (upper part, a) and refractive index (lower part, b) spectra for bulk GaAs at room temperature and plasma densities: (1) $2\cdot10^{18}$cm^{-3}, (2) $5\cdot10^{18}$cm^{-3}, (3) $1\cdot10^{19}$cm^{-3}. (From Ref. [11], where also the used material parameters are specified.)

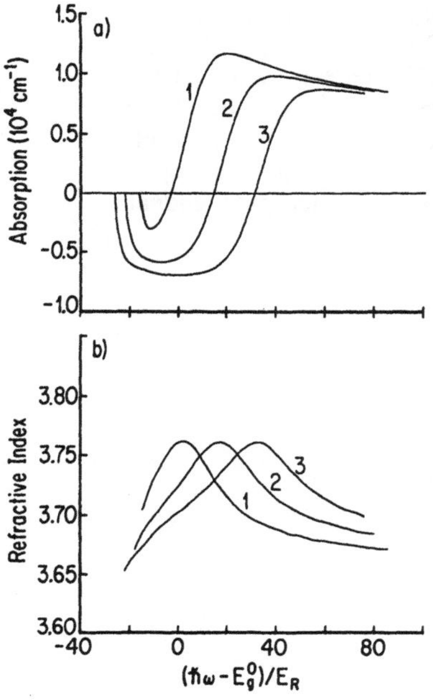

Fig. 3 Gain/absorption (upper part, a) and refractive index (lower part, b) spectra for quasi-2d GaAs at room-temperature for the plasma densities: (1) $2 \cdot 10^{12}$cm^{-2}, (2) $4 \cdot 10^{12}$cm^{-2}, (3) $6 \cdot 10^{12}$cm^{-2}. (From Ref. [11], where also the material parameters are specified.)

spectral position but not the width of the gain region. From the lower part of Fig. 2 we see that the peak of the corresponding refractive index spectra shifts to higher energies causing negative index changes in the gain regime.

The optical spectra for a quasi-2d GaAs quantum-well at T = 300K (Fig. 3) also exhibit the increasing chemical potential and the bandgap renormalization. The shape of the gain spectra (upper part of Fig. 3) deviates significantly from that of bulk GaAs, expressing the difference in the density of states in two and three dimensions. The more pronounced steepening of the spectra around the chemical potential and the small peak in the absorption region result from the stronger electron-hole Coulomb correlation in 2d. The combined influence of Coulomb enhancement and constant density of states leads to a gain spectrum with a broad flat maximum.

As we have also analyzed in Ref. [11], the temperature has a quite dramatic influence on the gain/absorption spectra of semiconductor lasers both in two and three dimensions. In Fig. 4 we compare the low-temperature gain/absorption spectra with and without the Coulomb enhancement factor, Eqs. (22) and (21), respectively. The dramatic differences seen in Fig. 4 are due to the pronounced maximum of $Q(\omega)$ at low temperatures which leads to sharp peaks in the optical spectra just below and above the chemical potential. Furthermore, in bulk material the spectral region of optical gain expands significantly with decreasing temperature, whereas the expansion is almost negligible in 2d due to the much weaker temperature dependence of the quasi-chemical potential in 2d.

A similar comparison as in Fig. 4 but for room-temperature conditions is presented in Ref. [11]. There we show that the Coulomb enhancement causes relatively small modifications in the gain regime of bulk GaAs but is still quite significant for the refractive index and for quantum-well material.

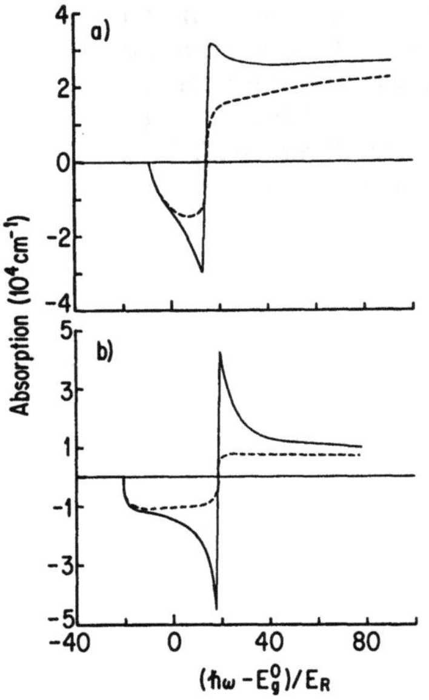

Fig. 4 Gain/absorption spectra for bulk (a) and quasi-2d GaAs (b) at T = 10 K for the plasma densities $2\cdot 10^{18}$ cm^{-3} (a) and $4\cdot 10^{-12}$ cm^{-2} (b), respectively. The dashed lines show the results without Coulomb enhancement but including bandgap renormalization and density-dependent chemical potential.

In summary, we now have a comprehensive many-body theory of the electron-hole plasma in a semiconductor laser. In this paper, we summarize the theory and its evaluation to obtain gain/absorption and refractive index spectra. In the following paper by Sargent et al. [14] we discuss the application of the theory to multiwave interactions and sidemode instabilities in semiconductor laser media.

Acknowledgements: This work and the visits of H. Haug and C. Ell in Tucson are supported by the Optical Circuitry Cooperative at the Optical Sciences Center, Tucson. Partial support through the U. S. Army Research Office and the U. S. Air Force Office of Scientific Research (JSOP), through NSF (grant INT-8713068), NATO (travel grant 0736/87), and through a grant for computer time at the John von Neumann Computer Center, Princeton, is gratefully acknowledged.

References

1. See, e.g., the articles in *Optical Nonlinearities and Instabilities in Semiconductors*, ed. H. Haug (Academic Press, N. Y. 1988)
2. S.W. Koch and H. Haug, *Quantum Theory of the Electro-Optical Properties of Semiconductors*, (World Scientific, Singapore, to be published)
3. J.F. Müller, R. Mewis and H. Haug, Z. Physik B69, 231 (1987)
4. M. Lindberg and S.W. Koch, Phys. Rev. B38, 3342 (1988)
5. S.W. Koch, N. Peyghambarian and M. Lindberg, J. Phys. C: Solid State 21, 5229 (1988).
6. A. Stahl and I. Balslev, *Electrodynamics of the Semiconductor Band-Edge*, Springer Tracts in Modern Physics 110, Springer, Berlin (1986)
7. S. Schmitt-Rink and D. S. Chemla, Phys. Rev. Lett. 57, 2752 (1986); S. Schmitt-Rink, D. S. Chemla, and H. Haug, Phys. Rev. B 37, 941 (1988).

8. R. Zimmerman, phys. stat. sol. b146, 371 (1988).
9. M.P. Kessler and E.P. Ippen, Appl. Phys. Letters 51, 1765 (1987)
10. H. Haug and S.W. Koch, Phys. Rev. A39, February 1989
11. C. Ell, H. Haug, and S.W. Koch, Opt. Lett. (April 1989)
12. for a discussion of laser theory using two-level systems see, e.g., M. Sargent III, M. O. Scully, and W. E. Lamb, Jr., *Laser Physics*, Addison-Wesley, Reading, MA (1977).
13. see, e.g., G.D. Mahan, *Many Particle Physics*, Plenum Press, New York (1981)
14. M. Sargent III, Fenglei Zhou, Sungyuck An, M. Lindberg, and S.W. Koch, this volume.

Theory of Semiconductor Lasers

M. Sargent III, F. Zhou, S. An, M. Lindberg, and S.W. Koch

Optical Sciences Center, University of Arizona, Tucson, AZ 85721, USA

This paper presents a simple theory of a semiconductor laser based on the many-body theory discussed in the preceding paper [1]. We show how to modify an inhomogeneously broadened two-level model to create a reasonable model of a semiconductor laser gain medium. We then include the cavity boundary conditions and find the laser steady-state oscillation condition. We analyse the stability of single-mode laser operation, obtaining relaxation-oscillation equations similar to those for a homogeneously broadened two-level atom laser in the rate equation limit. Further questions of instability lead us to a theory of multiwave mixing in the semiconductor medium. This theory is also useful for saturation spectroscopy and phase conjugation using such media. We finish by discussing the "spark" for sidemode buildup in a semiconductor laser, namely the resonance fluorescence spectrum.

1. Introduction

The semiconductor laser has become the most important laser due to its high efficiency (50%), low cost ($10), small size (like a pencil eraser), and long life (one or more years). Nevertheless compared to other lasers such as the He-Ne laser, the semiconductor laser has had relatively primitive theoretical modeling, typically limited to two-level rate equation analysis. While such analysis yields very useful predictions and has had remarkable success [2, 3], it requires the ad hoc specification of a variety of "constants" that should be derivable from a more fundamental theory. The present paper gives a theory based on a quasi-equilibrium model of the semiconductor described in the preceeding paper [1]. Although we neglect a number of the refinements of many-body theory, our theory is substantially more realistic than a two-level rate equation theory and is still simple enough to be used with the aid of a personal computer. The model is similar to that given in Ref. [4] and is based on the generalized Bloch equations of Ref. [5]. Background two-level models are described in Refs. [6, 7].

In the remainder of this introduction, we describe the two-band model of a semiconductor and introduce some notation. Section 2 derives the polarization of the semiconductor medium. Section 3 solves for the steady-state laser oscillation condition. Section 4 studies the relaxation oscillations toward this steady state, and Sec. 5 summarizes a theory of multiwave mixing that can be used to analyse the possible build-up of side modes.

The semiconductor has an "upper-level" band called the conduction band, where electrons can flow. It also has a lower-level band, called the valence band, where holes can flow. Holes are the absence of electrons. When we say electron, we mean a conduction electron, although there are valence electrons, and when we say hole, we mean a valence hole. As such holes and electrons are free charge carriers, and we often refer to them together simply as carriers. The two bands are diagrammed in Fig. 1. The electron and hole energies are given by

Springer Proceedings in Physics, Vol. 41 **Quantum Optics V**
Editors: J.D. Harvey and D.F. Walls

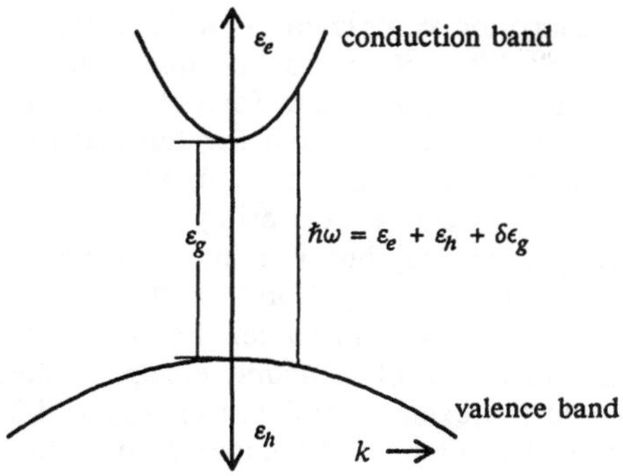

Fig. 1. Energy band diagram ε versus momentum k. The electron energy ε_e increases upward, and the hole energy ε_h increases downward. Optical transitions are vertical on this graph, since photons have negligible momentum compared to electrons. The semiconductor parameters illustrate GaAs, a common laser material. The bandgap energy ε_g = 1.462 meV, the dielectric constant ϵ = 12.35 giving an index of 3.5, the electron effective mass m_e = 1.127 m, where m is the reduced mass, and the hole effective mass m_h = 8.82 m. As seen from Eqs. (1) and (2), the curvatures of the bands are inversely proportional to their respective effective masses.

$$\varepsilon_e(k) = \frac{\hbar^2 k^2}{2m_e} + \frac{\varepsilon_g}{2}, \tag{1}$$

$$\varepsilon_h(k) = \frac{\hbar^2 k^2}{2m_h} + \frac{\varepsilon_g}{2}, \tag{2}$$

where m_e and m_h are the electron and hole effective masses and ε_g is the zero-field bandgap energy. If an electron in the valence band absorbs light, it is excited into the conduction band leaving a hole behind in the valence band. The energy of the photon inducing a transition at the carrier momentum k is given by

$$\hbar\omega = \varepsilon_e(k) + \varepsilon_h(k) + \delta\varepsilon_g = \varepsilon + \varepsilon_g + \delta\varepsilon_g , \tag{3}$$

where ε is the reduced mass energy

$$\varepsilon = \frac{\hbar^2 k^2}{2m}, \tag{4}$$

and m is the reduced mass defined by

$$\frac{1}{m} = \frac{1}{m_e} + \frac{1}{m_h}, \tag{5}$$

and $\delta\varepsilon_g$ is a bandgap renormalization energy that depends on the carrier density. Specifically the Coulomb potential between the carriers gives a net reduction in energy and there is a negative energy contribution given by the Fermion exchange correlation. Both of these effects are enhanced by increasing the carrier density. $\delta\varepsilon_g$ is typically about -10 to -30 meV. For our present purposes, we express our results relative to the renormalized band gap, so that we don't need to know its value. For further discussion, see Refs. [1, 8].

We wish to derive the semiconductor complex absorption coefficient as a function of temperature, current, decay rates, and effective masses. We can fix up the inhomogeneously broadened two-level theory for gain media to find an appropriate formula for ρ_{ab}. This provides the polarization of the medium, which we use as a source for the electromagnetic field in Maxwell's equations. In addition to the bandgap reduction $\delta\varepsilon_g$ of Eq. (3), the major new mechanism we need to include is carrier-carrier scattering. This has a counterpart in gas lasers known as velocity changing collisions, but is a much stronger effect in semiconductors and has consequences in addition to reequilibrating the carrier distributions. The carriers are charged and interact via the Coulomb force. For low carrier densities, this leads to excitons, H-like "atoms" consisting of a bound electron-hole pair. The exciton Bohr radius in GaAs is 124.3 Å, and the exciton Rydberg energy is 4.2 meV, which is tiny compared with 13.6 eV for the H atom and small compared to room-temperature k_BT = 25 meV. As carrier density increases, the Coulomb potential becomes increasingly screened, and for densities above 10^{16} the excitons are completely ionized. For densities high enough to get gain (2×10^{18}), the excitons are ionized and two main effects remain from the carrier-carrier scattering: 1) The intraband carrier distributions can each be described by Fermi-Dirac distributions provided external forces like light fields vary little in the carrier-carrier scattering time of .1 picoseconds or less. 2) The radiation recombination is proportional to the product of electron and hole occupation probabilities, while the two-level case is proportional to the probability of upper-level occupation alone.

The Fermi-Dirac distribution for the carrier α (= e or h) is given by

$$f_\alpha = \frac{1}{e^{\beta[\varepsilon_\alpha(k)-\mu_\alpha]} + 1}, \tag{6}$$

where $\beta = 1/k_BT$, k_B is Boltzmann's constant, T is the absolute temperature, and μ_α is the carrier chemical potential that yields a self-consistent total carrier density N. From Eq. (6), we see that μ_α is the carrier energy ε_α for which f_α is precisely $\frac{1}{2}$. For intrinsic semiconductors, there are as many electrons as holes, i.e.,

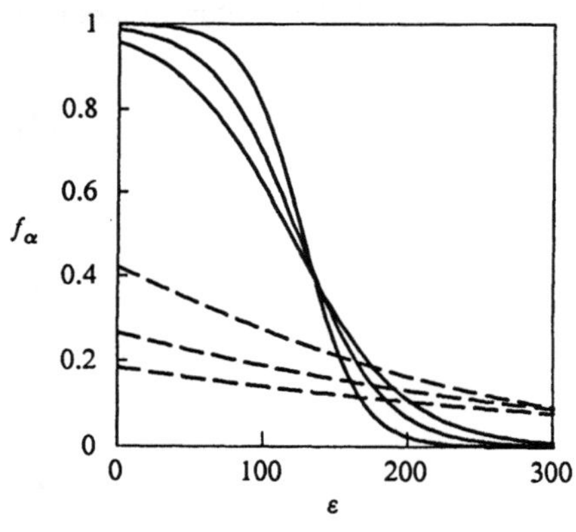

Fig. 2. Fermi-Dirac distributions given by Eq. (6) versus the "reduced" energy $\varepsilon = \hbar^2k^2/2m$ for electrons ($\alpha = e$, solid lines) and holes ($\alpha = h$, dashed lines) for the temperatures (high to low) T = 200, 300, and 400 K. The chemical potentials μ_α in Eq. (6) are chosen so that the total carrier density N of Eq. (7) is 3×10^{18} per cm^3 for all curves. This gives μ_e = 112.4, 109.4, and 104.8 meV and μ_h = -5.63, -26.5, and -51.5 meV for T = 200, 300, and 400 K, respectively.

$$N = V^{-1}\Sigma_k f_e = V^{-1}\Sigma_k f_h \,. \tag{7}$$

We use a computer table lookup to find the chemical potential μ_α that corresponds to the desired carrier density N. The Fermi-Dirac distribution has neat properties such as

$$\frac{1}{e^x + 1} + \frac{1}{e^{-x} + 1} = 1 \,, \tag{8}$$

which we use to determine the width [Eq. (27)] of the laser gain region. Figure 2 plots electron and hole Fermi-Dirac distributions for several temperatures. Note that the hole distributions all have negative chemical potentials and look very much like the decaying tails of Maxwell-Boltzmann distribtions.

2. Polarization of Semiconductor Gain Media

We consider a single-mode electric field of the form

$$E(z,t) = \tfrac{1}{2}\mathcal{E}(t)e^{i(Kz - \nu t)} + \text{c.c.} \,, \tag{9}$$

where $\mathcal{E}(t)$ is a complex amplitude that varies little in an optical frequency period, K is the wave number, ν is the oscillation frequency. This field induces a polarization of the medium of the form

$$P(z,t) = \tfrac{1}{2}\mathcal{P}(t)e^{i(Kz - \nu t)} + \text{c.c.} \,, \tag{10}$$

where $\mathcal{P}(t)$ is a slowly-varying complex polarization. The electric field obeys the slowly-varying Maxwell equation

$$\dot{\mathcal{E}} = -\frac{\nu}{2Q}\mathcal{E} + \frac{i\nu}{2\epsilon}\mathcal{P} = -\frac{\nu}{2Q}\mathcal{E} + \alpha\mathcal{E} \,, \tag{11}$$

where $\nu/2Q$ is the cavity linewidth, ϵ is the permittivity in the absence of free carriers, and α is the gain coefficient $i(\nu/2\epsilon)\mathcal{P}/\mathcal{E}$. The polarization (10) can be written in terms of a density matrix $\rho(k,z,t)$ according to

$$P(z,t) = V^{-1}\Sigma_k \wp \rho_{ab} + \text{c.c.}, \tag{12}$$

where $\wp$ is the electric-dipole matrix element, which we take to be independent of k, V is the volume of the medium, ρ_{ab} is the off-diagonal element of the two-band density matrix whose elements are functions of the carrier wave vector k. The subscript a refers to the conduction band and b to the valence band. The volume V cancels out since it appears in the Σ_k. Combining Eqs. (10) and (12), we find the slowly-varying complex polarization

$$\mathcal{P} = 2\wp e^{-i(Kz - \nu t)} V^{-1}\Sigma_k \rho_{ab} \,. \tag{13}$$

The equation of motion for ρ_{ab} couples it to the probabilities that the electron is in the conduction or valence bands. It is convenient to use the probability of a hole instead of the probability for valence electron. In this "electron-hole" picture, we define n_e to be the probability of having a conduction electron with momentum k, and n_h to be the probability of having a valence hole with momentum k. When a valence electron absorbs light, *both* a (conduction) electron and a hole are created. However due to carrier-carrier scattering, the probabilities that the electron and hole remain in the states of their creation decreases from unity to the values given by the electron and hole Fermi-Dirac distributions. Hence instead of two states for each k, we have the four states $|0_k\ 0_k\rangle$, $|0_k\ 1_k\rangle$, $|1_k\ 0_k\rangle$, and $|1_k\ 1_k\rangle$, where the first number is the occupation of the electron with momentum k and the second is that of the hole. Optically $|0_k\ 0_k\rangle$ is connected to $|1_k\ 1_k\rangle$, but carrier-carrier scattering connects them all. The states $|0_k\ 1_k\rangle$ and $|1_k\ 0_k\rangle$ do not occur in the two-level atom, since there's no way to take away the built-in hole of an excited two-level atom.

In principle we need to define a 4x4 density matrix in this basis, however for semiclassical problems, we can reduce it to a 2x2 density matrix. Specifically we identify $\rho_{aa} = \rho_{11,11} \equiv \langle 1\ 1|\rho|1\ 1\rangle$, $\rho_{bb} = \rho_{00,00}$, and $\rho_{ab} = \rho_{11,00}$. The commutator $[\mathcal{H}, \rho]$ tells us that ρ_{ab} is driven by $\rho_{aa} - \rho_{bb}$. Since $\mathrm{tr}\{\rho\} = 1$, i.e., $\rho_{11,11} + \rho_{10,10} + \rho_{01,01} + \rho_{00,00} = 1$, we have that

$$\rho_{aa} - \rho_{bb} = [\rho_{11,11} + \rho_{10,10}] + [\rho_{11,11} + \rho_{01,01}] - 1 = n_e + n_h - 1\ , \tag{14}$$

where n_e is the probability of an electron with momentum k *independent* of whether there's a hole of momentum k, and n_h is the corresponding probability of a hole. Hence ρ_{ab} is simply coupled to $n_e + n_h - 1$, and vice versa, and we don't have to treat a complicated four-level problem for a semiclassical theory. For spontaneous emission to occur, however, ρ_{aa} alone is involved, since the Pauli exclusion principle requires that both a hole and an electron of the same momentum be occupied. We approximate ρ_{aa} for this purpose by $n_e n_h$, which is the value given by many-body theory.

From Eq. (14), we see that $\dot{n}_e$ is given by $\dot{\rho}_{aa} + \dot{\rho}_{10,10}$ and that $\dot{n}_h$ is given by $\dot{\rho}_{aa} + \dot{\rho}_{01,01}$. The optical contributions to $\dot{\rho}_{aa}$ are identical to those for the two-level system, since the commutator in the Schrödinger equation is the same. $\dot{\rho}_{10,10}$ and $\dot{\rho}_{01,01}$ result from carrier-carrier scattering alone. Including phenomenological terms to describe decay processes and the carrier-carrier scattering, we have the density matrix equations of motion

$$\dot{\rho}_{ab} = -(i\omega + \gamma)\rho_{ab} + i\mathcal{V}_{ab}(z,t)(n_e + n_h - 1) \tag{15}$$

$$\dot{n}_e = \lambda_e - \gamma_{nr} n_e - \Gamma n_e n_h - \dot{n}_e|_{c\text{-}c} - (i\mathcal{V}_{ab}\rho_{ba} + \text{c.c.}) \tag{16}$$

$$\dot{n}_h = \lambda_h - \gamma_{nr} n_h - \Gamma n_e n_h - \dot{n}_h|_{c\text{-}c} - (i\mathcal{V}_{ab}\rho_{ba} + \text{c.c.})\ , \tag{17}$$

where λ_α, $\alpha = e$ or h, is the pump rate due to an injection current, γ_{nr} is the nonradiative decay constant for the electron and hole probabilities, Γ is the radiative recombination rate constant, and $\dot{n}_\alpha|_{c\text{-}c}$ is the carrier-carrier scattering contribution. This scattering drives the distribution n_α toward the Fermi-Dirac distributions of Eq. (6). In fact, the carrier-carrier scattering contribution found using many-body techniques vanishes when n_α is given by a Fermi-Dirac distribution. While rapidly suppressing deviations from the Fermi-Dirac distribution, the scattering does not change the total carrier density N of Eq. (7). Hence summing either Eq. (16) or (17), we find the equation of motion

$$\dot{N} = \lambda - \gamma_{nr} N - \frac{\Gamma}{V}\sum_k n_e n_h - \left[\frac{i\mathcal{V}_{ab}}{V}\sum_k \rho_{ba} + \text{c.c.}\right], \tag{18}$$

where the injection current pump λ is given by $\lambda = \eta J/ed$, η is the efficiency that the injected carriers reach the active region, J is the current density, e is the charge of an electron, and d is the thickness of the active region. Equations (15) through (18) are also derivable from the generalized Maxwell-Bloch equations of Ref. [5,1].

The interaction energy matrix element $\mathscr{V}_{ab}$ for the field of Eq. (9) is given in the rotating-wave approximation by

$$\mathscr{V}_{ab} = -\frac{\wp\mathscr{E}}{2\hbar}e^{i(Kz - \nu t)} , \tag{19}$$

where we assume that the electric-dipole matrix element $\wp$ varies little over the range of k values that interact. We suppose that the dipole decay rate γ is large compared to variations in $\mathscr{V}_{ab}$ (rate equation approximation). For this we can solve Eq. (15) as

$$\rho_{ab} = -\frac{i\mathscr{D}\wp\mathscr{E}}{2\hbar} e^{i(Kz - \nu t)}d_0 , \tag{20}$$

where the complex Lorentzian denominator $\mathscr{D}$ is defined by

$$\mathscr{D} = \frac{1}{\gamma + i(\omega - \nu)} , \tag{21}$$

and the probability difference d_0 is given by

$$d_0 = f_e + f_h - 1 . \tag{22}$$

We call this probability difference d_0 in view of Sec. 5, which is based on nonzero-frequency components ($d_{\pm1}$) associated with the probability response to beat frequencies (population pulsations). Substituting Eqs. (19) and (20) into Eq. (18), we find the carrier-density equation of motion

$$\dot{N} = \lambda - \gamma_{nr}N - \frac{\Gamma}{V}\sum_k f_e f_h - \frac{|\wp\mathscr{E}/\hbar|^2}{2\gamma V}\sum_k \mathscr{L}d_0 , \tag{23}$$

where $\mathscr{L}$ is the dimensionless saturator Lorentzian

$$\mathscr{L} = \frac{\gamma^2}{\gamma^2 + (\omega - \nu)^2} . \tag{24}$$

Inserting Eq. (20) into Eq. (13), we have the complex polarization

$$\mathscr{P} = -\frac{i\wp^2\mathscr{E}}{\hbar V}\Sigma_k \mathscr{D}d_0 . \tag{25}$$

Substituting Eq. (25) into Eq. (11), we have the gain coefficient

$$\alpha = \frac{K\wp^2}{2\epsilon\hbar\gamma V}\Sigma_k \gamma\mathscr{D}d_0 . \tag{26}$$

To evaluate α, we need to know the Fermi-Dirac distributions f_α of Eq. (6). These are implicit functions of N, since their chemical potentials μ_α must be chosen to satisfy the closure relation (7). We can determine N by solving Eq. (23) self-consistently using numerical techniques. Alternatively for any given N, we know the f_α and can evaluate Eq. (26) accordingly.

For a gain medium, Re{α} of Eq. (26) must be positive. For simplicity suppose the linewidth function $\mathscr{L}$ = Re{$\gamma\mathscr{D}$} is a δ function. Gain then occurs for $d_0 > 0$. We can easily show that $d_0 > 0$ provided the carrier reduced mass energy

$$0 < \varepsilon < \mu_e + \mu_h \; . \tag{27}$$

To see that $\mu_e+\mu_h$ is in fact the upper limit, note the identity given by Eq. (8). Hence the condition $d_0 = f_e + f_h - 1 = 0$ is satisfied for energies for which the sum of the electron and hole exponential arguments vanishes. Accordingly adding the exponential arguments implicit in Eq. (22) with the energy value $\varepsilon = \mu_e + \mu_h$ and using the reduced mass of Eq. (5), we find

$$\left[\frac{\mu_e + \mu_h}{m_e} - \mu_e\right] + \left[\frac{\mu_e + \mu_h}{m_h} - \mu_h\right] = 0 \; .$$

The *total chemical potential* $\mu_e + \mu_h$ is a crucial parameter in semiconductor laser theory since it defines the upper limit of the gain region. To the extent that the width $2\hbar\gamma$ of Lorentzian $\mathcal{L}$ is small compared to $\mu_e + \mu_h$, this sum in fact gives the width of the gain region. In general the gain width is somewhat smaller than $\mu_e + \mu_h$, since for sufficiently high energies $\mathcal{L}$ samples a medium with $d_0 < 0$. As we see in the next section, if the gain exceeds the cavity losses, laser action can take place.

3. Single-Mode Semiconductor Laser Theory

In its simplest form, the semiconductor laser consists of a diode, which is a p-type semiconductor joined to an n-type semiconductor. By itself the p-type material is neutral in charge, but contains doping atoms that need extra electrons to fill their outermost valence shell. By itself the n-type semiconductor is also neutral but contains doping atoms that have extra electrons outside their last complete shell. When the two media are brought into contact with one another, the extra electrons in the n-type medium transfer across the junction to the p-type medium, and the holes from the p-type medium transfer to the n-type medium. The Coulomb attraction prevents the carriers from penetrating into the other medium very far. If a positive voltage is applied across junction from the the p-type to the n-type side, a current can flow and the conduction electrons recombine in the junction with the holes, emitting light. If this light can propagate along the junction and be partially reflected at the end facits, there may be enough gain to overcome the losses, and the device lases. To simplify our discussion we assume that the junction consists of a thin layer of an intrinsic semiconductor, that is, an undoped semiconductor.

Steady-state operation occurs when Eq. (11) gives $\dot{\mathcal{E}} = 0$ for the gain coefficient of Eq. (26). Above the laser threshold, this implies the steady-state oscillation condition

$$\alpha_r = \frac{\nu\wp^2}{2\epsilon\gamma\hbar V}\sum_k \mathcal{L}d_0 = \frac{\nu}{2Q} \, , \tag{28}$$

i.e., the real part of the saturated gain equals the cavity losses. According to Eq. (28), the difference d_0 and hence the total carrier density N must remain at their respective laser-threshold values, a feature known as *gain clamping*. In particular, Eq. (28) determines the steady-state single-mode laser carrier density N_0 and the threshold pump value λ_{th} as functions of the cavity loss rate constant $\nu/2Q$ and temperature T. Since the intensity determines the total carrier density independent of the momentum k, we say that the semiconductor laser medium *saturates homogeneously* in spite of the linear gain profile being inhomogeneously broadened.

In terms of the gain α_r, the carrier-density equation of motion (23) becomes

$$\dot{N} = \lambda - \lambda_{th} - \alpha_r n \,, \tag{29}$$

where the photon number density $n(t)$ (note that this is a classical quantity in spite of the name we give it) is defined by

$$n(t) = \epsilon|\mathcal{E}|^2/\hbar\nu \,, \tag{30}$$

and the threshold pump value λ_{th} (that for $n = 0$) is given by

$$\lambda_{th} = \gamma_{nr} N_0 + \frac{\Gamma}{V}\sum_k f_e f_h \,, \tag{31}$$

such that $V^{-1}\Sigma_k f_\alpha = N_0$, for $\alpha = e$ or h. Multiplying the field amplitude equation of motion (11) by $2\epsilon\mathcal{E}^*/\hbar\nu$, we have the intensity equation of motion

$$\dot{n} = 2n(\alpha_r - \nu/2Q) \,. \tag{32}$$

The steady-state photon number is determined by the condition $\dot{n} = \dot{N} = 0$, which gives

$$n = (\mathcal{N} - 1)n_s \,, \tag{33}$$

where the relative excitation $\mathcal{N}$ is defined by the ratio

$$\mathcal{N} = \frac{\lambda}{\lambda_{th}} \,, \tag{34}$$

and the "saturation photon number" n_s is given by

$$n_s = \frac{\lambda_{th}}{\alpha_r} \,. \tag{35}$$

Equation (33) shows that the laser intensity increases as a linear function of pump rate with a slope proportional to n_s. The unidirectional ring laser with a homogeneously broadened two-level active medium has the dimensionless intensity (intensity in units of the saturation intensity I_s) $I_n = (\mathcal{N}\mathcal{L} - 1)/\mathcal{L}$, which has a similar form, although it displays its tuning dependence explicitly. For central tuning ($\mathcal{L} = 1$) and a relative excitation $\mathcal{N} = 2$, the two-level dimensionless intensity $I_n = 1$, i.e., the laser intensity equals the saturation intensity I_s. In this spirit, we call n_s the semiconductor-laser saturation photon number because it is the n value corresponding to the relative excitation $\mathcal{N} = 2$. We see that the semiconductor-laser steady-state oscillation intensity has similarities with the corresponding homogeneously broadened two-level laser intensity. However it differs in important ways, such as its dependences on carrier density, temperature and effective masses, and on the cavity tuning. The semiconductor-laser n_s differs from the two-level I_s significantly in various respects, not the least of which is that n_s is negative, i.e., unphysical, for an absorbing medium.

To evaluate the complex gain given by Eq. (26), we convert $V^{-1}\Sigma_k$ into an integral by introducing the density of states. The optical gain coefficient of Eq. (28) is given by

$$\alpha_r = \frac{\nu\wp^2}{2\epsilon\gamma\hbar(2\pi^2 a_0^3 E_R^{\,3/2})}\int_0^\infty \frac{\sqrt{\varepsilon}\,d\varepsilon}{1 + (\varepsilon-\hbar\delta)^2/(\hbar\gamma)^2}\left[\frac{1}{e^{\beta(\varepsilon/\bar{m}_e - \mu_e)} + 1} + \frac{1}{e^{\beta(\varepsilon/\bar{m}_h - \mu_h)} + 1} - 1\right] , \tag{36}$$

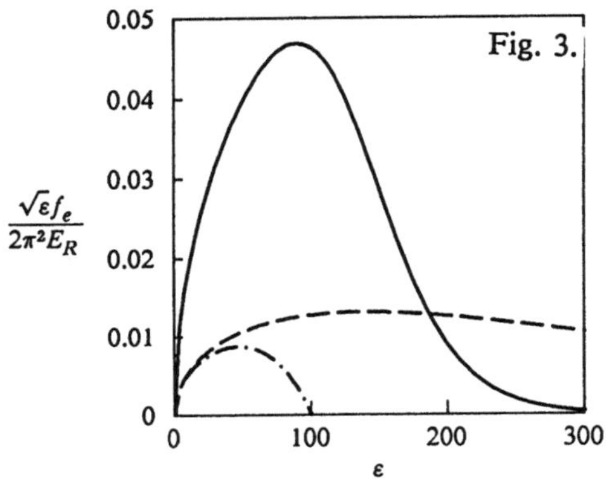

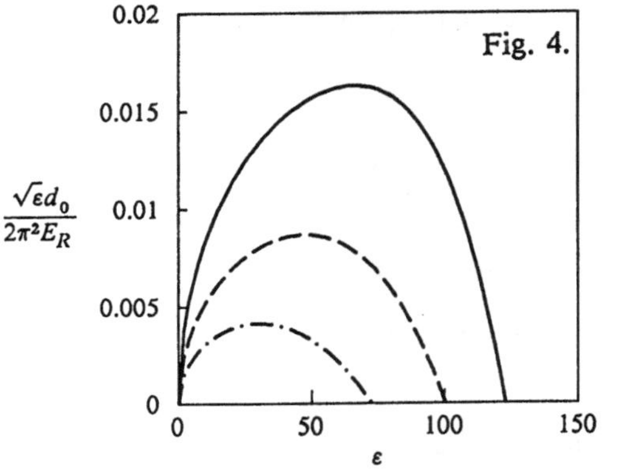

Fig. 3. $\sqrt{\varepsilon} f_e$ (solid line), $\sqrt{\varepsilon} f_h$ (dashed line), and $\sqrt{\varepsilon}(f_e + f_h - 1)$ (dot-dashed line) all in units of $2\pi^2 E_R$ versus energy ε in meV for $T = 300K$ and a carrier density of 3.5×10^{18} carriers/cm^3, $k_B = .086164$ meV/K, and carrier parameters appropriate for GaAs, namely, $a_0 = 1.243\times10^{-6}$ cm, $E_R = 4.2$ meV, $m_e/m = 1.127$, $m_h/m = 8.82$. The gain formula used is Eq. (28) with the chemical potentials determined by the closure relation (7) to be $\mu_e = 4.73 k_B T$ and $\mu_h = -.86 k_B$T.

Fig. 4. $\sqrt{\varepsilon} d_0 = \sqrt{\varepsilon}(f_e + f_h - 1)$ versus reduced-mass energy ε for the temperatures $T = 200K$ (solid line), $300K$ (dashed line), and $400K$ (dot-dashed line) and other parameters as in Fig. 3.

where the exciton Bohr radius $a_0 = \hbar^2\epsilon/me^2$ and Rydberg energy $E_R = \hbar^2/2ma_0^2$, ϵ is the permittivity in the medium, the effective mass ratio $\tilde{m}_\alpha \equiv m_\alpha/m$, the energy detuning $\hbar(\omega - \nu) = (\hbar\omega - \varepsilon_g - \delta\varepsilon_g) - (\hbar\nu - \varepsilon_g - \delta\varepsilon_g) = \varepsilon - \hbar\delta$, and the laser detuning relative to the renormalized bandgap is given by $\hbar\delta = \hbar\nu - \varepsilon_g - \delta\varepsilon_g$. To get a feeling for this gain function, we plot $\sqrt{\varepsilon} f_e$ (solid line) $\sqrt{\varepsilon} f_h$ (dashed line) and $\sqrt{\varepsilon}(f_e + f_h - 1)$ (dot-dashed line) in Fig. 3. The dot-dashed curve gives the gain spectrum for an optical probe for zero linewidth; linewidths ($2\hbar\gamma$) larger than zero sample a range of values, changing the gain spectrum somewhat. For the parameters chosen, the gain disappears below $N_0 \simeq 1.21\times10^{18}$ carriers/cm^3.

An increase in temperature reduces the gain by spreading the carriers out. This is shown in Fig. 4 for the three temperatures $T = 200$, 300, and $400K$. The gain peak is defined by the condition $\partial\alpha_r/\partial\delta = 0$. The laser threshold occurs for the minimum carrier density N_0 and the detuning δ_0 that allow this condition and Eq. (28) to be satisfied simultaneously. N_0 and δ_0 can be found by using a Newton-Raphson numerical procedure.

4. Linear Stability Analysis of the Single-Mode Steady-State

To determine if the single-mode photon number density of Eq. (38) is stable, we expand the photon-number and carrier densities about their steady-state values $\mathscr{n}_0$ and N_0 as

$$\mathscr{n}(t) = \mathscr{n}_0 + \delta\mathscr{n}(t) \tag{37}$$
$$N(t) = N_0 + \delta N(t)\ . \tag{38}$$

Substituting Eqs. (37) and (38) into the equations of motion (29) and (32) and using Eq. (34), we have

$$\frac{d}{dt}\delta n(t) = 2n_0\alpha_r{}'\delta N(t) \tag{39}$$

$$\frac{d}{dt}\delta N(t) = -\Gamma_1{}'\delta N(t) - \frac{\nu}{2Q}\delta n(t) , \tag{40}$$

where the rate of change of the gain with respect to the carrier density is

$$\alpha_r{}' = \frac{\partial\alpha_r}{\partial N} = \frac{\nu \wp^2}{2\epsilon\gamma\hbar V}\sum_k \mathcal{L}(g_e + g_h) , \tag{41}$$

the derivatives g_α of the Fermi-Dirac distributions are given by

$$\begin{aligned} g_\alpha &= \frac{\partial f_\alpha}{\partial N} = \frac{\partial f_\alpha}{\partial \mu_\alpha}\frac{\partial \mu_\alpha}{\partial N_0} \\ &= \frac{f_\alpha(1-f_\alpha)}{V^{-1}\Sigma_k f_\alpha(1-f_\alpha)} , \end{aligned} \tag{42}$$

the zero-field relaxation rate of the carrier density

$$\Gamma_1 = \frac{\partial\lambda_{th}}{\partial N} = \gamma_{nr} + \Gamma V^{-1}\Sigma_k(f_e g_h + f_h g_e) , \tag{43}$$

and the power-broadened relaxation rate

$$\Gamma_1{}' = \Gamma_1 + \alpha_r{}' n_0 . \tag{44}$$

The eigenvalues of the coupled set of equations (39) and (40) are

$$-\frac{\Gamma_1{}'}{2} \pm \sqrt{\frac{\Gamma_1{}'^2}{4} - \frac{\alpha_r{}' n_0 \nu}{Q}} . \tag{45}$$

Since $\alpha_r{}' n_0 \nu/Q$ is positive, both roots have negative real parts, implying that single-mode operation is stable. The solutions are oscillatory yielding linear relaxation oscillations provided

$$\frac{\alpha_r{}' n_0 \nu}{Q} > \frac{\Gamma_1{}'^2}{4} . \tag{46}$$

Nonlinear relaxation oscillations can be studied by direct integration of the coupled equations of motion (29) and (32).

Near the laser operating point, the long-T_1 two-level (like ruby) and semiconductor lasers are both describable by simple rate equation formalisms. However while the two-level laser rate constants are really or nearly *constant*, the semiconductor rate "constants" are *functions* of temperature, tuning, and carrier density. A definition of a saturation intensity for the semiconductor is really only valuable in defining the slope of the intensity versus relative-excitation, since it vanishes and then goes negative in passing from gain to absorption. In contrast the two-level saturation intensity is the same for both gain and absorption and determines the amount of power broadening.

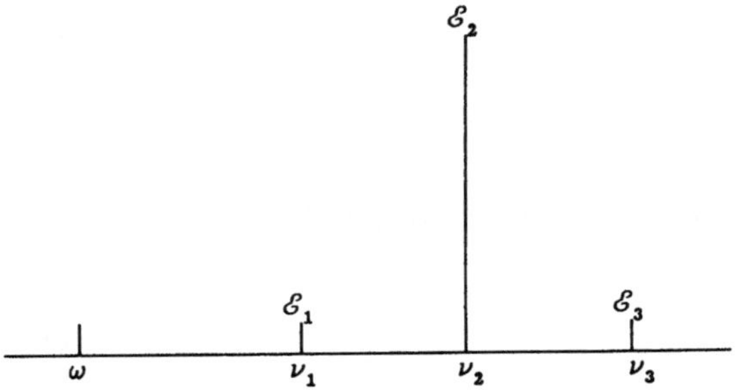

Fig. 5. Spectrum of three-mode field. Waves with frequencies ν_1 and ν_3 are usually taken to be weak (nonsaturating), while the ν_2 wave is allowed to be arbitrarily intense.

5. Sidemode Gain and Multiwave Mixing

An important question is whether other modes can build up in the laser. This section provides a preliminary step in answering this question by summarizing a nondegenerate theory of multiwave mixing in a semiconductor medium. We label the sidemodes by the indices 1 and 3 and the saturator mode by 2 as shown in Fig. 5. Our electric field has the form

$$E(z,t) = \frac{1}{2}\sum_n \mathcal{E}_n(t)\, e^{i(K_n z - \nu_n t)} + \text{c.c.}, \tag{47}$$

where the mode amplitudes $\mathcal{E}_n(t)$ are in general complex and K_n are the wave propagation vectors. For simplicity we take mode functions appropriate for a unidirectional ring laser. The mode index equals 1, 2, or 3.

One might guess that the sidemode polarization is given by a sidemode Lorentzian integrated over the sum of weighted Fermi-Dirac distributions saturated by the saturator wave. However an additional contribution enters due to pulsations in the total carrier density. Specifically the nonlinear probabilities respond to the superposition of the modes to give pulsations at the beat frequency $\Delta = \nu_2 - \nu_1$. Since we suppose the sidemode does not saturate, the pulsations occur only at $\pm\Delta$, a point proved below. These pulsations act as modulators (or like Raman "shifters"), putting sidebands onto the medium's response to the ν_2 mode. One of these sidebands falls precisely at ν_1, yielding a contribution to the sidemode absorption coefficient. The other sideband influences the polarization of the sidemode placed symmetrically on the other side of the strong mode, namely at the frequency $\nu_2 + \nu_2 - \nu_1$. The polarization of the medium can be analysed using Fourier techniques successful in the corresponding treatment of two-level media. In this way, Ref. [8] finds the coupled-mode equations

$$\dot{\mathcal{E}}_1 = [\alpha_1 - \nu/2Q_1 - i(\Omega_1 - \nu_1)]\mathcal{E}_1 + \chi_1 \mathcal{E}_3^*, \tag{48}$$

$$\dot{\mathcal{E}}_3^* = [\alpha_3 - \nu/2Q_3 + i(\Omega_3 - \nu_3)]^* \mathcal{E}_3^* + \chi_3^* \mathcal{E}_1, \tag{49}$$

where the sidemode gain coefficient α_1 is given by

$$\alpha_1 = \frac{\alpha_0}{V}\sum_k \gamma \mathcal{D}_2 \left[d_0 - \frac{(g_e + g_h) 2\gamma^{-1} |\wp\mathcal{E}_2/2\hbar|^2 V^{-1} \Sigma_k \mathcal{L}_2 d_0}{\Gamma_1' + i\Delta} \right] \tag{50}$$

with $d_0 \simeq f_{e0} + f_{h0} - 1$ and the coupling coefficient χ_1 is given by

$$\chi_1 = -\frac{\alpha_0}{V}\sum_k \gamma\mathscr{D}_2 \frac{(g_e + g_h)2\gamma^{-1}(\wp\mathscr{E}_2/2\hbar)^2\, V^{-1}\Sigma_k \mathscr{L}_2 d_0}{\Gamma_1' + i\Delta}. \tag{51}$$

where the coefficients α_3 and χ_3 are given by α_1 and χ_1, respectively, by interchanging the subscripts $_1$ and $_3$ (note this implies replacing Δ by $-\Delta$), g_α is given by Eq. (42), the complex denominator $\mathscr{D}_2$ and Lorentzian $\mathscr{L}_2$ are given by Eqs. (21) and (24), respectively, with ν replaced by ν_2, and the carrier-density decay constant Γ_1' is given by Eq. (44).

This carrier-density decay constant is *not* a function of k. The remaining sums over k produce various multiplicative constants. The α_1 and χ_1 of Eqs. (50) and (51) lead to coherent dips (see Fig. 6) in pump-probe and modulation spectroscopy similar to those that might be found in two-level media. They furthermore lead to narrow-band retroreflection in phase conjugation by nondegenerate four-wave mixing. The major distinguishing features of the semiconductor case is that the dips are asymmetric due to the product of the complex Lorentzian denominators integrated over an asymmetric weighted distributions. We see that one side of the sidemode gain curves is higher than the average gain, and hence a sidemode might build up on that side. The resonance fluorescence spectrum calculated with the same model, but with substantially greater algebraic complexity [9] reveals similar asymmetric dips. This fluoresence provides the "spark" for the build up of sidemodes.

In summary we have presented a theory of the semiconductor laser based on a quasiequilibrium model of the semiconductor medium. This model is motivated as an extension of the inhomogeneously broadened two-level system, but is nevertheless justified also as an approximation to a proper many-body theory. We have derived the steady-state oscillation condition and analysed both the approach through relaxation oscillations to the steady state as well as addressed the question as to whether sidemodes can build in the presence of a single oscillating mode. Substantial work remains to be done in this area, since multimode laser operation is commonplace and must be understood to maximize the utility of the lasers. In particular our current approach requires that the sidemode frequencies be held well within a reciprocal carrier-carrier scattering time of the main mode. In practice,

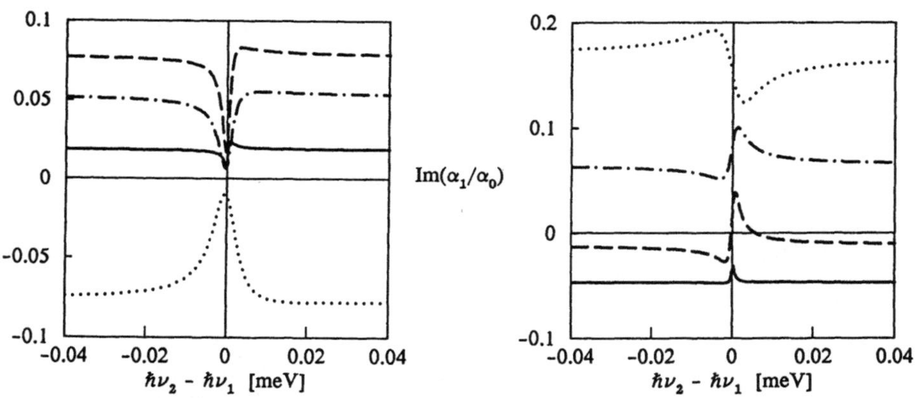

Fig. 6. Real and imaginary parts of the sidemode gain coefficient α_1 of Eq. (50) *vs* pump-probe detuning $\hbar\Delta$ for pump detunings $\hbar\nu_2 - (\varepsilon_g + \delta\varepsilon_g) = 0$ (solid line, at band gap), 33 meV (dashed line, above band gap), 66 meV (dot-dashed line), and 99 meV (dotted line, above gain region). The Rabi frequency $|\wp\mathscr{E}| = 0.5$ meV, which corresponds to a pump intensity of $\simeq 1.3$ MW/cm^2 (from Ref. [8]).

multimode operation can break this requirement. We have also assumed that the electron and hole Fermi-Dirac distributions are always described by the same temperature, namely that of the lattice. High frequency mode modulations could violate this approximation even before violating the establishment of quasiequilibrium Fermi-Dirac distributions with different temperatures. In the quest to figure out the reasons for multimode operation, we may well find better ways to understand the dynamics of the light-semiconductor interactions themselves. In any event, the present theory provides a substantially better way to understand single-mode operation than the typical rate equation theories used commonly today.

Acknowledgements

Work supported in part by the U. S. Army Research Office, in part by the U. S. Air Force Office of Scientific Research, and in part by the U. S. Office of Naval Research.

References

1. S. W. Koch, H. Haug, C. Ell, M. Lindberg, and M. Sargent III, preceeding paper.
2. G. A. Agrawal and N. K. Dutta, *Long-Wavelength Semiconductor Lasers* Van Nostrand Reinhold Co., New York (1986).
3. A. Yariv, *Quantum Electronics*, 3rd Edition, John-Wiley, New York (1988).
4. W. W. Chow, W. W., G. C. Dente, and D. Depatie, IEEE J. Quant. Electron. E-23 1314 (1987).
5. M. Lindberg and S. W. Koch, Phys. Rev. *B*38, 3342 (1988).
6. M. Sargent III, M. O. Scully, and W. E. Lamb, Jr., *Laser Physics*, Addison-Wesley Pub. Co., Reading, MA (1977).
7. R. W. Boyd and M. Sargent III, J. Opt. Soc. *B*, (1987).
8. M. Sargent III, F. Zhou, and S. W. Koch, Phys. Rev. *A*38, 4673 (1988).
9. M. Lindberg, S. An, S. W. Koch, and M. Sargent III, to be published in Phys. Rev. A.

Poster Papers

Absence of Laser Induced Continuum Structure in Third Harmonic Generation in Sodium
Philip B. Chappel, Kenneth G.H. Baldwin and Hans-A. Bachor
Australian National University, Canberra, Australia

Experimental Investigation of Squeezing Using Four Wave Mixing in Barium
H.A. Bachor, P.T.H. Fisk, P.J. Manson, D.E. McClelland and R.J. Sandemann
Australian National University, Canberra, Australia

Compton effect, and two-photon pair production, in an intense electromagnetic field (laser field), using the Volkov solution: a complete calculation
A. Hartin, H.S. Perlman, G.J. Troup
Monash University, Victoria, Australia

Towards a Bell's Inequality Experiment using Time and Energy
Lloyd M. Davis
University of Tennessee Space Institute, Tullahoma, USA

Laser Cavity with a Nonlinear Mirror
R. McDuff and N.R. Heckenberg
University of Queensland, St Lucia, Queensland, Australia

Theory of Photon Counting for Phase Dependent Three Level Detector Systems
B.J. Dalton and P.L. Knight
University of Queensland, St Lucia, Queensland, Australia

A Quantum Optical Fredkin Gate
G.J. Milburn
University of Queensland, St Lucia, Queensland, Australia

Dynamics of Nonlinear Optical Media
Reto Holzner, Nick Tufillaro and Tim Molteno
University of Otago, Dunedin, New Zealand

Quantum Tunnelling in a Kicked Nonlinear Top
B.C. Sanders and G.J. Milburn
University of Queensland, St Lucia, Queensland, Australia

Quantum Propagation in Optical Fibres
S.J. Carter and P.D.Drummond
University of Auckland, Auckland, New Zealand

Switching Times in Optically Bistable Devices
P. Kinsler and P.D. Drummond
University of Auckland, Auckland, New Zealand

Experimental Observation of Polarisation Separation and Self Focussing in Sodium Vapour
J.C. Wang, H.-A. Bachor and D.E. McClelland
Australian National University, Canberra, Australia

Complementarity in a Quantum Nondemolition Measurement
B.C. Sanders and G.J. Milburn
University of Queensland, St Lucia, Queensland, Australia

Are there Non-negligible Quantum Corrections to Reflection Ellipsometry?
John Lekner
Victoria University of Wellington, Wellington, New Zealand

Quantum Propagation in Isotropic and Anisotropic Nonlinear Media
T.A.B. Kennedy and S. Wabnitz
University of Auckland, Auckland, New Zealand

The Effect of Continuum-continuum Transitions on Laser Induced Continuum Structures
M.E.St.J. Dutton and B.J. Dalton
University of Queensland, St Lucia, Queensland, Australia

An Ultra-fast Phase Sensitive Photodetector
M.J. Fernée and P.D. Drummond
University of Queensland, St Lucia, Queensland, Australia

Honours Projects in Quantum Optics
Tim Molteno, Nick Mulgan and Andrew Wilson
University of Otago, Dunedin, New Zealand

Violations of Bell's Inequalities with Parametric Down Conversion
S.M. Tan and D.F. Walls
University of Auckland, Auckland, New Zealand

Inhibition of Atomic Phase Decays by Squeezed Light in a Microcavity
S. Parkins and C.W. Gardiner
University of Waikato, Hamilton, New Zealand

Instability and Chaos in Lasers with Delayed Feedback
Y. Cho
Institute of Scientific and Industrial Research, Osaka University, Osaka, Japan

Time resolved coherent Raman spectroscopy in the femtosecond time domain
R. Leonhardt, W. Zinth and W. Kaiser
University of Auckland, Auckland, New Zealand

Index of Contributors

MIX
Papier aus verantwortungsvollen Quellen
Paper from responsible sources
FSC® C105338

If you have any concerns about our products,
you can contact us on
ProductSafety@springernature.com

In case Publisher is established outside the EU,
the EU authorized representative is:
Springer Nature Customer Service Center GmbH
Europaplatz 3, 69115 Heidelberg, Germany

Printed by Libri Plureos GmbH
in Hamburg, Germany